Principles of Biochemistry and Biophysics

Principles of Biochemistry and Biophysics

Shiv Naz

RANDOM

RANDOM PUBLICATIONS

NEW DELHI (INDIA)

Principles of Biochemistry and Biophysics

ISBN 978-93-5111-206-8

Published in 2014 in India by

RANDOM PUBLICATIONS

4376-A/4B, Gali Murari Lal, Ansari Road
New Delhi-110 002
Phone : +91-11-43580356, +91-11-23289044
e-mail: randomexports@gmail.com, sales@randompublications.com, info@randompublications.com

Reprinted 2021

Type Setting by : Keystoneprintads, Delhi-110051
Digitally Printed at: Replika Press Pvt. Ltd.

Preface

Biochemists and biophysicists seek to understand life processes in terms of chemical and physical principles. Biochemistry and biophysics provide the basis for much of modern biotechnology.

Biochemistry and biophysics is focused on numerous processes central to understanding life. Several groups use biochemical and structural approaches to address the basic principles governing protein folding, function and biological recognition. Using in vitro approaches, the central steps in biological information transfer are being analyzed, from maintenance of the genome to protein synthesis, sorting and processing.

Biochemistry and biophysics are closely related fields. Biochemistry focuses on the interconversion of compounds in the many complex reactions of life, on the mechanisms whereby enzymes catalyze and regulate these reactions, and the function and structure of the molecular components of living organisms.

Biophysics is principally concerned with processes of energy conversion, information transmission, and the structure and properties of materials in biological systems, as explored with methods of physics. Biochemical and biophysical research is advancing the frontiers of research in the basic life sciences and making possible advances in more applied fields such as medicine and agriculture.

Biochemists are interested, for example, in mechanisms of brain function, cellular multiplication and differentiation, communication within and between cells and organs, and the chemical bases of inheritance and disease. The biochemist seeks to determine how specific molecules such as proteins, nucleic acids, lipids, vitamins and hormones function in such processes. Particular emphasis is placed on regulation of chemical reactions in living cells.

Biophysics is that branch of knowledge that applies the principles of physics and chemistry and the methods of mathematical analysis and computer modeling to biological systems, with the ultimate goal of understanding at a fundamental level the structure, dynamics, interactions, and ultimately the function of biological systems. Biophysics seeks to explain biological function in terms of the physical properties of specific molecules. The present book has been written in simple language in such a way that a student with very little chemistry background can easily follow the various aspects of biochemistry presented in the book.

I thank all members of my team who have helped in the preparation of the book. My special thanks go to "Random Publications" who have published the book.

– ***Shiv Naz***

Contents

1

Structure of Membrane Proteins

The lipid bilayer constitutes the fundamental structural unit of all biological membranes. Proteins, in contrast, carry out essentially all of the active functions of membranes, including transport activities, receptor functions, and other related processes. As suggested by Singer and Nicolson, most membrane proteins can be classified as peripheral or integral. The peripheral proteins are globular proteins that interact with the membrane mainly through electrostatic and hydrogen-bonding interactions with integral proteins.

Although peripheral proteins are not discussed further here, many proteins of this class are described in the context of other discussions throughout this textbook. Integral proteins are those that are strongly associated with the lipid bilayer, with a portion of the protein embedded in, or extending all the way across, the lipid bilayer. Another class of proteins not anticipated by Singer and Nicolson, the lipid-anchored proteins, are important in a variety of functions in different cells and tissues. These proteins associate with membranes by means of a variety of covalently linked lipid anchors.

PROTEINS

Proteins are any of a group of complex organic compounds containing carbon, hydrogen, nitrogen and sulfur. Proteins, the principle constituents of the protoplasm of all cells, are of high molecular weight and consist of alpha-amino acids joined by peptide linkages. Different amino acids are commonly found in proteins, each protein having a unique, genetically defined amino-acid sequence, which determines its specific shape and function. They serve as enzymes, structural elements, hormones, immunoglobulins, etc. and are involved in oxygen transport, muscle contraction, electron transport, and other activities.

The importance of adequate protein intake to proper immune function has been extensively studied. The most severe effects of"Protein-Calorie Malnutrition" (PCM) are on cell-mediated immunity, although all facets of immune function are ultimately affected. PCM is not, however, a single nutrient deficiency. It is normally associated with multiple nutrient deficiencies, and some immune dysfunctions attributed to PCM are most likely due to these other factors. Partial deficiencies of dietary vitamins produce a comparatively greater depression on immune functions than do partial protein deficiencies. Non-the less, adequate protein is essential for optimal immune function. High protein diets are not recommended for individuals with kidney or liver disease. There are no other known side effects.

A macromolecule made from amino acids; some consist of one, long polypeptide chain while others are made from several intertwined polypeptides (proteios = first place) Proteins are about 50 per cent of the dry weight of most cells, and are the most structurally complex macromolecules known. Each type of protein has its own unique structure and function.

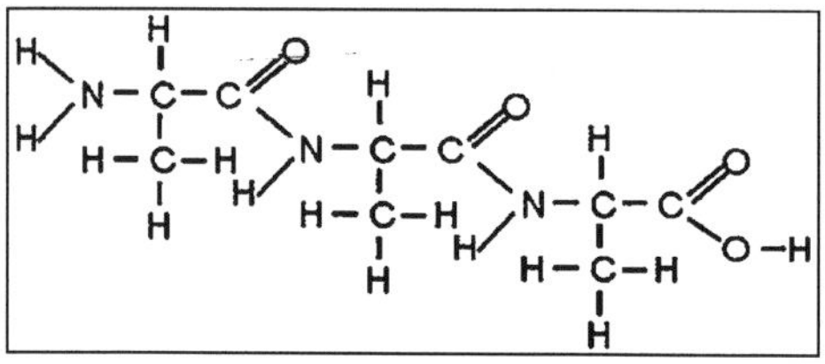

POLYMER

A large molecule made of many, smaller, repeating, identical or similar subunits (poly = many; mer = a part) Polymers are any kind of large molecules made of repeating identical or similar subunits called monomers. The starch and cellulose we previously discussed are polymers of glucose, which in that case, is the monomer. Proteins are polymers of about 20 amino acids (the monomer).

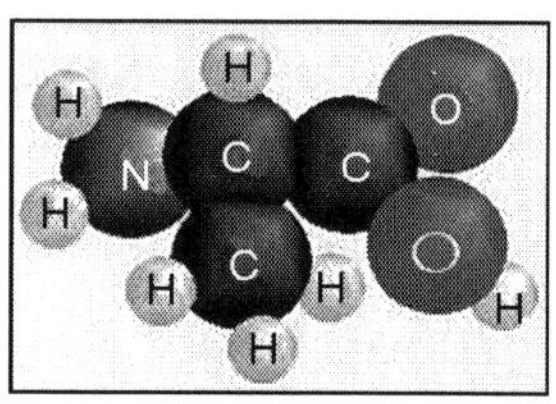

Amino acids are built from a central carbon bonded to four different groups. (run your mouse over the amino acid to see the names of the parts).

- Hydrogen (–H),
- Amino group (–NH_2),
- Carboxyl group (–COOH), and some side chain symbolized by"R".

In the approximately-20 amino acids found in our bodies, what varies is the side chain. Some side chains are hydrophilic while others are hydrophobic. Since these side chains stick out from the backbone of the molecule, they help determine the properties of the protein made from them. Most naturally-occurring amino acids are the l- form, whereas synthetically-produced amino acids give a 50:50 mixture. Notice that these molecules are mirror images of each other, thus there is no way you can rotate one molecule to make it look like the other.

To form protein, the amino acids are linked by dehydration synthesis to form peptide bonds. The chain of amino acids is also known as a polypeptide. Some proteins contain only one polypeptide chain while others, such as hemoglobin, contain several polypeptide chains all twisted together. The sequence of amino acids in each polypeptide or protein is unique to that protein, so each protein has its own, unique 3-D shape or native conformation.

If even one amino acid in the sequence is changed, that can potentially change the protein's ability to function. For example, sickle cell anemia is caused by a change in only one nucleotide in the DNA sequence that causes just one amino acid in one of the hemoglobin polypeptide molecules to be different. Because of this, the whole red blood cell ends up being deformed and unable to carry oxygen properly.

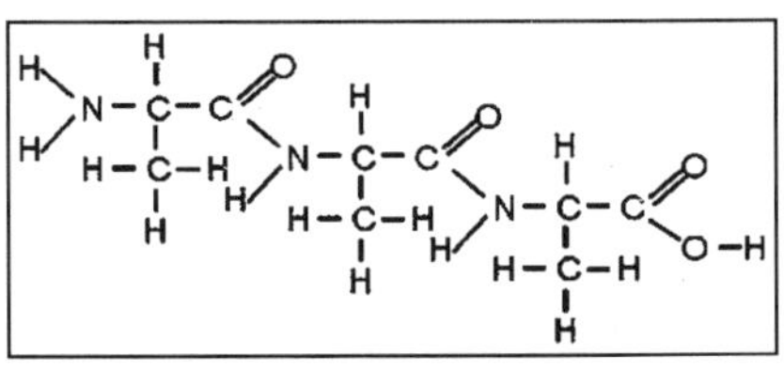

As a polypeptide chain forms, it naturally twists and bends into its native conformation. One of the things that helps determine the native conformation of a protein is the side chains of all the amino acids involved. Remember some amino acid side chains are hydrophobic while others are hydrophilic. In this case, the"likes" attract: all the hydrophobic side chains (here represented by yellow beads) try to"get together" in the Centre of the molecule, away from the watery environment, while the hydrophilic side chains are attracted to the outside of the molecule, near the watery environment. Additionally, some of the hydrophilic side chains have groups of atoms attached that make them acidic, while others have groups attached that make them basic. Side chains with acidic ends are attracted to side chains with basic ends, and can form ionic bonds.

Thus, the side chains interacting with each other help to hold the protein in its native conformation.

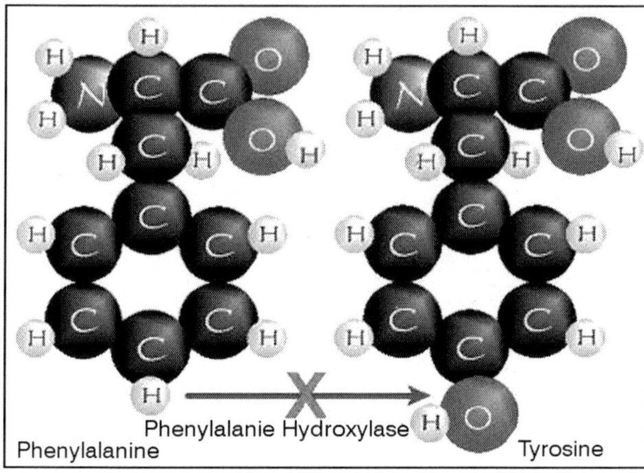

Denaturation is when a protein loses its native conformation. For example, egg white (also called albumen) contains a protein called albumin which is water-soluble. However, if heated, albumin becomes denatured and loses its ability to be water-soluble. There are several possible things that can denature proteins including changing the temperature (adding heat), changing the pH or salt concentration of the solution, or putting the protein into a hydrophobic solvent. In a hydrophobic solvent, the amino acids with hydrophobic side chains (the yellow beads) would all try to go to the outside of the molecule, and all those with hydrophilic side chains would cluster in the Centre of the molecule. If a protein remains water-soluble when denatured (unlike albumin), it can return to its native conformation if/when placed back into a"normal" environment.

PKU or phenylketonuria is a genetic disorder for which, by law, all newborns must be tested. People with this disorder are lacking an enzyme called phenylalanine hydroxylase needed to turn the amino acid, phenylalanine into tyrosine so it can be eliminated via the urine. Since phenylalanine cannot be eliminated, it builds up and in children whose nervous systems are still developing, causes mental retardation.

According to the Merck Manual, the body is slowly able to turn it into several other compounds which can be eliminated via the urine, but these chemical reactions are too inefficient to get rid of the excess phenylalanine. If PKU is detected shortly after birth, the child is put on a special diet low in phenylalanine and has normal mental development. Since phenylalanine is an amino acid that is needed by the body for normal growth and development, the goal is to limit intake to just what the child needs without any excess that could build up in his/her system. The Merck Manual says that, unfortunately, most natural protein sources, including milk, are too high in phenylalanine for children with PKU, so these babies/children must be put on a special formula from which most, if not all, of the phenylalanine has been removed.

As the child grows, some low-protein, natural foods such as fruits, vegetables, and some kinds of grain are OK to eat. The needed phenylalanine is supplied via carefully-measured amounts of natural protein plus the residual in the formula. One area of"debate" is how long the special diet must be continued. Some people think the diet may be discontinued after the brain finishes developing at about age 5. Others have noted some Behaviour problems and learning difficulties that appear to be related to going off the diet, and recommend that people with PKU stay on the diet for life.

Staying on the special diet is especially important if a woman who has PKU is even thinking about trying to become pregnant. Because any excess phenylalanine in her system could cross the placenta and could cause mental retardation in the baby, especially in the first few weeks while the nervous system is forming, she has to make sure that she adheres to the special diet from before conception. Note that this is not a major concern for a normal woman carrying a baby with PKU (she wouldn't know that until after the baby was born and tested) because the woman's body can digest phenylalanine, including any excess from the baby's body that is sent back, via the placenta, to her body.

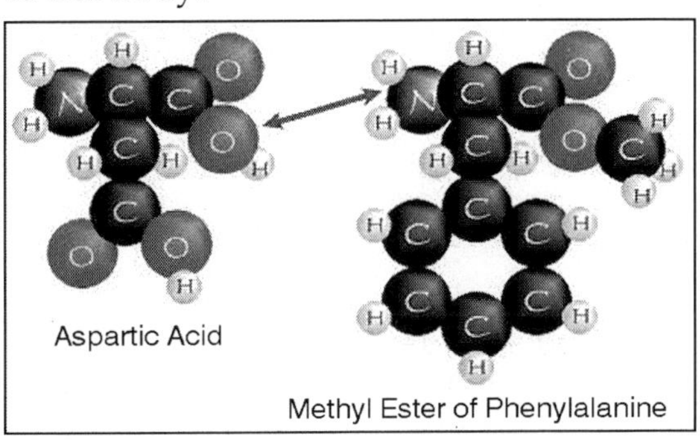

Nutrasweet has the chemical name aspartame and is made of two amino acids: l-aspartic acid and the methyl ester of phenylalanine, thus people with PKU can't have it. Look at a can of diet soft drink that is sweetened with aspartame. Somewhere on the can, there's always a cautionary label warning people with PKU not to drink it.

But what about the rest of us? That methyl group on the phenylalanine can break off pretty easily, and at temperatures as"low" as 86° F (= 30° C - lower than body temperature or a summer day), aspartame starts to disintegrate, releasing methanol (which in larger quantity causes blindness, etc. - it's toxic). As the liver tries to deal with the methanol using normal metabolic pathways, it ends up turning the methanol into formaldehyde, which is a known carcinogen. Interestingly, because of this and several other concerns, originally there was controversy over FDA approval of aspartame. Even still, its safety is controversial. More recently, some people have suggested a possible link between aspartame consumption and an increase in brain tumors in this country.

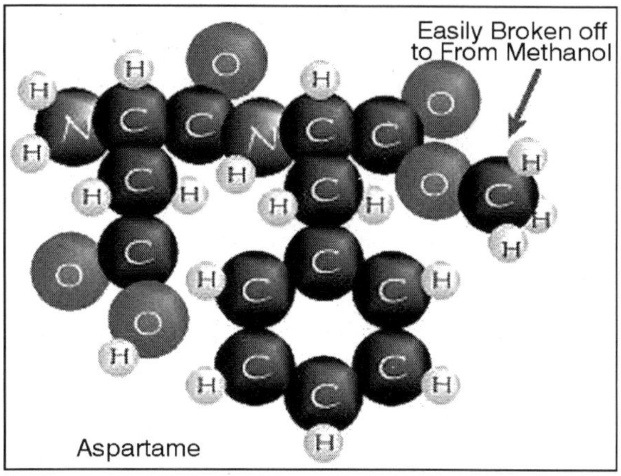

Equal® is not the same as Nutrasweet, but does contain some of it. There's an interesting psychological thing here: in our culture, people think they need sweetener by the teaspoon, but Nutrasweet is so much sweeter than sugar that only a tiny pinch would be needed to gain the equivalent sweetness. Artificial sweetner manufacturers know that if they don't make a product such that one teaspoon is as sweet as one teaspoon of the sugar people are used to using, people will use one teaspoon anyway (a tiny pinch just doesn't look like enough), then say it was too sweet and stop buying it. Thus, artificial sweetners are designed to take up about a teaspoon of space, and Equal® is mostly glucose (which diabetics and hypoglycemics shouldn't have) with a little Nutrasweet added, so in theory, not quite as much of the sugar is needed (also remember that glucose isn't as sweet as sucrose).

To make the sweetness of one teaspoon of sweetner powder equivalent to that of one teaspoon of sucrose, various fillers (things like cellulose) are added. These products also contain a number of other chemicals, including silicon dioxide ($SiO2$)"to prevent caking." This is the chemical name for sand or glass (although, from what I've heard, a bit of silicon is needed for healthy fingernails)! Pre-mixed, pre-packaged items, such as artificially-sweetened soft drinks, can contain all aspartame as their sweetener because the manufacturers of these products do not have to be concerned with the"one teaspoon" psychological factor.

Remember, because aspartame degrades into methanol at warm temperatures, it should not be used in anything hot or anything that will be cooked (coffee sweetener, cake mixes, cocoa, etc.), and bottles of artificially-sweetened soft drinks that have been sitting out in the hot sun at the local gas station are suspect.

THE DIVERSITY OF INTEGRAL MEMBRANE PROTEINS

Despite the diversity of integral membrane proteins, most fall into two general classes. One of these includes proteins attached or anchored to the membrane by only a small hydrophobic segment, such that most of the protein extends out into the water solvent on one or both sides of the membrane. The other class includes those proteins that are more or less globular in shape and more totally embedded in the membrane, exposing only a small surface to the water solvent outside the membrane. In general, those structures of integral membrane protein within the nonpolar core of the lipid bilayer are dominated by a-helices or ß-sheets because these secondary structures neutralize the highly polar N —H and C = O functions of the peptide backbone through H-bond formation.

PROTEIN WITH A SINGLE TRANSMEMBRANE SEGMENT

In the case of the proteins that are anchored by a small hydrophobic polypeptide segment, that segment often takes the form of a single a-helix. One of the best examples of a membrane protein with such an a-helical structure is glycophorin. Most of glycophorin's mass is oriented on the outside surface of the cell, exposed to the aqueous milieu.

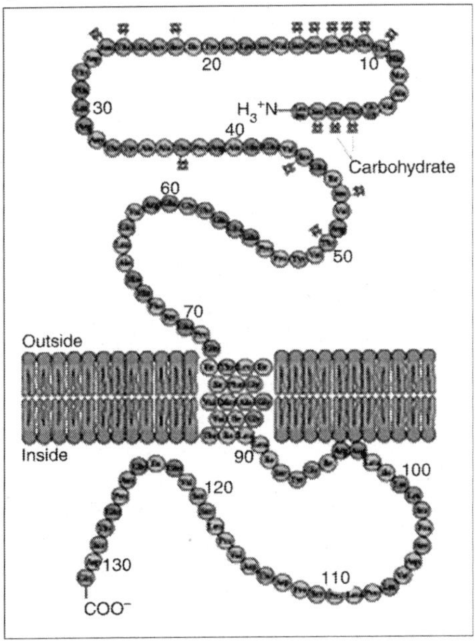

Fig. Glycophorin A Spans the Membrane of the Human Eerythrocyte via a Single a-helical Transmembrane Segment.

The C-terminus of the peptide, whose sequence is shown here, faces the cytosol of the erythrocyte; the N-terminal domain is extracellular. Points of attachment of carbohydrate groups are indicated.

A variety of hydrophilic oligosaccharide units are attached to this extracellular domain. These oligosaccharide groups constitute the ABO and MN blood group antigenic specificities of the red cell. This extracellular portion of the protein also serves as the receptor for the influenza virus. Glycophorin has a total molecular weight of about 31,000 and is approximately 40 per cent protein and 60 per cent carbohydrate. The glycophorin primary structure consists of a segment of 19 hydrophobic amino acid residues with a short hydrophilic sequence on one end and a longer hydrophilic sequence on the other end. The 19-residue sequence is just the right length to span the cell membrane if it is coiled in the shape of an a-helix. The large hydrophilic sequence includes the amino terminal residue of the polypeptide chain.

Numerous other membrane proteins are also attached to the membrane by means of a single hydrophobic a-helix, with hydrophilic segments extending into either the cytoplasm or the extracellular space. These proteins often function as receptors for extracellular molecules or as recognition sites that allow the immune system to recognize and distinguish the cells of the host organism from invading foreign cells or viruses. The proteins that represent the major transplantation antigens H2 in mice and human leukocyte associated (HLA) proteins in humans are members of this class. Other such proteins include thesurface immunoglobulin receptors on B lymphocytes and the spike proteins of many membrane viruses. The function of many of these proteins depends primarily on their extracellular domain, and thus the segment facing the intracellular surface is often a shorter one.

Bacteriorhodopsin: A 7-Transmembrane Segment Protein

Membrane proteins that take on a more globular shape, instead of the rodlike structure previously described, are often involved with transport activities and other functions requiring a substantial portion of the peptide to be embedded in the membrane. These proteins may consist of numerous hydrophobic a-helical segments joined by hinge regions so that the protein winds in a zig-zag pattern back and forth across the membrane. A well-characterized example of such a protein is bacteriorhodopsin, which clusters in purple patches in the membrane of the bacterium Halobacterium halobium.

The name Halobacterium refers to the fact that this bacterium thrives in solutions having high concentrations of sodium chloride, such as the salt beds of San Francisco Bay. Halobacterium carries out a light-driven proton transport by means of bacterio-rhodopsin, named in reference to its spectral similarities to rhodopsin in the rod outer segments of the mammalian retina. When this organism is deprived of oxygen for oxidative metabolism, it switches to the capture of energy from sunlight, using this energy to pump protons out of the cell. The proton gradient generated by such light-driven proton pumping represents potential energy, which is exploited elsewhere in the membrane to synthesize ATP.

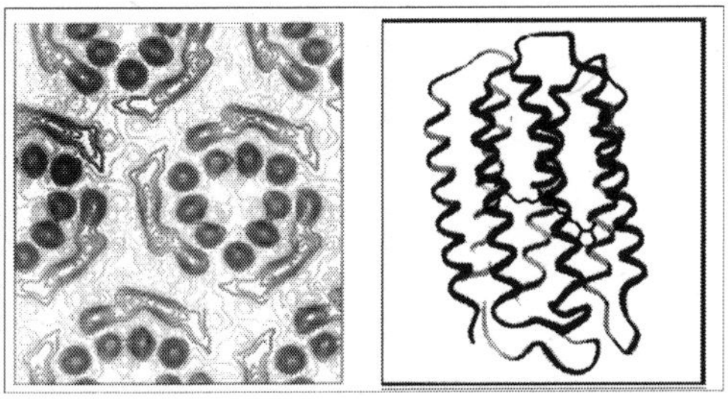

Fig. An Electron Density Profile Illustrating the Three Centers of Threefold Symmetry in Arrays of Bacterio-Rhodopsin in the Purple Membrane of Halobacterium Halobium, Together with a Computer-generated Model Showing the Seven a-helical Transmembrane Segments in Bacteriorhodopsin.

Bacteriorhodopsin clusters in hexagonal arrays in the purple membrane patches of Halobacterium, and it was this orderly, repeating arrangement of proteins in the membrane that enabled Nigel Unwin and Richard Henderson in 1975 to determine the bacteriorhodopsin structure. The polypeptide chain crosses the membrane seven times, in seven a-helical segments, with very little of the protein exposed to the aqueous milieu. The bacteriorhodopsin structure has become a model of globular membrane protein structure. Many other integral membrane proteins contain numerous hydro-phobic sequences that, like those of bacteriorhodopsin, could form a-helical transmembrane segments. For example, the amino acid sequence of the sodium - potassium transport ATPase contains ten hydrophobic segments of length sufficient to span the plasma membrane. By analogy with bacterio-rhodopsin, one would expect that these segments form a globular hydrophobic core that anchors the ATPase in the membrane. The helical segments may also account for the transport properties of the enzyme itself.

NATURE OF GLYCOPROTEINS

Many proteins found in nature are glycoproteins because they contain covalently linked oligo - and polysaccharide groups. The list of known glycoproteins includes structural proteins, enzymes, membrane receptors, transport proteins, and immunoglobulins, among others. In most cases, the precise function of the bound carbohydrate moiety is not understood. Carbohydrate groups may be linked to polypeptide chains via the hydroxyl groups of serine, threonine, or hydroxylysine residues (in O-linked saccharides) or via the amide nitrogen of an asparagine residue (in N-linked saccharides). The carbohydrate residue linked to the protein in O-linked saccharides is usually an N-acetylgalactosamine, but mannose, galactose, and xylose residues linked to protein hydroxyls are also found.

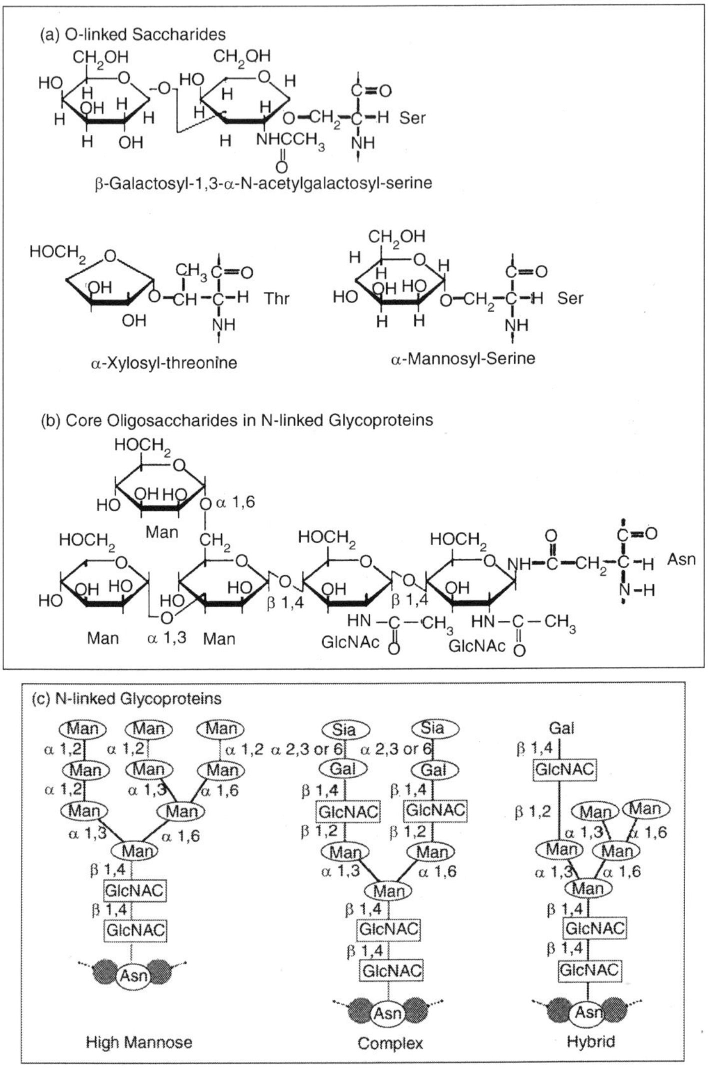

Fig. The Carbohydrate Moieties of Glycoproteins may be Linked to the Protein via (a) Serine or Threonine Residues (in the O-linked Saccharides) or (b) Asparagine Residues (in the N-linked Saccharides). (c) N-Linked Glycoproteins are of Three Types: high Mannose, Complex, and Hybrid, the Latter of which Combines Structures found in the High Mannose and Complex Saccharides.

Oligosaccharides O-linked to glycophorin involve N-acetylgalactosamine linkages and are rich in sialic acid residues. N-linked saccharides always have a unique core structure composed of two N-acetylglucosamine residues linked to a branched mannose triad. Many other sugar units may be linked to each of the mannose residues of this branched core. O-Linked saccharides are often found in cell surface glycoproteins and in mucins, the large glycoproteins that coat and protect mucous membranes in the respiratory and gastrointestinal tracts in the body. Certain viral glycoproteins also contain O-linked sugars. O-Linked saccharides in glycoproteins are often found clustered in richly glycosylated domains of the polypeptide chain. Physical studies on mucins show that they adopt rigid, extended structures so that an individual mucin molecule (Mr =107) may extend over a distance of 150 to 200 nm in solution. Inherent steric interactions between the sugar residues and the protein residues in these cluster regions cause the peptide core to fold into an extended and relatively rigid conformation.

This interesting effect may be related to the function of O-linked saccharides in glycoproteins. It allows aggregates of mucin molecules to form extensive, intertwined networks, even at low concentrations. These viscous networks protect the mucosal surface of the respiratory and gastrointestinal tracts from harmful environmental agents.

There appear to be two structural motifs for membrane glycoproteins containing O-linked saccharides. Certain glycoproteins, such as leukosialin,are O-glycosylated throughout much or most of their extracellular domain.

Leukosialin, like mucin, adopts a highly extended conformation, allowing it to project great distances above the membrane surface, perhaps protecting the cell from unwanted interactions with macromolecules or other cells. The second structural motif is exemplified by the low density lipoprotein (LDL) receptor and by decay accelerating factor (DAF).These proteins contain a highly O-glycosylated stem region that separates the transmembrane domain from the globular, functional extracellular domain. The O-glycosylated stem serves to raise the functional domain of the protein far enough above the membrane surface to make it accessible to the extracellular macromolecules with which it interacts.

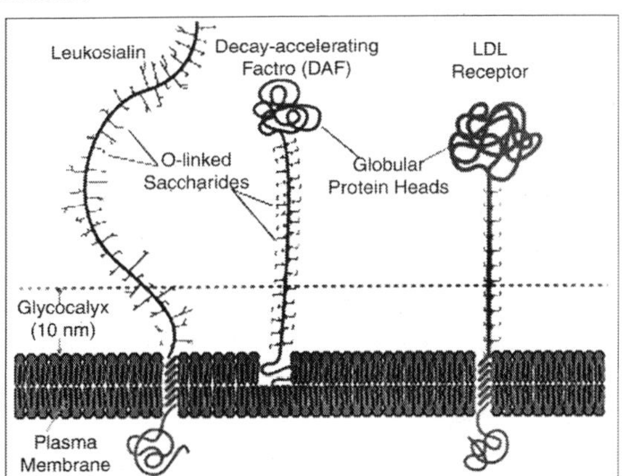

Fig. The O-linked Saccharides of Glycoproteins Appear in many Cases to Adopt Extended Conformations that Serve to Extend the Functional Domains of these Proteins above the Membrane Surface.

ANTIFREEZE GLYCOPROTEINS

A unique family of O-linked glycoproteins permits fish to live in the icy seawater of the Arctic and Antarctic regions where water temperature may reach as low as -1.9°C. Antifreeze glycoproteins (AFGPs) are found in the blood of nearly all Antarctic fish and at least five Arctic fish. These glycoproteins have the peptide structure

$$[\text{Ala-Ala-Thr}]_n\text{-Ala-Ala}$$

where n can be 4, 5, 6, 12, 17, 28, 35, 45, or 50. Each of the threonine residues is glycosylated with the disaccharide β-galactosyl -(1 → 3)-a-N-acetylgalactosamine. This glycoprotein adopts a flexible rod conformation with regions of threefold left-handed helix. The evidence suggests that antifreeze glycoproteins may inhibit the formation of ice in the fish by binding specifically to the growth sites of ice crystals, inhibiting further growth of the crystals.

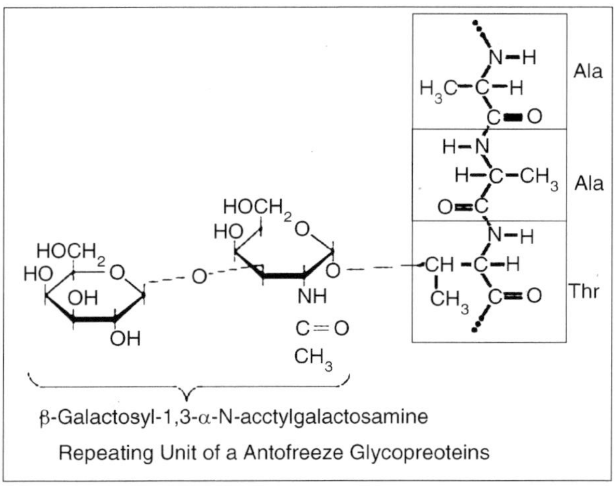

Fig. The Structure of the Repeating unit of Antifreeze Glycoproteins, a Disaccharide Consisting of β -galactosyl-(1e3)-a-N-acetylgalactosamine in Glycosidic Linkage to a Threonine Residue.

N-LINKED OLIGOSACCHARIDES

N-Linked oligosaccharides are found in many different proteins, including immunoglobulins G and M, ribonuclease B, ovalbumin, and peptide hormones. Many different functions are known or suspected for N-glycosylation of proteins. Glycosylation can affect the physical and chemical properties of proteins, altering solubility, mass, and electrical charge.

Carbohydrate moieties have been shown to stabilize protein conformations and protect proteins against proteolysis. Eukaryotic organisms use posttranslational additions of N-linked oligosaccharides to direct selected proteins to various intracellular organelles.

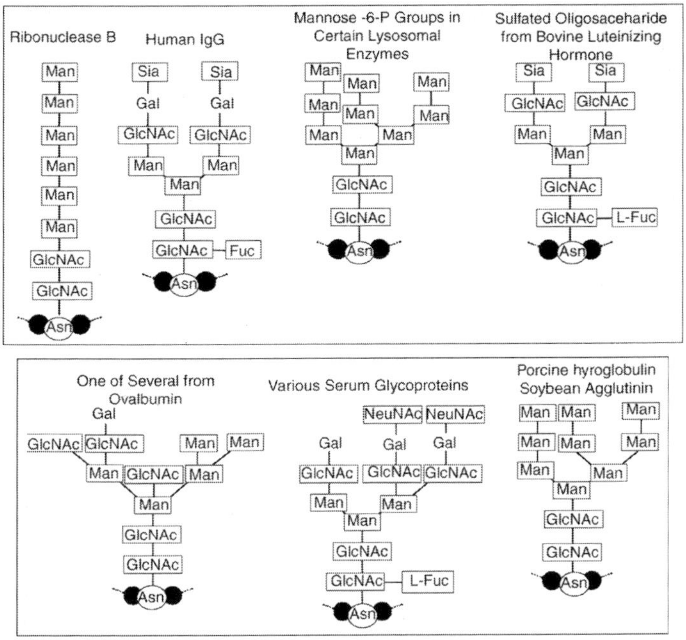

Fig. Some of the Oligosaccharides Found in N-linked Glycoproteins.

OLIGOSACCHARIDE CLEAVAGE AS A TIMING DEVICE FOR PROTEIN DEGRADATION

The slow cleavage of monosaccharide residues from N-linked glycoproteins circulating in the blood targets these proteins for degradation by the organism. The liver contains specific receptor proteins that recognize and bind glycoproteins that are ready to be degraded and recycled. Newly synthesized serum glycoproteins contain N-linked triantennary (three-chain) oligosaccharides having structures similar to those in Figure, in which sialic acid residues cap galactose residues.

As these glycoproteins circulate, enzymes on the blood vessel walls cleave off the sialic acid groups, exposing the galactose residues. In the liver, the asialoglycoprotein receptor binds the exposed galactose residues of these glycoproteins with very high affinity ($K_D = 10^{-9}$ to 10^{-8}M). The complex of receptor and glycoprotein is then taken into the cell by endocytosis, and the glycoprotein is degraded in cellular lysosomes. Highest affinity binding of glycoprotein to the asialoglycoprotein receptor requires three free galactose residues.

Oligosaccharides with only one or two exposed galactose residues bind less tightly. This is an elegant way for the body to keep track of how long glycoproteins have been in circulation. Over a period of time, anywhere from a few hours to weeks, the sialic acid groups are cleaved one by one. The longer the glycoprotein circulates and the more sialic acid residues are removed, the more galactose residues become exposed so that the glycoprotein is eventually bound to the liver receptor.

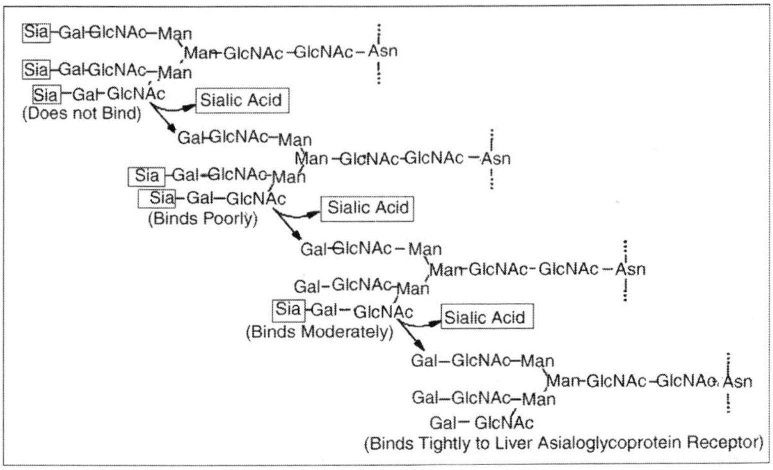

Fig. Progressive Cleavage of Sialic Acid Residues Exposes Galactose Residues. Binding to the Asialoglycoprotein Receptor in the Liver becomes Progressively more Likely as more Gal Residues are Exposed.

MODELS FOR THE ALLOSTERIC BEHAVIOUR OF PROTEINS

In 1965, Jacques Monod, Jeffries Wyman, and Jean-Pierre Changeux proposed a theoretical model of allosteric transitions based on the observation that allosteric proteins are oligomers. They suggested that allosteric proteins can exist in (at least) two conformational states, designated R, signifying "relaxed," and T, or "taut," and that, in each protein molecule, all of the subunits have the same conformation (either R or T). That is, molecular symmetry is conserved. Molecules of mixed conformation (havi ng subunits of both R and T states) are not allowed by this model.

In the absence of ligand, the two states of the allosteric protein are in equilibrium:

$$R_0 \leftrightharpoons T_0$$

(Note that the subscript "0" signifies "in the absence of ligand.") The equilibrium constant is termed L: L = T_0/R_0. L is assumed to be large; that is, the amount of the protein in the T conformational state is much greater than the amount in the R conformation. Let us suppose that L = 104.

The affinities of the two states for substrate, S, are characterized by the respective dissociation constants, KR and K_T. The model supposes that $K_T \gg K_R$. That is, the affinity of R_0 for S is much greater than the affinity of T0 for S. Let us choose the extreme where $K_R/K_T = 0$ (that is, K_T is infinitely greater than K_R). In effect, we are picking conditions where S binds only to R. (If K_T is infinite, T does not bind S.) Given these parameters, consider what happens when S is added to a solution of the allosteric protein at conformational equilibrium. Although the relative [R0] concentration is small, S will bind "only" to R0, forming R1. This depletes the concentration of R0, perturbing the T_0/R_0 equilibrium.

To restore equilibrium, molecules in the T_0 conformation undergo a transition to R0. This shift renders more R_0 available to bind S, yielding R1, diminishing $[R_0]$, perturbing the T_0/R_0 equilibrium, and so on. Thus, these linked equilibria are such that S-binding by the R_0 state of the allosteric protein perturbs the T_0/R_0 equilibrium with the result that S-binding drives the conformational transition, $T_0 \rightarrow R_0$.

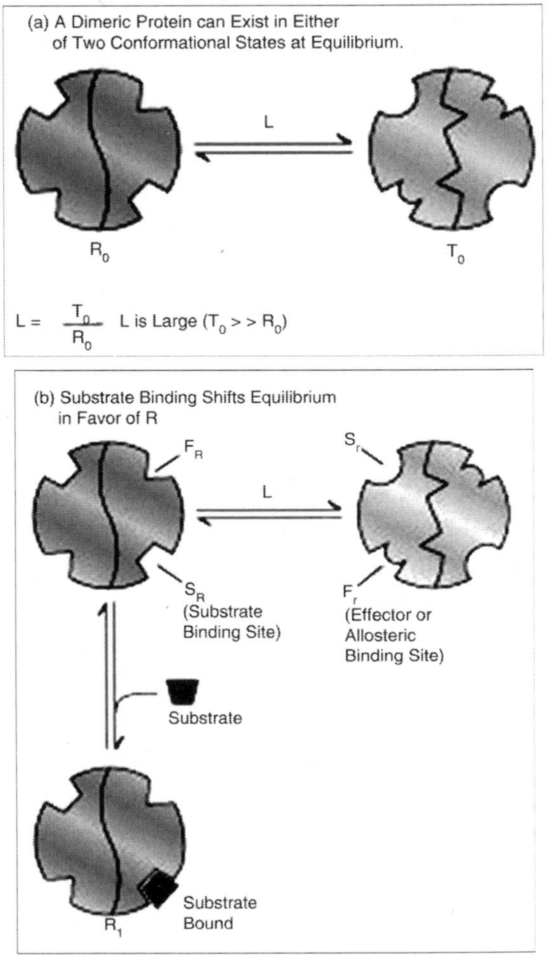

Fig. Monod-Wyman-Changeux (MWC) Model for Allosteric Transitions.

Consider a dimeric protein that can exist in either of two conformational states, R or T. Each subunit in the dimer has a binding site for substrate S and an allosteric effector site, F. The promoters are symmetrically related to one another in the protein, and symmetry is conserved regardless of the conformational state of the protein. The different states of the protein, with or without bound ligand, are linked to one another through the various equilibria. Thus, the relative population of protein molecules in the R or T state is a function of these equilibria and the concentration of the various ligands, substrate (S), and effectors (which bind at F_R or F_T). As [S] is increased, the T/R equilibrium shifts in favour of an increased proportion of R-conformers in the total population (that is, more protein molecules in the R conformational state).

In just this simple system, cooperativity is achieved because each subunit has a binding site for S, and thus, each protein molecule has more than one binding site for S. Therefore, the increase in the population of R conformers gives a progressive increase in the number of sites available for S. The extent of cooperativity depends on the relative T0/R0 ratio and the relative affinities of R and T for S. If L is large (that is, the equilibrium lies strongly in favour of T0) and if KT >> KR, as in the example we have chosen, cooperativity is great. Ligands such as S here that bind in a cooperative manner, so that binding of one equivalent enhances the binding of additional equivalents of S to the same protein molecule, are termed positive homotropic effectors. (The prefix "homo" indicates that the ligand influences the binding of like molecules.)

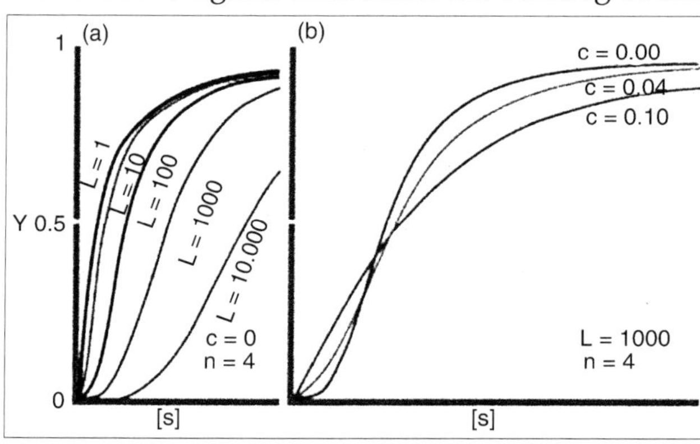

Fig. The Monod - Wyman - Changeux Model.

Graphs of allosteric effects for a tetramer (n = 4) in terms of Y, the saturation function, versus [S]. Y is defined as [ligand-binding sites that are occupied by ligand]/[total ligand-binding sites]. (a) A plot of Y as a function of [S], at various L values. (b) Y as a function of [S], at different c, where c = K_R/K_T. (When c = 0, K_T is infinite.)

HETEROTROPIC EFFECTORS

This simple system also provides an explanation for the more complex substrate-binding responses to positive and negative effectors. Effectors that influence the binding of something other than themselves are termed heterotropic effectors. For example, effectors that promote S binding are termed positive heterotropic effectors or allosteric activators. Effectors that diminish S binding are negative heterotropic effectors or allosteric inhibitors. Feedback inhibitors fit this class. Consider a protein composed of two subunits, each of which has two binding sites: one for the substrate, S, and one to which allosteric effectors bind, the allosteric site. Assume that S binds preferentially ("only") to the R conformer; further assume that the positive heterotropic effector, A, binds to the allosteric site only when the protein is in the R conformation, and thenegative allosteric effector, I, binds at the allosteric site only if the protein is in the T conformation. Thus, with respect to binding at the allosteric site, A and I are competitive with each other.

POSITIVE EFFECTORS

If A binds to R0, forming the new species R1(A), the relative concentration of R0 is decreased and the T0/R0 equilibrium is perturbed. As a consequence, a relative $T_0 \rightarrow R_0$ shift occurs in order to restore equilibrium. The net effect is an increase in the number of R conformers in the presence of A, meaning that more binding sites for S are available. For this reason, A leads to a decrease in the cooperativity of the substrate saturation curve, as seen by a shift of this curve to the left. Effectively, the presence of A lowers the apparent value of L.

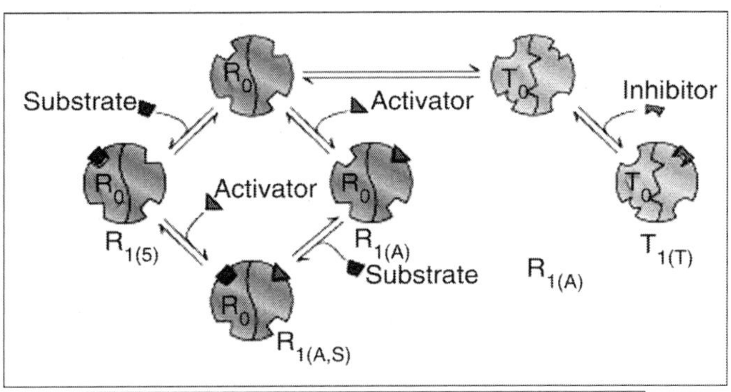

A Dimeric Protein which can Exist in Either of Two States R_0 and T_0.
This Protein can Bind 3 LIgands:

1. Substrate (S) : A Positive Homotropic Effector that Binds Only to R at Site S
2. Activator (A) : A Positive Heterotropic Effector that Binds Only to R at Site F
3. Inhibitor (I) : A Negative Heterotropic Effector that Binds Only to T a Site F

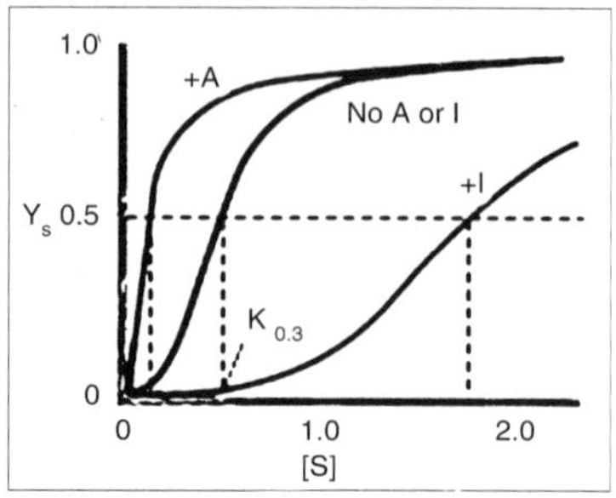

Fig. Heterotropic Allosteric Effects: A and I binding to R and T, Respectively.

Note: Effects of A: $A + R_0 \rightarrow R_{1(A)}$

Increase in number of R-conformers shifs $R_0 \rightleftharpoons T_0$ so that $T_0 \rightarrow R_0$

1. More binding sites for S made available
2. Decrease in cooperativity of substrate saturation curve. Effector A lowers the apparent value of L.

   Effects of I: I + T0 → T1(T)

Increase in number of T-conformers (Decrease in R_0 as $R_0 \rightarrow T_0$ to restore equilibrium). Thus , I inhibits association of S and A with R by lowering R_0 level. I Increase cooperativity of substrate saturation curve. I raises the apparent value of L.

The linked equilibria lead to changes in the relative amounts of R and T and, therefore, shifts in the substrate saturation curve. This behaviour, depicted by the graph, defines an allosteric "K" system. The parameters of such a system are: (1) S and A (or I) have different affinities for R and T and (2) A (or I) modifies the apparent K0.5 for S by shifting the relative R versus T population.

NEGATIVE EFFECTORS

The converse situation applies in the presence of I, which binds "only" to T. I-binding will lead to an increase in the population of T conformers, at the expense of R0. The decline in [R0] means that it is less likely for S (or A) to bind. Consequently, the presence of I increases the cooperativity (that is, the sigmoidicity) of the substrate saturation curve, as evidenced by the shift of this curve to the right. The presence of I raises the apparent value of L.

K SYSTEMS AND V SYSTEMS

The allosteric model just presented is called a K system because the concentration of substrate giving half-maximal velocity, defined as K0.5, changes in response to effectors. Note that Vmax is constant in this system.

An allosteric situation where $K_{0.5}$ is constant but the apparent Vmax changes in response to effectors is termed a V system. In a V system, all v versus S plots are hyperbolic rather than sigmoid. The positive heterotropic effector A activates by raising Vmax, whereas I, the negative heterotropic effector, decreases it. Note that neither A nor I affects $K_{0.5}$.

This situation arises if R and T have the same affinity for the substrate, S, but differ in their catalytic ability and their affinities for A and I. A and I thus can shift the relative T/R distribution. Acetyl-coenzyme A carboxylase, the enzyme catalyzing the committed step in the fatty acid biosynthetic pathway, behaves as a V system in response to its allosteric activator, citrate.

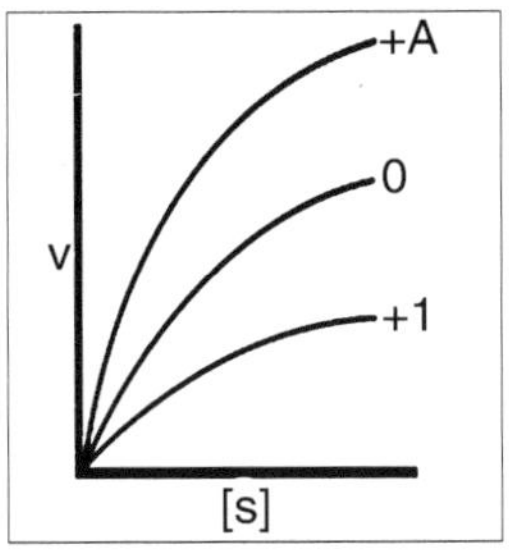

Fig. v Versus [S] Curves for an Allosteric "V" System.

The V system fits the model of Monod, Wyman, and Changeux, given the following conditions:

(1) R and T have the same affinity for the substrate, S.

(2) The effectors A and I have different affinities for R and T and thus can shift the relative T/R distribution. (That is, A and I change the apparent value of L.) Assume as before that A binds "only" to the R state and I binds "only" to the T state.

(3) R and T differ in their catalytic ability. Assume that R is the enzymatically active form, whereas T is inactive. Because A perturbs the T/R equilibrium in favour of more R, A increases the apparent Vmax. I favours transition to the inactive T state.

K SYSTEMS AND V SYSTEMS FILL DIFFERENT BIOLOGICAL ROLES

The K and V systems have design features that mean they work best under different physiological situations. "K system" enzymes are adapted to conditions in which the prevailing substrate concentration is rate-limiting, as when [S] in vivo ? K0.5. On the other hand, when the physiological conditions are such that [S] is usually saturating for the regulatory enzyme of interest, the enzyme conforms to the "V system" mode in order to have an effective regulatory response.

HUMAN BIOCHEMISTRY

Allergies represent overreactions of the immune system caused by exposure to foreign substances referred to as allergens. The inhalation of allergens, such as pollen, pet dander, and dust, can cause a variety of allergic responses, including itchy eyes, a runny nose, shortness of breath, and wheezing. Allergies can also be caused by food, drugs, dyes, and other chemicals. The visible symptoms of such an allergic response are caused by the release of histamine by mast cells, a type of cell found in loose connective tissue. Histamine dilates blood vessels, increases the permeability of capillaries (allowing antibodies to pass from the capillaries to surrounding tissue), and constricts bronchial air passages. Histamine acts by binding to specialized membrane proteins called histamine H1 receptors.

These integral membrane proteins possess seven transmembrane a-helical segments, with an extracellular amino terminus and a cytoplasmic carboxy terminus. When histamine binds to the extracellular domain of an H1 receptor, the intracellular domain undergoes a conformation change that stimulates a GTP-binding protein, which in turn activates the allergic response in the affected cell.

A variety of highly effective antihistamine drugs are availablefor the treatment of allergy symptoms. These drugs share the property of binding tightly to histamine H1 receptors, without eliciting the same effects as histamine itself.

They are referred to as histamine H1 receptor antagonists because they prevent the binding of histamine to the receptors. The structures of Allegra (made by Hoechst Marion Roussel, Inc.), Claritin (by Schering-Plough Corp.), and Zyrtec (by Pfizer) are all shown at right.

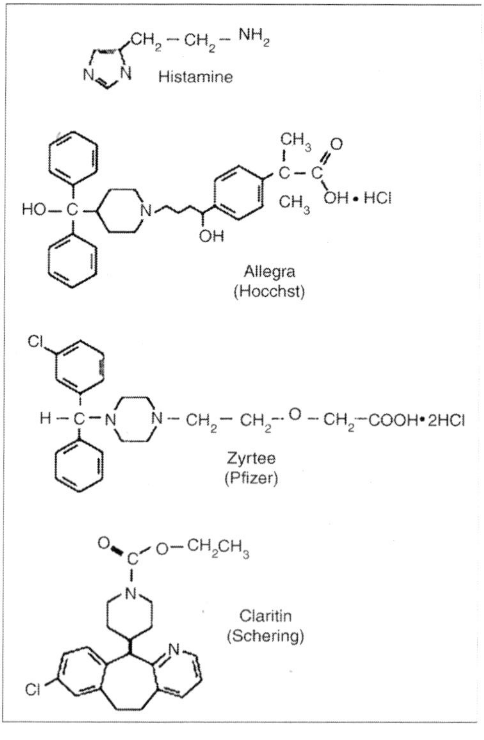

Porins — A β-Sheet Motif for Membrane Proteins

The β-sheet is another structural motif that provides extensive hydrogen bonding for transmembrane peptide segments. Porin proteins found in the outer membranes (OM) of Gram-negative bacteria such as Escherichia coli, and also in the outer mitochondrial membranes of eukaryotic cells, span their respective membranes with large ß-sheets. A good example is maltoporin, also known as LamB protein orlambda receptor, which participates in the entry of maltose and maltodextrins into E. coli. Maltoporin is active as a trimer. The 421-residue monomer is an aesthetically pleasing 18-strand β-barrel.

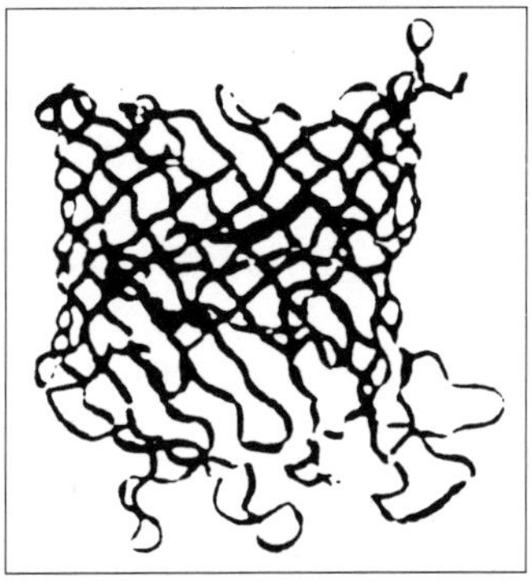

Fig. The Three-dimensional Structure of Maltoporin from E. coli.

The ß-strands are connected to their nearest neighbours either by long loops or by β-turns. The long loops are found at the end of the barrel that is exposed to the cell exterior, whereas the turns are located on the intracellular face of the barrel. Three of the loops fold into the center of the barrel.

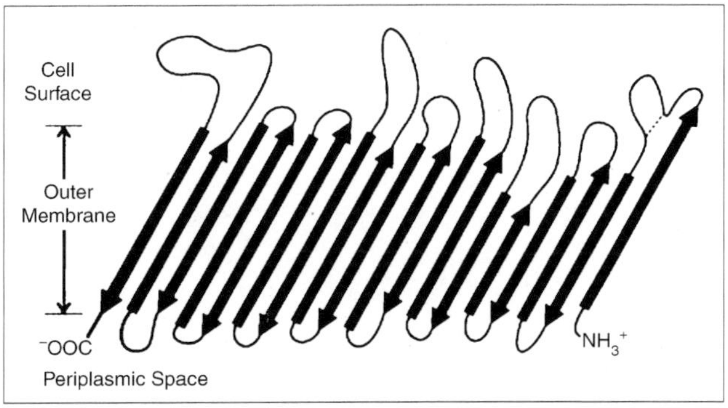

Fig. The Arrangement of the Peptide Chain in Maltoporin from E. coli.

The amino acid compositions and sequences of the β-strands in porin proteins are novel. Polar and nonpolar residues alternate along the ß-strands, with polar residues facing the central pore or cavity of the barrel and non-polar residues facing out from the barrel where they can interact with the hydrophobic lipid milieu of the membrane. The smallest diameter of the porin channel is about 5 Å. Thus, a maltodextrin polymer (composed of two or more glucose units) must pass through the porin in an extended conformation (like a spaghetti strand).

LIPID-ANCHORED MEMBRANE PROTEINS

Certain proteins are found to be covalently linked to lipid molecules. For many of these proteins, covalent attachment of lipid is required for association with a membrane. The lipid moieties can insert into the membrane bilayer, effectively anchoring their linked proteins to the membrane. Some proteins with covalently linked lipid normally behave as soluble proteins; others are integral membrane proteins and remain membrane-associated even when the lipid is removed. Covalently bound lipid in these latter proteins can play a role distinct from membrane anchoring. In many cases, attachment to the membrane via the lipid anchor serves to modulate the activity of the protein.

Control of biological pests, including mosquitoes, houseflies, gnats, and tree-consuming predators like the eastern tent caterpillar, is frequently achieved through the use of microbial membrane proteins.

For example, several varieties of Bacillus thurigiensis produce proteins that bind to cell membranes in the digestive systems of insects that consume them, creating transmembrane ion channels. Leakage of Na+, K+, and H+ ions through these membranes in the insect gut destroys crucial ion gradients and interferes with digestion of food. Insects that ingest these toxins eventually die of starvation.B. thurigiensis toxins account for more than 90 per cent of sales of biological pest control agents.

B. thurigiensis is a common Gram-positive, spore-forming soil bacterium that produces inclusion bodies, microcrystalline clusters of many different proteins. These crystalline proteins, called d-endotoxins, are the ion channel toxins that are sold commercially for pest control. Most such endotoxins are protoxins, which are inactive until cleaved to smaller, active proteins by proteases in the gut of a susceptible insect. One such crystalline protoxin, lethal to mosquitoes, is a 27 kD protein, which is cleaved to form the active 25 kD toxin in the mosquito. This toxin has no effect on membranes at neutral pH, but at pH 9.5 (the pH of the mosquito gut) the toxin forms cation channels in the gut membranes.

This 25 kD protein is not toxic to tent caterpillars, but a larger, 130 kD protein in the B. thurigiensis inclusion bodies is cleaved by a caterpillar gut protease to produce a 55 kD toxin that is active in the caterpillar. Remarkably, the strain of B. thurigiensis known asazawai produces a protoxin with dual specificity: in the caterpillar gut, this 130 kD protein is cleaved to form a 55 kD toxin active in the caterpillar. However, when the same 130 kD protoxin is consumed by mosquitoes or houseflies, it is cleaved to form a 53 kD protein (15 amino acid residues shorter than the caterpillar toxin) that is toxic to these latter organisms. Under-standing the molecular basis of the toxicity and specificity of these proteins and the means by which they interact with membranes to form lethal ion channels is a fascinating biochemical challenge with far-reaching commercial implications.

Another interesting facet of lipid anchors is that they are transient. Lipid anchors can be reversibly attached to and detached from proteins. This provides a "switching device" for altering the affinity of a protein for the membrane. Reversible lipid anchoring is one factor in the control of signal transduction pathways in eukaryotic cells.

Four different types of lipid-anchoring motifs have been found to date. These are amide-linked myristoyl anchors, thioester -linked fatty acyl anchors, thioether -linked prenyl anchors, and amide-linked glycosyl phosphatidylinositol anchors. Each of these anchoring motifs is used by a variety of membrane proteins, but each nonetheless exhibits a characteristic pattern of structural requirements.

AMIDE-LINKED MYRISTOYL ANCHORS

Myristic acid may be linked via an amide bond to the a-amino group of the N-terminal glycine residue of selected proteins. The reaction is referred to as N-myristoylation and is catalyzed by myristoyl - CoA:protein N-myristoyltransferase, known simply asNMT. N-Myristoyl -anchored proteins include the catalytic subunit of cAMP -dependent protein kinase, the pp60src tyrosine kinase, the phosphatase known as calcineurin B, the a-subunit of G proteins (involved in GTP-dependent transmembrane signaling events), and thegag proteins of certain retroviruses, including the HIV-1 virus that causes AIDS.

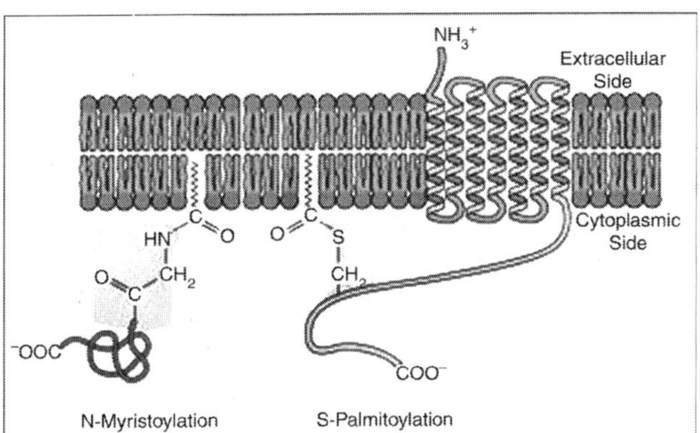

Fig. Certain Proteins are Anchored to Biological Membranes by Lipid Anchors.

Thioester -Linked Fatty Acyl Anchors

A variety of cellular and viral proteins contain fatty acids covalently bound via ester linkages to the side chains of cysteine and sometimes to serine or threonine residues within a polypeptide chain. This type of fatty acyl chain linkage has a broader fatty acid specificity than N-myristoylation. Myristate, palmitate, stearate, and oleate can all be esterified in this way, with the C16 and C18 chain lengths being most commonly found. Proteins anchored to membranes via fatty acyl thioesters include G-protein-coupled receptors, thesurface glycoproteins of several viruses, and the transferrin receptor protein.

Thioether -Linked Prenyl Anchors

As noted, polyprenyl (or simply prenyl) groups are long-chain polyisoprenoid groups derived from isoprene units. Prenylation of proteins destined for membrane anchoring can involve either farnesyl or geranylgeranyl groups. The addition of a prenyl group typically occurs at the cysteine residue of a carboxy -terminal CAAX sequence of the target protein, where C is cysteine, A is any aliphatic residue, and X can be any amino acid. As shown in Figure, the result is a thioether -linked farnesyl or geranylgeranyl group. Once the prenylation reaction has occurred, a specific protease cleaves the three carboxy -terminal residues, and the carboxyl group of the now terminal Cys is methylated to produce an ester.

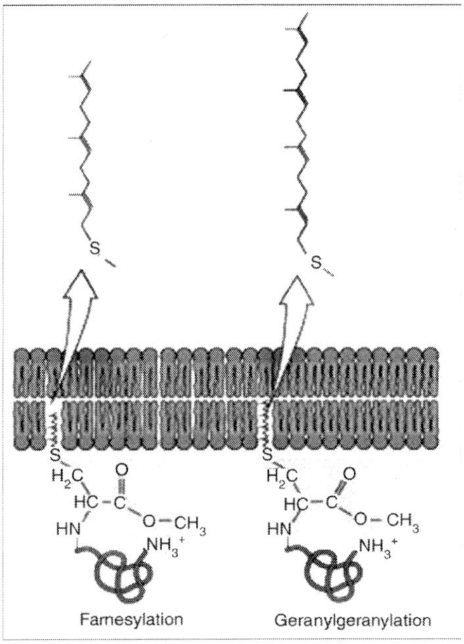

Fig. Proteins Containing the C-terminal Sequence CAAX can Undergo Prenylation Reactions that Place Thioether-linked Farnesyl or Geranylgeranyl Groups at the Cysteine side Chain.

Prenylation is accompanied by removal of the AAX peptide and methylation of the carboxyl group of the cysteine residue, which has become the C-terminal residue. All of these modifications appear to be important for subsequent activity of the prenyl -anchored protein. Proteins anchored to membranes via prenyl groups include yeast mating factors, the p21ras protein, and the nuclear lamins, structural components of the lamina of the inner nuclear membrane.

Glycosyl Phosphatidylinositol Anchors

Glycosyl phosphatidylinositol, or GPI, groups are structurally more elaborate membrane anchors than fatty acyl or prenyl groups. GPI groups modify the carboxy -terminal amino acid of a target protein via an ethanolamine residue linked to an oligosaccharide, which is linked in turn to the inositol moiety of a phosphatidylinositol. The oligosaccharide typically consists of a conserved tetras-accharide core of three mannose residues and a glucosamine, which can be altered by modifications of the mannose residues or addition of galactosyl side chains of various sizes, extra phosphoethanolamines, or additional N-acetylgalactose or mannosyl residues.

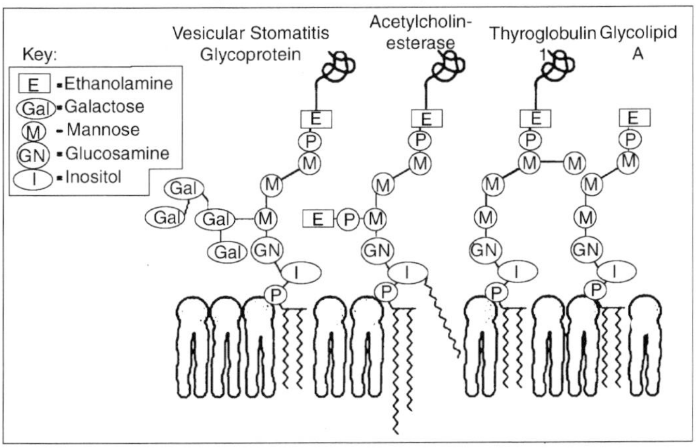

Fig. The Glycosyl Phosphatidylinositol (GPI) Moiety is an Elaborate Lipid-anchoring Group.

Note the core of three mannose residues and a glucosamine. Additional modifications may include fatty acids at the inositol and glycerol OOH groups.

The inositol moiety can also be modified by an additional fatty acid, and a variety of fatty acyl groups are found linked to the glycerol group. GPI groups anchor a wide variety of surface antigens, adhesion molecules, and cell surface hydrolases to plasma membranes in various eukaryotic organisms. GPI anchors have not yet been observed in prokaryotic organisms or plants.

STRUCTURE OF PROTEINS

This explains how amino acids combine to make proteins and what is meant by the primary, secondary and tertiary structures of proteins. Quaternary structure isn't covered. It only applies to proteins consisting of more than one polypeptide chain. There is a mention of quaternary structure on the IB chemistry syllabus, but on no other UK-based syllabus at this level.

Quaternary structure can be very complicated, and I don't know exactly what depth the IB syllabus wants for this (which is why I haven't included it). I suspect what is wanted is fairly trivial. IB students should ask the advice of their teacher or lecturer.

PRIMARY STRUCTURE OF PROTEINS

Drawing the Amino Acids

In chemistry, if you were to draw the structure of a general 2-amino acid, you would probably draw it like this:

$$\begin{array}{c} NH_2 \\ | \\ R-CH-COOH \end{array}$$

However, for drawing the structures of proteins, we usually twist it so that the"R" group sticks out at the side. It is much easier to see what is happening if you do that.

$$\begin{array}{l} R \\ | \\ NH2-CH-COOH \end{array}$$

That means that the two simplest amino acids, glycine and alanine, would be shown as:

$$\begin{array}{c} H \\ | \\ NH_2-CH-COOH \\ \text{Glycine} \end{array}$$

$$\begin{array}{c} CH_3 \\ | \\ NH_2-CH-COOH \\ \text{alanine} \end{array}$$

PEPTIDES AND POLYPEPTIDES

Glycine and alanine can combine together with the elimination of a molecule of water to produce a dipeptide. It is possible for this to happen in one of two different ways - so you might get two different dipeptides.

Either:

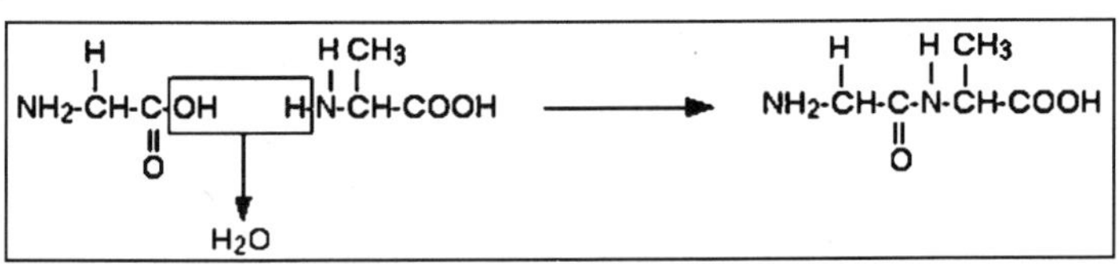

Or:

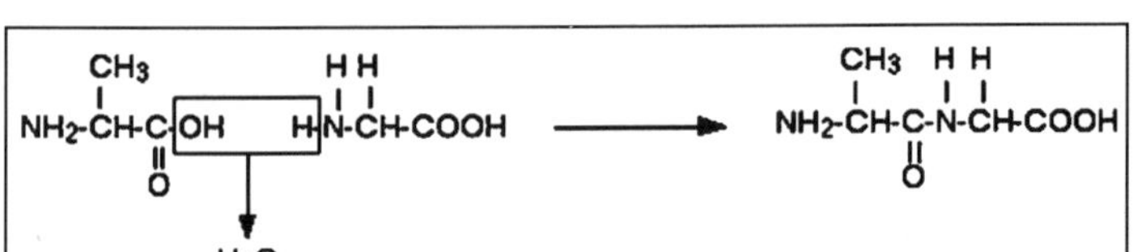

In each case, the linkage shown in blue in the structure of the dipeptide is known as a peptide link. In chemistry, this would also be known as an amide link, but since we are now in the realms of biochemistry and biology, we'll use their terms.

If you joined three amino acids together, you would get a tripeptide. If you joined lots and lots together (as in a protein chain), you get a polypeptide.

A protein chain will have somewhere in the range of 50 to 2000amino acid residues. You have to use this term because strictly speaking a peptide chain isn't made up of amino acids. When the amino acids combine together, a water molecule is lost. The peptide chain is made up from what is left after the water is lost - in other words, is made up of amino acid residues. By convention, when you are drawing peptide chains, the -NH_2group which hasn't been converted into a peptide link is written at the left-hand end. The unchanged –COOH group is written at the right-hand end. The end of the peptide chain with the –NH_2 group is known as theN-terminal, and the end with the –COOH group is the C–

Terminal

A protein chain (with the N-terminal on the left) will therefore look like this:

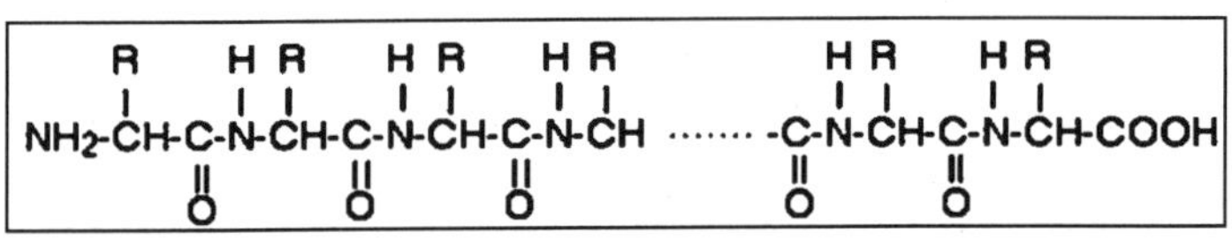

The"R" groups come from the 20 amino acids which occur in proteins. The peptide chain is known as the backbone, and the"R" groups are known as side chains. Note: In the case where the"R" group comes from the amino acid proline, the pattern is broken. In this case, the hydrogen on the nitrogen nearest the"R" group is missing, and the"R" group loops around and is attached to that nitrogen as well as to the carbon atom in the chain.

I mention this for the sake of completeness - not because you would be expected to know about it in chemistry at this introductory level.

At its simplest, the term is used to describe the order of the amino acids joined together to make the protein. In other words, if you replaced the"R" groups in the last diagram by real groups you would have the primary structure of a particular protein. This primary structure is usually shown using abbreviations for the amino acid residues. These abbreviations commonly consist of three letters or one letter.

Using three letter abbreviations, a bit of a protein chain might be represented by, for example:

Try.Gly.Lys.pro.Val.Gly. Lys.Lys.Arg.Arg.Pro.Val.Lys.
Tyr.Pro.Ala.Gly.Glu...

If you look carefully, you will spot the abbreviations for glycine (Gly) and alanine (Ala) amongst the others.

If you followed the protein chain all the way to its left-hand end, you would find an amino acid residue with an unattached $-NH_2$ group. The N-terminal is always written on the left of a diagram for a protein's primary structure - whether you draw it in full or use these abbreviations.

The wider definition of primary structure includes all the features of a protein which are a result of covalent bonds. Obviously, all the peptide links are made of covalent bonds, so that isn't a problem.

But there is an additional feature in proteins which is also covalently bound. It involves the amino acid cysteine.

$$\begin{array}{c} CH_2SH \\ | \\ NH_2-CH-COOH \end{array}$$

Cysteine

If two cysteine side chains end up next to each other because of folding in the peptide chain, they can react to form a sulphur bridge. This is another covalent link and so some people count it as a part of the primary structure of the protein.

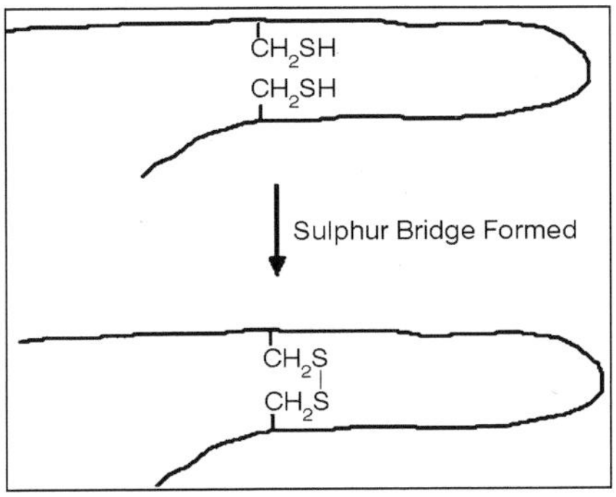

Because of the way sulphur bridges affect the way the protein folds, other people count this as a part of the tertiary structure. This is obviously a potential source of confusion!

Important: You need to know where your particular examiners are going to include sulphur bridges - as a part of the primary structure or as a part of the tertiary structure. You need to check your current syllabus and past papers. If you are studying a UK-based syllabus and haven't got these, follow this link to find out how to get hold of them.

Secondary Structure of Proteins

Within the long protein chains there are regions in which the chains are organised into regular structures known as alpha-helices (alpha-helixes) and beta-pleated sheets. These are the secondary structures in proteins.

These secondary structures are held together by hydrogen bonds. These form as shown in the diagram between one of the lone pairs on an oxygen atom and the hydrogen attached to a nitrogen atom:

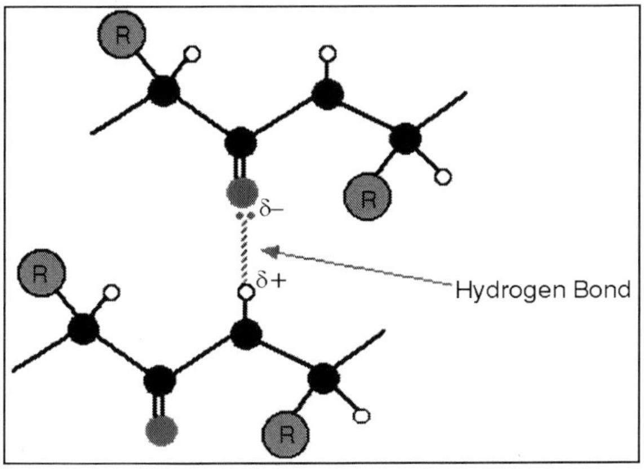

Important: If you aren't happy about hydrogen bonding and are unsure about what this diagram means, follow this link before you go on.

What follows is difficult enough to visualise anyway without having to worry about what hydrogen bonds are as well!

You must also find out exactly how much detail you need to know about this next bit. It may well be that all you need is to have heard of an alpha-helix and know that it is held together by hydrogen bonds between the C=O and N-H groups. Once again, you need to check your syllabus and past papers - particularly mark schemes for the past papers.

The Alpha-helix

In an alpha-helix, the protein chain is coiled like a loosely-coiled spring. The"alpha" means that if you look down the length of the spring, the coiling is happening in a clockwise direction as it goes away from you.

Note: If your visual imagination is as hopeless as mine, the only way to really understand this is to get a bit of wire and coil it into a spring shape. The lead on your computer mouse is fine for doing this!

The next diagram shows how the alpha-helix is held together by hydrogen bonds. This is a very simplified diagram, missing out lots of atoms. We'll talk it through in some detail after you have had a look at it.

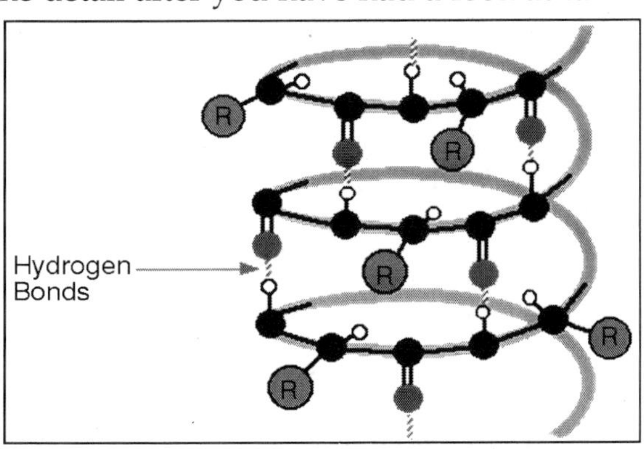

What's wrong with the diagram? Two things: First of all, only the atoms on the parts of the coils facing you are shown. If you try to show all the atoms, the whole thing gets so complicated that it is virtually impossible to understand what is going on. So, what do you need to notice? Notice that all the"R" groups are sticking out sideways from the main helix. Notice the regular arrangement of the hydrogen bonds. All the N-H groups are pointing upwards, and all the C=O groups pointing downwards. Each of them is involved in a hydrogen bond. And finally, although you can't see it from this incomplete diagram, each complete turn of the spiral has 3.6 (approximately) amino acid residues in it. If you had a whole number of amino acid residues per turn, each group would have an identical group underneath it on the turn below. Hydrogen bonding can't happen under those circumstances. Each turn has 3 complete amino acid residues and two atoms from the next one. That means that each turn is offset from the ones above and below, such that the N-H and C=O groups are brought into line with each other.

Beta-pleated Sheets

In a beta-pleated sheet, the chains are folded so that they lie alongside each other. The next diagram shows what is known as an"anti-parallel" sheet.

All that means is that next-door chains are heading in opposite directions. Given the way this particular folding happens, that would seem to be inevitable.

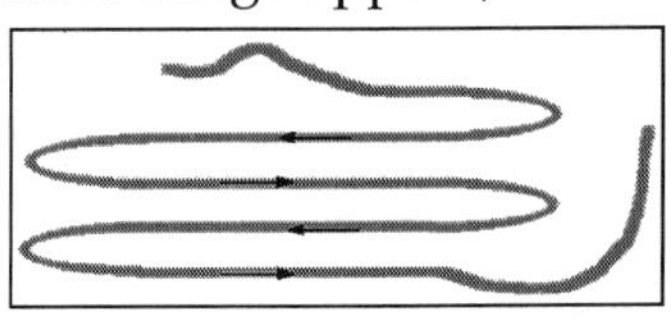

It isn't, in fact, inevitable! It is possible to have some much more complicated folding so that next-door chains are actually heading in the same direction. We are getting well beyond the demands of UK A level chemistry (and its equivalents) now. The folded chains are again held together by hydrogen bonds involving exactly the same groups as in the alpha-helix.

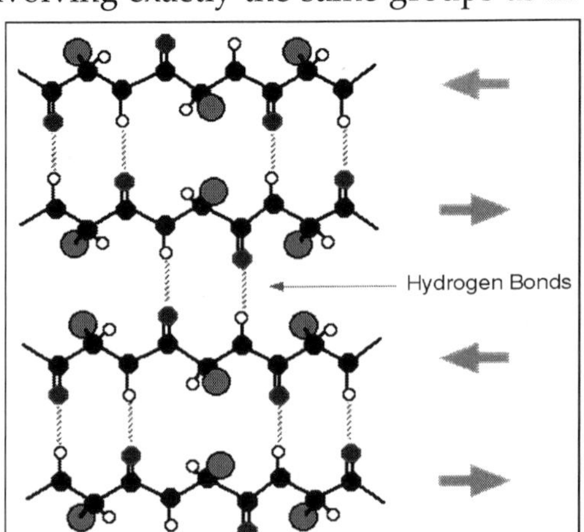

Note: Note that there is no reason why these sheets have to be made from four bits of folded chain alongside each other as shown in this diagram. That was an arbitrary choice which produced a diagram which fitted nicely on the screen!

TERTIARY STRUCTURE OF PROTEINS

Definition

The tertiary structure of a protein is a description of the way the whole chain (including the secondary structures) folds itself into its final 3-dimensional shape. This is often simplified into models like the following one for the enzyme dihydrofolate reductase. Enzymes are, of course, based on proteins.

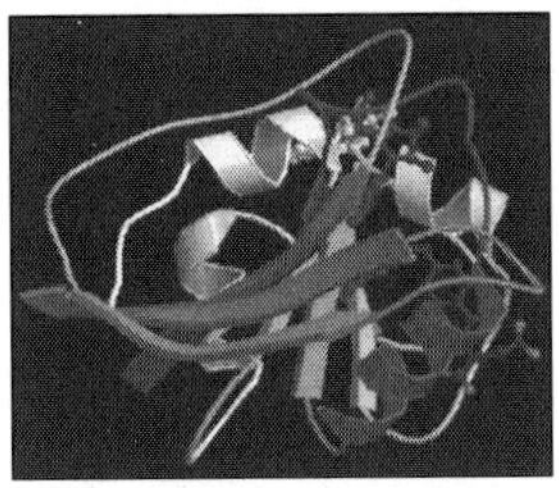

Note: This diagram was obtained from the RCSB Protein Data Bank. If you want to find more information about dihydrofolate reductase, their reference number for it is 7DFR. There is nothing particularly special about this enzyme in terms of structure. The model shows the alpha-helices in the secondary structure as coils of"ribbon". The beta-pleated sheets are shown as flat bits of ribbon ending in an arrow head. The bits of the protein chain which are just random coils and loops are shown as bits of"string".

The colour coding in the model helps you to track your way around the structure - going through the spectrum from dark blue to end up at red. You will also notice that this particular model has two other molecules locked into it (shown as ordinary molecular models). These are the two molecules whose reaction this enzyme catalyses.

Holding Protein into its Tertiary Structure

The tertiary structure of a protein is held together by interactions between the the side chains - the"R" groups. There are several ways this can happen.

Ionic Interactions

Some amino acids (such as aspartic acid and glutamic acid) contain an extra -COOH group. Some amino acids (such as lysine) contain an extra -NH_2 group. You can get a transfer of a hydrogen ion from the -COOH to the -NH2 group to form zwitterions just as in simple amino acids. You could obviously get an ionic bond between the negative and the positive group if the chains folded in such a way that they were close to each other.

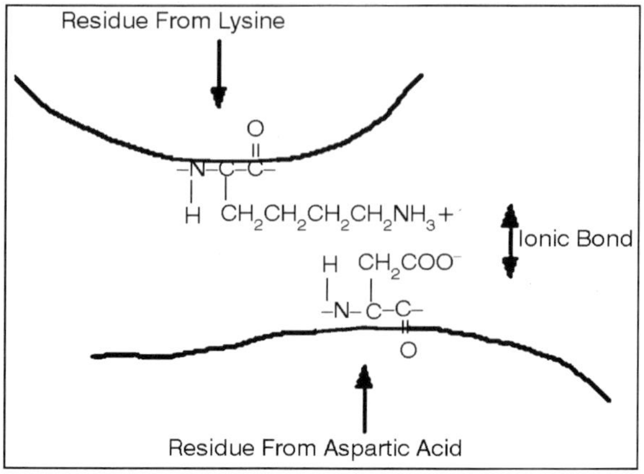

Hydrogen Bonds

Notice that we are now talking about hydrogen bonds between side groups - not between groups actually in the backbone of the chain.

Lots of amino acids contain groups in the side chains which have a hydrogen atom attached to either an oxygen or a nitrogen atom. This is a classic situation where hydrogen bonding can occur. For example, the amino acid serine contains an -OH group in the side chain. You could have a hydrogen bond set up between two serine residues in different parts of a folded chain.

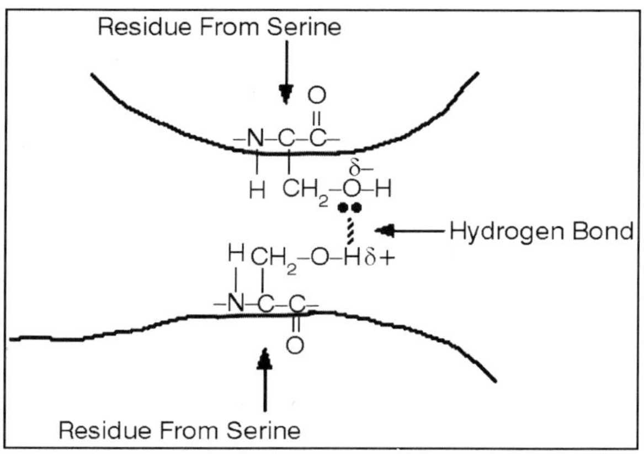

You could easily imagine similar hydrogen bonding involving -OH groups, or -COOH groups, or -CONH2 groups, or -NH2 groups in various combinations - although you would have to be careful to remember that a -COOH group and an -NH2 group would form a zwitterion and produce stronger ionic bonding instead of hydrogen bonds.

Van der Waals Dispersion Forces

Several amino acids have quite large hydrocarbon groups in their side chains. A few examples are shown below. Temporary fluctuating dipoles in one of these groups could induce opposite dipoles in another group on a nearby folded chain. The dispersion forces set up would be enough to hold the folded structure together.

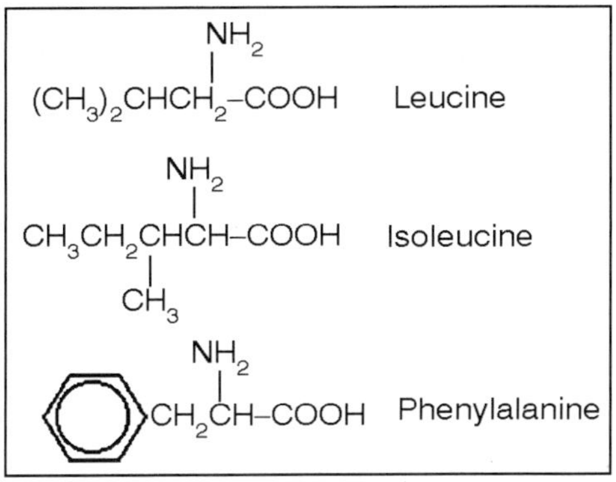

Sulphur Bridges

Sulphur bridges which form between two cysteine residues have already been discussed under primary structures. Wherever you choose to place them doesn't affect how they are formed.

PROTEIN DIGESTION AND METABOLISM

Ingested proteins are first split into smaller fragments by pepsin in the stomach or by trypsin or chymotrypsin from the pancreas. These peptides are then further reduced by the action of carboxypeptidase which hydrolyzes off one amino acid at a time beginning at the free carboxyl end of the molecule or by aminopeptidase which splits off one amino acid at a time beginning at the free amino end of the polypeptide chain. The free amino acids released into the digestive system are then absorbed through the walls of the gastro intestinal tract into the blood stream where they are then resynthesized into new tissue proteins or are catabolyzed for energy or for fragments for further tissue metabolism.

GROSS PROTEIN REQUIREMENTS

Gross protein requirements have been determined for a few species of fish. Simulated whole egg protein component of test diets contains an excess of indispensable amino acids. These diets were kept approximately isocaloric by adjusting total protein plus digestible carbohydrate components to a fixed amount as the protein diet treatments were varied over the ranges tested. Tests in feeding fry, fingerling and yearling fish have shown that gross protein requirements are highest in initial feeding fry and that they decrease as fish size increases. To grow at the maximum rate, fry must have a diet in which nearly half of the digestible ingredients consist of balanced protein; at 6-8 weeks this requirement is decreased to about 40 per cent of the diet for salmon and trout and to about 35 per cent of the diet for yearling salmonids raised at standard environmental temperature (SET).

Initially feeding fry require that about 50 per cent of the digestible components of the ration be protein and the requirement decreases with size. Some feeding trials with salmon have indicated direct relationships between changes in the protein requirements of young fish and changes in water temperature. Chinook salmon in 7 C water require about 40 per cent whole egg protein for maximum growth; the same fish in 15 C water require about 50 per cent protein.

Salmon, trout and catfish can use more protein than required for maximum growth because of efficiency in eliminating nitrogenous wastes in the form of soluble ammonia compounds through the gill tissue directly into the water environment. This system for eliminating nitrogen is more efficient than that available to fowl and mammals. Fowl and mammals consume energy to synthesize urea, uric acid, or other nitrogen compounds which are excreted through the kidney tissue and expelled in urine. Digestible carbohydrate and fat will spare excess protein in the diet as long as the protein requirement for maximum growth is met.

Table. Estimated Dietary Protein Requirement of Certain Fish

Species	Crude protein level in diet for optimal growth (g/kg)
Rainbow trout (Salmo gairdneri)	400-460
Carp (Cyprinus carpio)	380
Chinook salmon (Oncorhynchus tshawytscha)	400
Eel (Anguilla japonica)	445
Plaice (Pleuronectes platessa)	500
Gilthead bream (Chrysophrys aurata)	400
Grass carp (Ctenopharyngodon idella)	410-430
Brycon sp.	356
Red sea bream (Chrysophrys major)	550
Yellowtail (Seriola quinqueradiata)	550

Basically the fish must be given a diet containing graded levels of high quality protein and energy and adequate balances of essential fatty acids, vitamins and minerals over a prolonged period. From the resulting dose/response curve the protein requirement is usually obtained by an Almquist plot. These differences in apparent protein requirement are thought to be due to differences in culture techniques and diet composition.

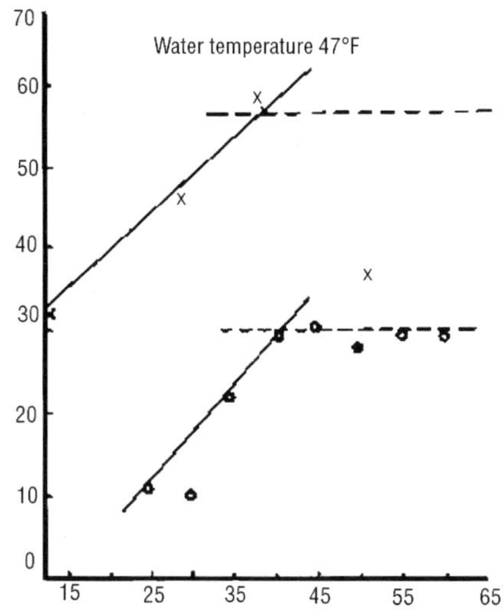

Fig. Protein Requirement of Chinook Salmon at 47°F. Top Curve: Initial Individual Average Weight of Fish, 1.5g. Bottom Curve: Initial Individual Average Weight of Fish, 5.6g.

The relatively high dietary protein levels required for maximal growth of certain fish such as grass carp, Ctenopharyngodon idella and Brycon spp. are surprising as these fish are omnivorous. Brycon spp. are grown on unwanted fruit and other plant material of low protein content and under these conditions there is presumably a substantial contribution to their protein intake from a natural food chain. Protein requirement of eurythaline fish such as the rainbow trout, Salmo gairdneri and the coho salmon, Oncorhynchus kisutch, reared in water of salinity 20 ppt is about the same as the requirement in freshwater. No data are available for the protein requirement of these species in full strength sea water.(35 ppt).

PROCEDURE OF PROTEIN SYNTHESIS

Initiation

Peptide synthesis always starts from methionine (Met). Therefore, the initial aminoacyl-tRNA is Met-tRNAiMet, where the subscript "iΔ specifies "initiation". In bacteria, the methionine of the initial aminoacyl-tRNA has been modified by the addition of a formyl group (HCO) to its amino group. The modified methionine is called formylmethionine (fMet), which is unique for bacteria. Thus, fMet is an obvious foreign substance in eukaryotes. It can elicit a strong immune response. In humans, the immune response elicited by the peptide "fMet-Leu-Phe" is about a thousand times greater than "Met-Leu-Phe".

Elongation

A ribosome contains two major tRNA-binding sites: A site and P site. After the large subunit joins the initiation complex, the initial Met-tRNAiMet enters the P site and the newly arrived aminoacyl-tRNA is always placed at the A site ("A" for "aminoacyl"). Then, methionine is transferred to the new aminoacyl-tRNA, forming a "peptidyl-tRNA" where a peptide is attached to the tRNA. Subsequently, the empty tRNA at the P site is ejected from the ribosome and the peptidyl-tRNA jumps to the P site ("P" for "peptidyl"). During this translocation step, the ribosome also moves one codon down the mRNA chain. Similar steps are repeated in the next cycles of elongation.

Termination

Protein synthesis will terminate when the ribosome arrives at one of three stop codons. The termination process is assisted by special proteins called termination factors which recognize the stop codons. Their association stimulates the release of the peptidyl-tRNA from the ribosome. Subsequently, the released peptidyl-tRNA divides into tRNA and a newly synthesized peptide chain. The ribosome also divides into the large and small subunits, ready for synthesizing another peptide.

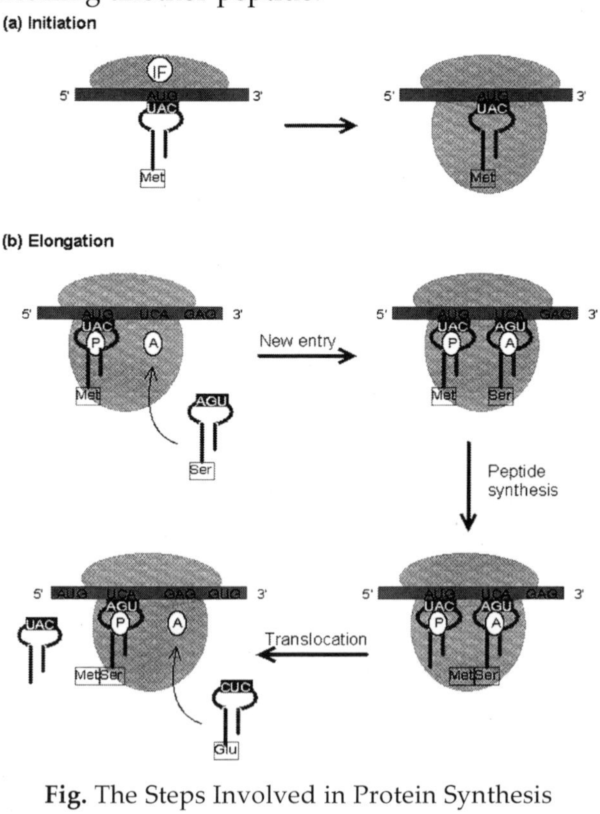

Fig. The Steps Involved in Protein Synthesis

- In the absence of mRNA, the large and small subunits of a ribosome are separated. At the beginning of peptide synthesis, initiation factors (IF) first assist the assembly of the small subunit, mRNA and the initial aminoacyl-tRNA. Then, the large subunit is recruited to join the complex.
- In the elongation process, one cycle involves the following steps:
  - *New entry.* A new aminoacyl-tRNA with a correct anticodon is brought to the A site. This step is catalyzed by elongation factors Tu and Ts in prokaryotes and by elongation factors EF1 and EF1b in eukaryotes..
  - *Peptide synthesis.* The peptide attached to the peptidyl-tRNA at the P site is transferred to the new aminoacyl-tRNA at the A site, generating a peptidyl-tRNA with a longer peptide. This step is catalyzed by peptidyl transferase.
  - *Translocation.* The empty tRNA at the P site is ejected from the ribosome and the peptidyl-tRNA generated at the A site takes over the vacant P site. In the mean time, the ribosome moves one codon down the mRNA chain. The A to P switch is catalyzed by the elongation factor G in bacteria and by the elongation factor EF2 in eukaryotes.

SORTING OF PROTEINS

As discussed in the previous section, proteins are synthesized on ribosomes which are located mainly in the cytosol. Only a small number of ribosomes are located in mitochondria and chloroplasts. Proteins synthesized on these ribosomes can be directly incorporated into the compartments within these organelles.

However, most mitochondrial and chloroplast proteins are encoded by nuclear DNA and synthesized on cytosolic ribosomes. These and all other proteins synthesized in the cytosol must be transported to appropriate locations in the cell. This is made possible by the specific signal sequence in the newly synthesized peptide.

- If the N terminus of the new peptide contains a stretch of hydrophobic residues, it is sent to the rough endoplasmic reticulum (ER) for further sorting. Otherwise, it goes to non-ER pathways.
- The new peptide is retained in the rough ER if its C-terminus contains the sequence "Lys-Asp-Glu-Leu" (KDEL in one-letter code). Otherwise, it will move to the Golgi apparatus.
- Proteins containing a specific transmembrane a helix will be localized in the Golgi apparatus.
- After glycosylation at the Golgi apparatus, the modified protein containing a mannose 6-phosphate (M6P) will be delivered to the lysosome.

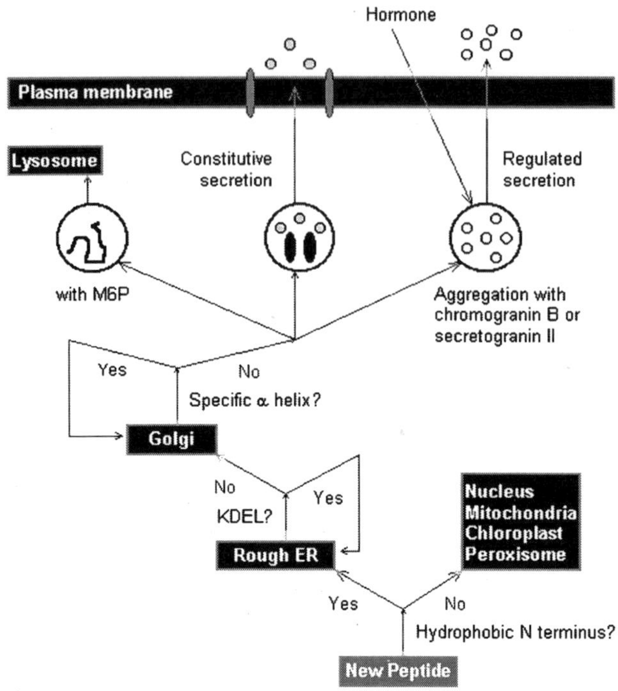

Fig. Protein Sorting

- Proteins that can aggregate with chromogranin B (secretogranin I) or secretogranin II will be sorted into regulated secretory vesicles where proteins are released upon specific stimulation. Otherwise, they are sorted into another type of vesicles which continuously move to the plasma membrane or outside the cell.

GENE AND PROTEIN STRUCTURES

Characteristics of the human and mouse members of the ACAT/DGAT gene family are shown in Table. The human ACAT1 gene encodes four mRNAs of 7.0, 4.3, 3.6, and 2.8 kilobases (kb) (12, 18-20); all contain the same translational reading frame but differ in the length of the untranslated regions.

The two shorter mRNAs are products of a proximal ACAT promoter. The 4.3-kb mRNA is derived from an unusual RNA recombination mechanism involving *trans*-splicing of two discontinuous precursor RNAs produced from chromosomes 1 and 7 (21). The origin of the 7.0-kb transcript is unknown. The human ACAT2 gene encodes a single mRNA of 2.2 kb and the human DGAT gene encodes two mRNAs of 2.0 and 2.4 kb. Regions of shared sequence identity are present throughout the ACAT and DGAT proteins, but the greatest similarity is found in their C termini. Human and mouse ACAT2 are ~40 per cent identical to human ACAT1 and DGAT is ~20 per cent identical to ACAT1.

Little is known about the relationship between protein structure and function in ACAT/DGAT family members. One motif, FYXDWWN (amino acids 403-409 of human ACAT1), is highly conserved in all family members and may be involved in fatty acyl-CoA binding.

Another motif, MKXXSF (amino acids 265-270 of human ACAT1), is conserved in ACAT family members. The serine of this motif is required for ACAT activity. All three family members have a potential tyrosine phosphorylation motif and at least one *N*-linked glycosylation site, both of unknown significance. ACAT1 and ACAT2 contain an N-terminal leucine zipper motif. Although ACAT1 functions as a homotetramer it is unknown whether this leucine zipper motif participates in tetramer formation or whether ACAT2 and DGAT act as multimers. ACAT1 and ACAT2 do not form hetero-oligomeric complexes.

All ACAT/DGAT proteins possess multiple hydrophobic regions predicted to serve as transmembrane domains, although there is not yet a consensus structural model for the proteins. Investigations of human ACAT1 topology have yielded different results. One study, performed by inserting epitopes at different positions in the expressed protein found seven transmembrane domains. Another study, performed by expressing truncated forms of the protein found five transmembrane domains.

In both studies, the N terminus of the protein localized to the cytoplasm and the C terminus to the ER lumen. Topologic studies of human ACAT2 performed with the truncation mutant method also found five transmembrane domains with the N terminus in the cytoplasm and the C terminus in the ER lumen. The latter study found that a serine residue essential for ACAT activity was located in the cytoplasm for ACAT1 and in the ER lumen for ACAT2, suggesting that the active sites of these enzymes reside on opposite sides of the ER membrane. Several putative transmembrane sequences are conserved in ACAT1 and ACAT2 but not in DGAT suggesting that these regions are involved in cholesterol binding. The topology of DGAT has not been reported, but a putative diacylglycerol binding motif has been identified

TISSUE AND CELLULAR DISTRIBUTIONS

The tissue distributions of ACAT1 and ACAT2 are largely complementary. ACAT1 mRNA is present in many tissues, with the highest levels in macrophages and in adrenal and sebaceous glands all of which store cholesterol esters in cytoplasmic droplets. ACAT1 is also highly expressed in human atherosclerotic lesions, particularly in macrophage foam cells. ACAT2 is expressed predominantly in the liver and small intestine. In humans, nonhuman primates and mice, ACAT2 appears to be the major ACAT in the small intestine and the predominant isozyme in hepatocytes of nonhuman primates and mice. Whether ACAT1 or ACAT2 is the predominant enzyme in human hepatocytes is controversial. Both are expressed in human hepatocytes and HepG2 cells. However, immunodepletion experiments found that ACAT1 accounts for ~90 per cent of ACAT activity in human liver, whereas ACAT2 accounts for most of the ACAT activity in adult small intestine and fetal liver but only 10-20 per cent of activity in adult liver.

It is unclear whether the hepatic ACAT1 activity resides in hepatocytes or macrophage-derived Kupffer cells. ACAT1 expression was detected immunohistochemically in both cell types, with stronger staining in Kupffer cells. In nonhuman primates, ACAT1 was expressed only in Kupffer cells in the liver. A significant proportion of the ACAT1 activity in human liver thus may be attributable to Kupffer cells. Another study found little evidence of ACAT2 expression in human hepatocytes.

DGAT mRNA and activity are ubiquitous in mouse and human tissues, with the highest levels in liver, small intestine and adipose tissue and somewhat lower levels in testis and adrenal gland. This wide range of tissue expression is consistent with the involvement of DGAT in the glycerol phosphate pathway of triglyceride synthesis, which is common to most cells. DGAT protein expression in tissues has not been studied because of the lack of suitable antibodies.

The intracellular localization of enzymes in this class has been examined only for ACAT1. In human melanoma cells and fibroblasts and in mouse macrophages, ACAT1 is located primarily in the ER. In mouse macrophages, a small portion of ACAT1 immunoreactivity localizes to a region near the *trans*-Golgi network. In macrophages, ACAT1 localization may change with different conditions

BIOCHEMICAL ACTIVITY AND REGULATION

ACAT and DGAT substrates have been identified by expressing the proteins in insect cells or yeast devoid of sterol esterification activity. In insect cells, mouse and human ACAT1 and ACAT2 utilize a variety of oxysterols in addition to cholesterol as substrates. Ergosterol and plant sterols are poor substrates. Both enzymes also utilize a wide variety of long chain fatty acyl-CoAs as substrates although arachidonyl-CoA appears to be a better substrate for ACAT1 than ACAT2. In insect cells, DGAT utilizes only diacylglycerol as the acyl acceptor. DGAT has a broad fatty acyl-CoA specificity.

ACAT1 is regulated primarily by post-translational mechanisms. In cholesterol-loaded cells, esterification increases without changes in the expression level of ACAT1 mRNA or protein. Purified ACAT1 is activated allosterically by cholesterol or oxysterols.

ACAT1 activity may also be regulated by its intracellular localization. For example, a portion of ACAT1 immunoreactivity in macrophages is found near the *trans*-Golgi network and the endocytic recycling compartment.

These cholesterol-rich organelles are involved in membrane traffic with the plasma membrane and the proximity of ACAT1 to these organelles may be important for cholesterol ester synthesis during the formation of macrophage foam cells. The observation that 30-40 per cent of ACAT1 redistributed from the ER to small vesicles during foam cell formation is consistent with this possibility. Regulation by phosphorylation has not been demonstrated.

ACAT1 is also transcriptionally regulated; however, only the expression levels of the shorter ACAT1 mRNAs (2.8 and 3.6 kb) change. In mice and rabbits a fat and cholesterol-rich diet increases ACAT1 mRNA levels 2-3-fold in the liver. In human HepG2 cells, free fatty acids, but not 25-hydroxycholesterol, increase ACAT1 mRNA levels 1.5-2-fold. ACAT1 mRNA and protein expression levels also increase during the differentiation of monocytes into macrophages. Two factors that induce gene expression during monocyte-macrophage differentiation1,25-dihydroxyvitamin D_3 and 9-*cis*-retinoic acid, also increased ACAT1 mRNA, protein and cholesterol esterification activity in THP-1 cells.

Little is known about the regulation of ACAT2 and DGAT. Levels of DGAT mRNA increase markedly in parallel with enzyme activity in NIH 3T3-L1 cells during their differentiation into adipocytes. This finding suggests that transcription factors involved in adipogenesis activate DGAT expression. Like other family members, DGAT possesses a putative tyrosine phosphorylation site. The use of this site in DGAT regulation has not been demonstrated, although two reports have indicated that the enzyme may be posttranslationally regulated by a tyrosine kinase.

FUNCTIONAL ANALYSIS OF ACAT1 AND ACAT2

Gene knockout experiments in mice have provided insights into the biological roles of ACAT1 and ACAT2. In ACAT1-deficient (ACAT1$^{/}$) mice, plasma cholesterol levels, hepatic cholesterol esterification activity and intestinal cholesterol absorption are normal, consistent with the presence of a second ACAT enzyme in the small intestine and liver of mice. ACAT1 deficiency has its greatest effect in the adrenal glands and macrophages.

Cholesterol esters and cholesterol esterification activity are virtually undetectable in the adrenal cortex of ACAT1$^{/}$ mice, although the depletion of cholesterol esters does not affect corticosterone synthesis. In addition, peritoneal macrophages from ACAT1$^{/}$ mice do not accumulate cholesterol esters when incubated with cholesterol-containing lipoproteins. This finding provides a unique tool for testing the role of ACAT1 in macrophage foam cell formation and atherosclerosis.

Another ACAT1 gene disruption study in mice confirmed general aspects of the ACAT1$^{/}$ phenotype and showed that ACAT1 deficiency reduces cholesterol esterification activity by >95 per cent in testis and ovary and results in atrophy of the meibomian glands of the eyelids. A naturally occurring mutation in the mouse ACAT1 gene is present in the AKR strain of inbred mice which are homozygous for the *adrenocortical lipid depletion* (*ald*) allele. The analysis of ACAT2-deficient mice (ACAT2$^{/}$) (34) complemented findings from ACAT1$^{/}$ mice. ACAT2$^{/}$ mice have nearly undetectable ACAT activity and lack cholesterol esters in the liver and small intestine. Because they cannot synthesize cholesterol esters in the small intestine, ACAT2$^{/}$ mice have a reduced capacity to obtain cholesterol from the diet. As a result, they do not develop hypercholesterolemia or form cholesterol gallstones when fed a diet high in fat, cholesterol ^{a}nd cholic acid. ACAT2 deficiency markedly decreases the cholesterol ester content of apolipoprotein (apo) B-containing lipoproteins in mice fed either a chow diet or the above diet. Thus, ACAT2 is the cholesterol esterification enzyme in mouse liver and small intestine and plays a major role in mediating the response to dietary cholesterol.

Gene disruption studies suggest that both ACAT enzymes synthesize cholesterol esters: ACAT1 for storage in cytosolic droplets and ACAT2 for secretion in apoB-containing lipoproteins. However, these results may reflect their different tissue distributions rather than inherent biochemical differences. Both ACAT1 and ACAT2, when expressed in a cell line deficient in cholesterol esterification activity (AC29 cells), synthesize cholesterol esters that are stored in cytoplasmic droplets (25). Also, overexpression of ACAT1 in mouse liver causes both cholesterol ester storage and secretion.

2

Vitamins

WATER AS SOLUBLE VITAMINS

This manuscript describes the lecture on vitamins contained in the core course entitled,"The Biochemical Basis of Drugs and Diseases. In the first year, our curriculum is designed to focus on systems and diseases. As such, this course acts in concert with both the Anatomy and Physiology course and the Pathophysiology course to present an integrated view of human diseases and systems. The second year of our curriculum is focused on learning about drugs classes and the mechanism of action.

The lecture on vitamins is the last topic of a course focused primarily on metabolism. Because many vitamins play major roles in metabolic cycles, this lecture allows for a brief review of much of the material covered throughout the course. Therefore, vitamin examples are primarily chosen to reinforce major metabolic cycles and also their role in human disease. There is more clinical information that is relevant to vitamins than what is presented in this lecture. However, since first year students in our curriculum have almost no knowledge of therapeutics, the focus of the lecture is on diseases and not clinical practice.

BIOCHEMISTRY AS CHEMISTRY OF WATER

We immediately see the evident truth in this statement when we consider the complex thermodynamics of protein and peptide structural stability and the prominent role of solvent in influencing this equilibrium. Proteins unfold from their native fold upon the addition of urea or other denaturants, peptide helicity increases upon addition of TFE, and alcohol is even known to induce helical structure in regions that otherwise adopt beta-sheet conformations.

While the helix inducing effects of alcohol upon peptide conformation have long been used to probe and enhance latent structure, thereby making it measurable by CD or NMR, a recent paper illustrates how this dependence of structure on solvent conditions may be harnessed to make a sort of'conformational switch'. Successful molecular design or peptide engineering applications will necessarily include an assessment of solvation effects for conformational stability. Turning from a macroscopic picture of the role of water and aqueous solutions in establishing the equilibria of proteinaceous substances to a more microscopic view, we can consider the surprising role played by discrete water molecules in antigen-antibody recognition. In a 1993 paper in Current Opinion in Immunology, Mariuzza and Poljak report a case in which interfacial water plays a critical role in determining the complementarity of the antigen and antibody surfaces.

This work was illuminating, since previous work was leading to the conclusion that the packing density was lower at protein interfaces than in protein interiors. It also appeared that polar atoms at these interfaces were apparently tolerating the absence of neutralizing partners - far more than they do in protein interiors. In this work, a 1.8 angstrom crystal structure of an immune complex was studied and revealed that the water network was an integral part of the interface.

As pointed out by these authors, the analysis of low resolution crystal structures may in fact lead to erroneous conclusions on the nature of the interface in these complexes. Furthermore, the evidence reported in this work points to the inescapable conclusion that the intermolecular complementarity essential for molecular recognition is not necessarily imparted solely by the large molecules.

And, contrary to what our notions of solvent entropy might lead us to expect, bound water may play a role in stabilizing the structure of single proteins as well as that of complexes. And, this bound water may, apparently, serve this function at the surface of the protein, and not solely while confined to the interior. That's just the tip of the iceberg as far as far as water and its impact for biological systems at what is considered physiological temperature and pressure; the realm normally explored by biochemists and pharmacologists. Much, however is being learned about the so-called"hot life", life found at thermal vents deep in the ocean floor. Reports have documented the existence of organisms at temperatures in excess of 110 deg. C. These organisms are referred to collectively as thermophiles. It is important to note the enormous temperature variation of sea water at these depths, with a documented range from 2 deg. C to 900 deg. C at depths of a few km where the high pressure precludes boiling. Hot life, or the existence of thermophiles, has been noted at temperature in excess of 100 deg. C at these elevated pressures, and it remains a matter of some dispute whether or not bacteria can exist at even higher temperatures still.

So, what are the properies of ambient sea water for organisms at deep sea thermal vents? We can expect these properties to influence the kinetics and thermodynamics of the biochemistry pertinent to this life, just as they influence the systems with which we are, perhaps, more familiar. We can begin to understand the fluid environment presented to thermophiles by considering the temperature dependence of a few properties of pure water. A quick look at the following data from Shock show the dramatic effect of temperature on water's dielectric. Water at 25 deg. C and 1 atm pressure has the familiar dielectric near 80; however near water's critical point the dielectric has fallen to less than 10. For comparison, ethanol has a dielectric of roughly 25 at STP.

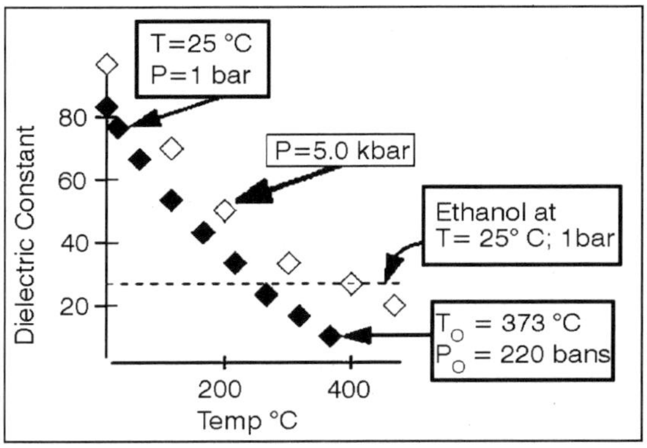

Fig. From Shock.

In this graph all white and blue diamonds pertain to water data points. The blue diamonds are for pressures below the critical point where water exists in two phases, while the white diamonds pertain to 5 kbar data. For the low pressure data, two key data points are indicated. The grey dashed line indicates the dielectric of ethanol at standard conditions. (This graph is drawn to approximate that shown in the original paper which should be consulted for quantitative work.)

The weak pressure dependence of this data should be noted. In order to understand the relevance of this sort of analysis to a complete understanding of ambient conditions at deep sea thermal vents, it should be understood that the pressure at these depths ranges from a few hundred bars to several kilobars. Shock should be consulted for additional details, data and a general groovy time. Well, if all this is true, then water must have a pretty different structure at these T&P's than we are used to. Right? Not surprisingly, this is what is being found. Some of these results are shown in this graph:

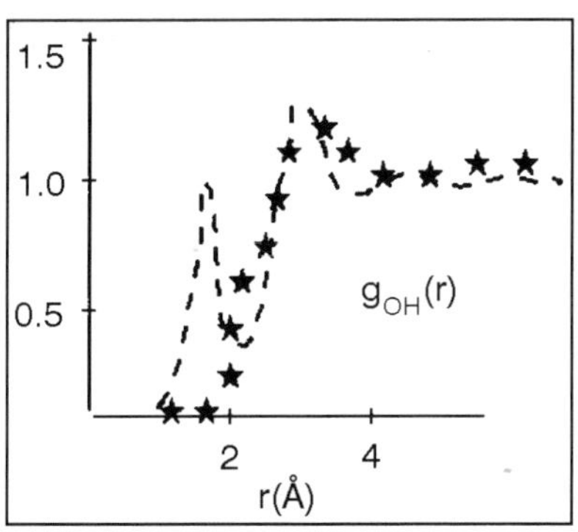

Fig. From Postorino

The pair distribution function for intermolecular OH atom pairs is shown as a function of interatomic distance for water at 25 deg. C (red dashed line) and 400 deg. C. (This graph is drawn to approximate that shown in the original paper which should be consulted for quantitative work.)

This particular pair distribution function tells us the likelihood with which a H atom will be found at any distance from an O atom (The intramolecular data isn't shown here, but the water molecule does have the same structure at 400 deg. C as at 25 deg. C.

This graph considers only intermolecular associations.) Immediately, then, we see that at room temperature it is relatively likely to find these atom pairs separated by 1.95 angstroms. This is the hydrogen bond. And then there are strong secondary structural phenomena, manifest as the damped oscillations that we see here out to beyond 5 angstroms, all due to the overwhelming hydrogen bonding Behaviour.

Looking at the high temperature data, it's clear that the hydrogen bonding peak, which dominates the room temperature data, is completely gone.

The OO correlation function, not reproduced here, shows a slight broadening of the nearest-Neighbour peak with temperature. The high temperature OO pair correlation function also yields a larger coordination number than is seen at room temperature. In other words, each water molecule is surrounded by more water molecules in its first solvation shell at 400 deg. C than at 25 deg. C. At 400 deg. C the packing is not restricted by the severe hydrogen-bonding geometrical constraints of the the room temperature fluid. This result is in line with the close packing Behaviour of most liquids.

These results taken together with those for the dielectric phenomena described above begin to paint a picture of high temperature water. Water at high temperature exhibits many of the properties we associate with less polar liquids under standard conditions. In order to understand something of the biochemistry that might pertain under extreme aqueous conditions, Shock has been estimating the stability of peptidic species at elevated temperatures and pressures. Specifically he has been asking questions like"How does the free energy of the peptide bond change with temperature and pressure?" A snapshot view of some of this work will serve to further the discussion of aqueous properties under extreme conditions while at the same time giving a hint for some of the very different biochemistry that will undoubtedly be revealed by further study in this area. Considering only Glycine here, the free energy of interest here is that which determines the following equilibrium:

$$\text{Glycine} + \text{Glycine} = \text{Glycylglycine} + H_2O$$

In other words, what is the free energy cost to form the peptide bond between two Glycine molecules? This equilibrium can be calculated from the free energies of formation for the individual species involved in this reaction, Glycine and Glycylglycine, and the temperature dependence of this free energy in the following way:

$$\Delta G^{\circ}_{PT} = \Delta G^{\circ}_{f} - S_{Pr,Tr}(T - Tr) + \int C^{\circ}_{P}$$

$$-T\int C^{\circ}_{P} d\ln T + \int V^{\circ} dP$$

This equation has the usual host of characters. G represents free energy, S is entropy, Cp is heat capacity, V is volume and P is pressure. If any of this is murky please consult the original paper by Shock. What we're basically doing is getting the free energy (OK, so it's really the standard partial molal Gibbs free energy) at any T&P from that at STP by integrating over the temperature and pressure dependence of the heat capacity and volume. In order to do this, we need to know something about the way heat capacity and volume depend on the thermodynamic variables of T and P, respectively. We'll look at this first:

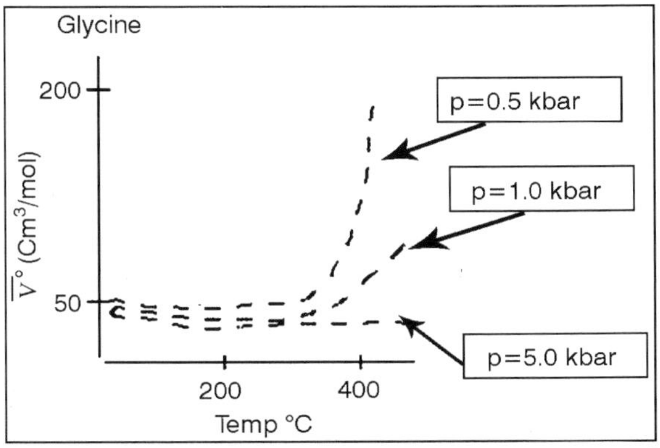

Standard partial molal volume for Glycine. From Shock. (This graph is drawn to approximate that shown in the original paper which should be consulted for quantitative work.) These results show the very large dependence of the molal volume on temperature at lower pressures, and the much weaker temperature dependence at extremely high pressures. Here are the results for the heat capacity:

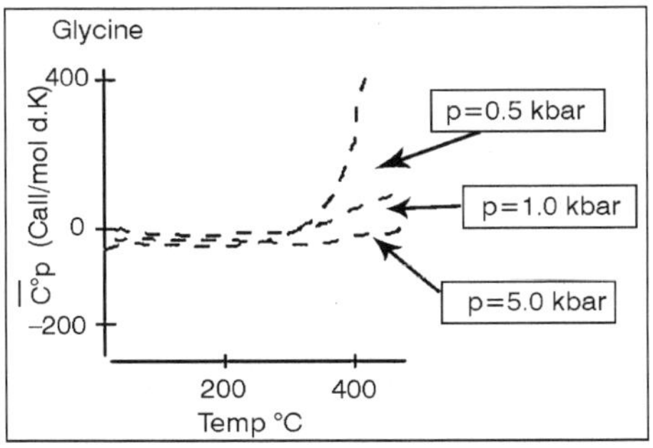

Standard partial molal heat capacity (constant pressure variety) for Glycine. From Shock. (This graph is drawn to approximate that shown in the original paper which should be consulted for quantitative work.) Pretty much the same thing. A large T dependence is witnessed at lower temperature, but this Behaviour is convincingly squashed at high pressure.

The results for Glycine have been shown here. Analogous results for Glycylglycine are obtained from these by calculating a regression equation that relates, for instance, the partial molal volume of Glycine to that of Glycylglycine. Putting it all together with some other necessary data (like the free energy of formation at STP, and the entropy at STP; you got it, Shock) the free energies of formation for both Glycine and Glycylglycine can be obtained at any temperature and pressure, and from these the free energy governing the equilibria may be directly obtained.

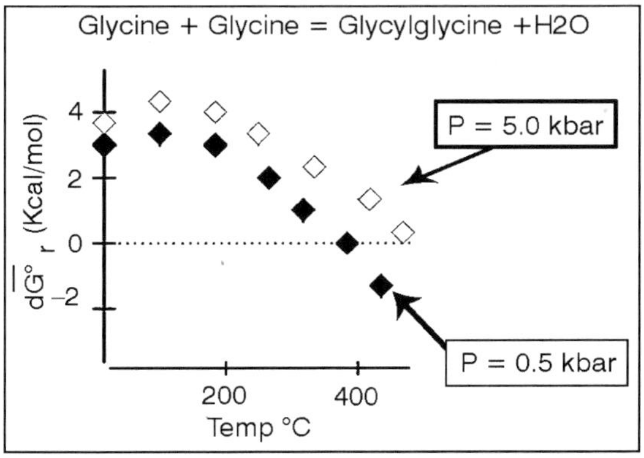

Free energy required to form the Glycylglycine dipeptide from the constituent Glycine moieties. From Shock. (This graph is drawn to approximate that shown in the original paper which should be consulted for quantitative work.) The results indicate that while it costs free energy to form the dipeptide at lower temperatures and pressures (at STP the free energy of this reaction is roughly 3.5 kcal/mol), at extremes of temperature and pressure the dipeptide is the thermodynamically favoured species.

In any event, these results are consistent with those for many other dipeptides not covered here and predict that peptide bonds become more stable at elevated temperatures. This Behaviour just begins to hint at the interesting biochemical phenomena that will be revealed for life in water under extreme conditions where water has very different properties from those to which we are accustomed.

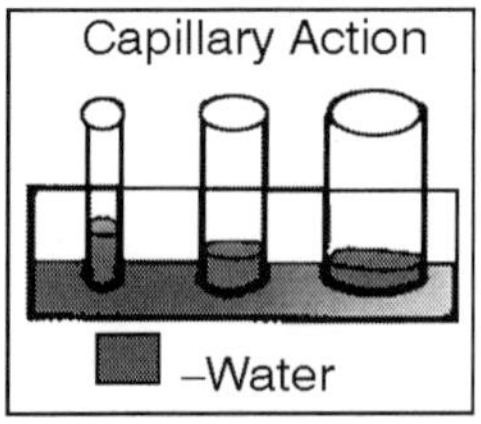

Water has a high specific heat capacity. It takes a lot of energy to change the temperature of water. This characteristic makes water a good insulator and a good coolant. When you sweat, your body is using water as a coolant. The eVapourating water removes heat with it. At night, oceans are a good insulator. The energy that the sun spent in heating the water all day is slowly released into the night.

Finally, water is the most dense at 4 degrees Celsius. Hydrogen bonding arranges water molecules into hollow"cells" when water freezes, making it less dense than liquid water. This characteristic of water is what allows fish in lakes and ponds to survive in the winter. When the water freezes, it becomes less dense and floats to the surface -- leaving the bottom of the lake or pond unfrozen.

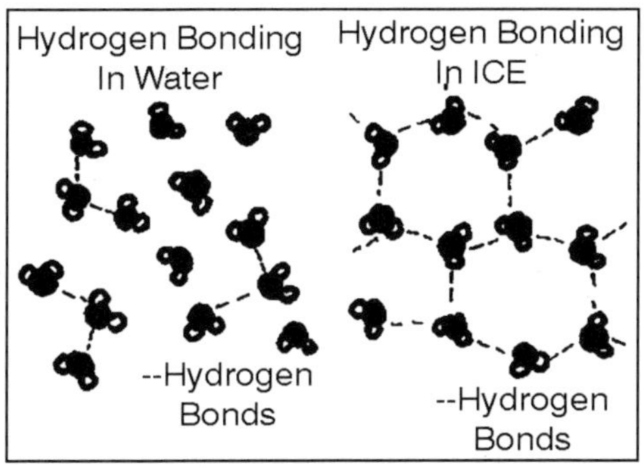

WEAK INTERACTIONS IN AQUEOUS SYSTEMS

Water is a good solvent for polar solutes with which it forms hydrogen bonds. Non-polar compounds dissolve poorly in water. They can not form a hydrogen bond with the solvent. Numerous weak non covalent interactions influence the folding of macromolecules. The physical properties of aqueous solutions are strongly influenced by the concentration of solutes. Ex: The tendency for water to move across a semi permeable membrane is the osmotic pressure.

IONIZATION OF WATER

Pure water is slightly ionized. pH is defined as the negative logarithm of the molar hydronium ion concentration. An acid that ionizes completely in water is a strong acid. An acid which is partially ionized in water is a weak acid. pk provides the measure of the ease of the dissociation of the proton. The lower the number the stronger is the acid.

BUFFERING AGAINST PH

Buffers are aqueous systems that tend to resist in changes in pH when small amounts or acids or base are added. It consists of a weak acid and its conjugate base. As an example a mixture of equal concentrations of acetic acid and acetate ion is a buffer system. Every life form is extremely sensitive to slight pH changes. Human blood for example needs to remain within the pH range of 7.38 to 7.42. The relationship between pH and pk is described by Henderson - Hasselbach equation. It is used to calculate the pH of biological fluids. Two especially important biological buffers are the phosphate and bicarbonate systems. The property of a solution that depends on the concentration of a dissolved solute but not on its chemical identity is known as colligative properties. Substances that do not ionize in water are non electrolytes. Vapour pressure, boiling point, freezing point and osmosis are colligative properties.

WATER AS A REACTANT

Water is very often a direct participant in the chemical reactions of living cells. Formation of ATP from ADP and inorganic phosphate is an example of a condensation reaction in which the elements of water are eliminated. Two amino acid molecules are covalently joined through a peptide bond with a loss of water molecule.

FITNESS OF AQUEOUS ENVIRONMENT FOR LIVING ORGANISMS

The high specific heat of water is useful to cells and organisms because it allows water to act as a'heat buffer', keeping the temperature of an organism relatively constant as the temperature of the surroundings fluctuates.

DESCRIPTION ABOUT VITAMINS

It is a complex of biological"wheels within wheels" consisting of enzymes, coenzymes, antioxidants and trace mineral activators. Enzymes, being protein, contain amino acids and trace minerals. Every mineral and amino acid needed by living cells may be found in a natural assemblage of vitamin concentrates obtained from wholesome foods. Identifying a vitamin in terms of a single chemical structure is self-defeating--you need the whole complex to get the vitamin function.

The natural vitamin complex carries trace mineral activators and without them, the vitamin fails in its metabolic function. Copper in its organic, biologically active form tyrosinase, is the trace element activator of the vitamin C-complex. Likewise, manganese activates the B complex. In turn, selenium is the trace mineral activator of the vitamin E-complex. Enzyme activators, destroyed by heat, steam sterilization and pasteurization, include Manganese, Cobalt, Molybdenum, Zinc, Selenium and Vanadium...and are effective only in the organic state.

By reducing a vitamin complex to a single chemical, chemists can claim to duplicate it in a Labouratory and easily mass produce it and sell it as an additive to food or a pill. This wouldn't be possible if they were charged with keeping all the synergistic components intact. As such, vitamins are given single chemical names and single chemical structures for the purpose of selling synthetic vitamins. The failure of the pharmaceutical approach to vitamins originated in the chemists' synthetic isolation of aelements, uncombined with its biological matrix. The discovery of the vitamins coincided with the boom in industry, agriculture, and chemistry, which were crowding their way into all facets of life. But in the attempted duplication of living things, like artificial fertilizers or vitamins, chemistry had its most influence.

The biological activity of synthetic and natural vitamins is not identical. Anyone who has studied anatomy, physiology, neurology and biochemistry will readily admit that the human body is a neuro-biochemical plant--a composite of tissues, fluids and organs that exist in a living state by virtue of neuroelectric and biochemical reactions.

Being biochemical in nature, the body operates on biochemicals—Live Chemicals—in contrast to dead or lifeless chemicals. Those biochemical factors are derived from whole, natural, uncooked and unprocessed foods. Cooking and processing alters biochemicals to dead chemicals. The human body cannot properly function if it is fueled with dead chemicals rather than live, complex biochemicals.

The difference between the live biochemical food factor and the dead chemical (counterfeit food) is dependent upon two primary requisites:

- Unaltered proteins.
- Viable enzyme systems.

Dr. Casmir Funk, who discovered the method for concentrating vitamin B, and the one who coined the term, Vitamine, said,"The synthetic product is less effective and more toxic." Every process used to extract vitamins from foods, destroys the vitamin complex. What you end up with is similar to a banana peel without the banana. When you take these incomplete supplements, they rob from the body's nutritional reserve what is missing (the banana), to try to complete the complex.

This creates what is called a rebound deficiency. If you find the shed skin of a snake, you haven't found a snake. "Synthetic vitamins are reverses conformationally or mirror images of natural vitamins. However, synthetic vitamins are not vitamins at all but synthesized fractions of a vitamin complex...a mirror image duplicate of just a portion of the real, biologically active and physiologically-precise, nutritional complex. There is no way that a fraction of a vitamin can be called a vitamin." He continues,"Textbooks record that vitamin B_1 was first isolated in the year 1926. About 10 years later, scientists were able to synthesize the mirror image of Thiamine from coal-tar.

About the beginning of WWII enrichment of refined, devitalized flour became compulsory. Bleached wheat flour has more than 30 known nutrients removed with 4 synthetics added back in. Today, the majority of people in the civilized world are becoming victims of degenerative disease resulting from chronic poisoning and malnutrition. Even though well fed, they are starving to death for the most essential micronutrients. Since World War II, the American people and people of other countries as well, have had a daily ration of a genetic poison in most of the bread, flour products, cereals and other food items that are forced, by law, to be enriched with synthetic vitamins.

Then, after there develops a suspicion that these'foods' are inadequate to support life, modern advertising science steps in to propagandize the people into believing that there is nothing wrong with counterfeit or devitalized foods, that they are products of value...and...the confused public is totally unable to arrive at any conclusion of fact. In turn, they continue to buy the rubbish that is killing them off, years prematurely."

Gilbert Levin discuss optical isomers or asymmetrical molecules with atomic arrangements that are mirror images of each other, states:"Because its structure is reversed, a left-handed molecule can't take part in chemical reactions meant for a right-handed molecule any more than a left hand can fit in a right-handed glove--its odd geometry prevents it from being metabolized by the body." In turn, the human physiology cannot properly utilize synthetic (mirror-image) fractions in the way natural complexes find their way into the biological reactions that are essential to tissue repair and the sustenance of life. A synthetic vitamin fraction can only be utilized for a drug or pharmacologic effect. Drugs are poisons, by definition. All pharmacologic therapies induce side effects and high-dose nutritional therapies are no exception. Only synthetic supplements achieve drug status.

The effect of a drug is palliative not curative. Palliation means a masking or covering-over of symptoms, while the disease process remains unchanged or progressively gets worse for lack of attention."It is better to have no vitamins at all than a mixture of crystalline-pure vitamin fractions, with one or more members of the complex missing."

Natural complexes differ from synthetic or crystalline vitamins in many ways:

- Natural complexes are colloidal, protein in nature and usually exist as enzymes or coenzymes.
- The crystalline vitamin can't be separated from its protein part without destroying its biological activity.
- The natural"live" complex carries trace mineral activators...without which the enzyme fails as a biochemical catalyst.
- Crystalline vitamins, taken into the body, must recombine (if possible) with their natural required components (robbed from the body's reserve) before they can function as a vitamin. Most, if not all of the crystalline component is lost through the kidneys.

Vitamins are not meant to be eliminated in urine any more than silverware is to be thrown out with the leftovers. Since most poisons stimulate circulation in the human body for the purpose of moving blood through the kidneys, more often to filter out the said toxin as quickly a possible...the speeded-up circulation also delivers more blood and more oxygen to the brain with a created sense of euphoria. Natural and low-potency vitamin complexes do not and will not produce euphoria, any more than eating a (non-poison-sprayed) carrot. A synthetically-derived substance may cause a reaction in a chemically susceptible person when the same material of natural origin is tolerated, despite the two substances having identical chemical structures. The layman who doesn't understand the biochemical nature of true nutritional changes, is easily deceived by the pseudo-nutritional therapist and believes the initial response (drug effect), to the massive dose of one or more prescribed megavitamins, represents that what is being taken is exactly what is needed.

Nutritional therapists know after a time on synthetic vitamin fractions, the biochemical reaction is reversed, the euphoria ends and the original symptoms return; except amplified by vitamin poisoning. Also, the diagnosis and evaluation of symptoms are now clouded by the weeks, months or years of improper supplementation. Why the great mystery about all of this? Why don't doctors know about vitamins or nutrition in general? One big answer was revealed in the Journal of the Medical Association. It honestly admitted:"Nutrition has been neglected by the medical profession." Most medical schools devote less than three hours of total instruction in nutritional deficiency and therapy.

In short, Physicians in the United States are not required to understand nutrition to be licensed to practice medicine. It's not hard to understand why, when medical experiments are performed with synthetic vitamin fractions, these counterfeits fail to produce impressive results. But a reductionist seeks to understand the whole by dismantling its parts. This precludes the wholistic principle of synergism, that the whole is greater than the sum of its parts. For example, what if we tried to understand the properties of water (H_2O) by testing hydrogen separately and oxygen separately. Presumably, we would learn enough about each to understand what their combined characteristics would be. Combined as 2 parts hydrogen and 1 part oxygen, H_2O will extinguish fire. Separately, the elements are among the most flammable and explosive elements in the universe.

The functions are exactly the reverse in their isolated state from their organically combined state. Similarly,vitamins function as biological mechanisms only when whole and complete, combined with their synergists, as in whole food. Isolated into synthetic chemicals, they fail as catalysts.

CATEGORIES OF SUPPLEMENTS

The following is a summarization of the basic differences between the four categories of supplements available today.

Pure, Hypoallergenic or Natural

This is the most prevalent category of professional supplements available today. Contrary to popular belief, the vitamins and minerals used to create supplements in this category are not derived from consumable food or botanical sources. Instead, they are synthetically produced, fractionated or isolated from refined raw materials such as crushed rock, petroleum by-products and organic solvents to produce a pure, crystalline vitamin or mineral analogue. Although the final compound is an analogous structure to a specific vitamin or mineral, it is no longer attached or associated with any of the synergistic co-factors that the nutrient is inherently found within foods. As a result, pure vitamins and minerals have significantly reduced biological activity and a lower rate of utilization in the body.

Food-Based

This category of supplements is created from a physical mix of pure, fractionated nutrients with varying amounts of food powders or extracts. There are several different factors that need to be considered when determining the absolute nutritional value of a food-based formula. The primary factors that affect the quality of a food-based supplement are the amount, variance and form of the foods included in a particular formula.

A premium food-base formula is one that provides a broad range of unadulterated whole food or plant concentrates (preferred over powders) in at least a 3:1 ratio of food to nutrients in a formula will significantly improve the bioavailability and biological activity of the vitamins and minerals. Delivering isolated vitamins and minerals in a premium food base is an energy-saving process, as the body will not waste vital energy stores to transform the vitamins and minerals into absorbable nutrients. Ascorbic acid is a good example of a chemically synthesized vitamin that has been vastly studied; it has demonstrated enhanced utilization when the nutrient is delivered with its natural co-factors, such as bioflavonoids, rutin and tyrosinase, which always co-exist with vitamin C in food. The utilization factor can be increased by as much as 3 to 5 times (from 10 per cent for vitamins in the first category, to 3 – 50 per cent for food-based).

To determine if a supplement has a food-base, review the label for chemical names, as well as a combination of food extracts or powders. To determine the ratio of nutrients to foods, add up the total milligrams of the listed nutrients and for the food extracts.

Biocultured Supplements

Biocultured supplements are a physical mix of food, pure isolated nutrients and may have one or more strains of yeast and/or probiotics. Once combined, this mixture is allowed to ferment in a controlled environment. The process of fermentation, also known as culturing, biologically alters the biochemistry and nutrient content of the food. Cultured foods have been shown to have numerous health benefits and have played a role in the diet of many cultures for centuries; however, the body cannot sustain optimal health on cultured foods alone. This is due to the loss of substances, nutrients or compounds that are protective of health, such as vitamin c. Depending on the strains used during the culturing process, the generation or inactivation of toxic substances may also occur. In addition, many proteins and enzymes are denatured during the culturing of biological activity of the nutrients.

Supplements that fall into this category will denote the foods on which the yeast and/or probiotics feed. Not all cultured supplements list the isolated vitamins and minerals used in a particular formula.

WHOLE-FOOD SUPPLEMENTS

Many companies tout that their supplements are made from whole-foods; however, there are only a few companies that actually can substantiate this claim. Authentic whole food supplements are made with vitamins and minerals delivered in whole-food concentrates. The food concentrates selected have been grown in a controlled environment where budding plants are fed a nourishing broth enriched with bioactive peptide carriers. The peptide carriers act as chaperones, which govern a vitamin's or mineral's capacity to store the nutrient within its cells is successfully achieved, the growing process is complete, and harvesting is commenced using proteolytic enzymes to break down the cells' walls, and release the intra-cellular material.

The whole-food extract or concentrate produced is rich in a particular vitamin or mineral and is delivered in a synergistic state, with all of the plant's inherent food attachments intact and unadulterated. These attachments are vital co-factors facilitating the body's natural ability to recognize and utilize the nutrients present within the concentrate. Whole-food nutrients are recognized by the body as food, and therefore are better utilized and retained by the body when compared to any other supplemental form. Food is the complex form in which all nutrients are found in nature.

Supplements in this category will list the nutrient names--vitamin C, folate, calcium, etc.--and the mg. dosage of each, butthey will not denote a fractionated chemical name. Nutrients as they exist in food cannot be listed as one identifiable organic compound in this way, for the reason that they always exist within an interconnected weave of parts, and never as one isolated element. Nutrients in a whole-food state will be listed with their food source in the supplement facts panel such as vitamin C.

Bioavailability Spectrum

Let's take a closer look at the bioavailability of supplements on a spectrum, from least bioavailable to mostbioavailable. On the low end of the spectrum you would have an inexpensive, store bought, hard-packed tablet. Something brightly Coloured that you can't break with your fingers. Such supplements have many fillers and binders, including ingredients that can be found in wood putty and tar. You can't take these on an empty stomach, or else you'll get an upset stomach.

These are difficult to digest and breakdown, and often times your body doesn't at all. And these have been known to have been passed whole! For that reason you can understand why they have a bioavailability of 0-25 per cent. So, just as a reminder, when you look at the daily values of nutrients on the label of these products, your body is not getting those Percentages of nutrition, but rather up to 25 per cent tops.

Moving up the bioavailability spectrum we come to the capsule delivery system. They have no fillers or binders, and you usually have to take larger quantities of these because they are not very concentrated. The bioavailability is a little better, but still only 40-60 per cent. Getting higher up the bioavailability spectrum brings us to liquid supplements. These are better then the previous two in bioavailability and range between 80-90 per cent, but many liquid herbs and oils do not taste very good.

Finally we arrive to isotonic-capable supplements. Isotonic supplements are the most bioavailable delivery of supplements and are up to 95 per cent absorbable.

Isotonic-Capable Supplements

It is the proper ratio of nutrients to water so that it can be seamlessly integrated into the blood stream. An isotonic capable supplement works similarly. It is the proper measurement of nutrients to water to create an isotonic capable solution. When it is taken on an empty stomach, it is recognized as the same pressure as the stomach. The pyloric valve is signaled to open, and the full concentration of nutrients is delivered to the small intestine (our absorption site) where it is up to 95 per cent absorbable within 5-15 minutes of consumption. It is important to note that not all liquids are isotonic capable, again it is a specific ratio of medicine or nutrients to water which makes it isotonic. Everything that you ingest and put into your stomach has to be digested down to an isotonic state before it is released to the small intestine. When you eat some food and take a standard pill or capsule vitamin, it all remains in your stomach digesting for anywhere from 40 minutes to 4 hours.

At that point a certain amount of dilution of nutrients has taken place. Some low quality pills do not even break down due to harsh fillers and binders, but still pass to the small intestine where none of the content of the supplement can be absorbed. This is why the bioavailability of pills and capsules are so significantly lower than isotonic capable supplements.

FAT-SOLUBLE VITAMINS

Fat-soluble vitamins A, D, E, and K differ from the water-soluble vitamins in their accumulative action. Little evidence has been recorded for hypervitaminosis with the water-soluble vitamins since these compounds are rapidly metabolized and excreted when intake exceeds liver or tissue storage capacity, but hypervitaminosis is a common occurrence in fish and other animals when large quantities of any one of the fat-soluble vitamins are ingested. Toxicity symptoms involving vitamins A and D are indistinguishable from deficiency symptoms for the same vitamins. On the other hand, symptoms of excess vitamins E and K intake are more discrete. Fish rations may often be enriched with fish oils to increase caloric density of the ration resulting in excessive intake of the fat-soluble vitamins.

VITAMIN A

A fat-soluble factor that promoted growth in rats was described by Hopkins and by Osborne and Mendel at the beginning of this century. McCollum and Simmonds cured xerophthalmia, a disease of the eye, with this material. The chemical structure of vitamin A and its relationship to -carotene was shown by Von Euler in 1928. Active vitamin A was synthesized in the mid-thirties.

Chemistry

The relationship of the vitamin A alcohols to naturally occurring -carotene is as follows:

Retinene, the aldehyde form of vitamin A, has been isolated from the retina of dark adapted eyes and is involved in vision in dim light. Retinoic acid, which is the oxidized form of vitamin A alcohol, has been shown to have some vitamin A activity. Vitamin A1 is found in saltwater fish, whereas vitamin A is more abundant in freshwater fish. Interconversion of the two forms in living fish tissue has been demonstrated.

Fish oils contain vitamin A as free alcohols or esters. Vitamin A alcohol occurs as a light coloured viscous oil which is heat labile and subject to air oxidation. Beta-carotene occurs as an orange, crystalline compound which is more stable to heat and oxidation. Vitamin A is water insoluble but is soluble in fat and organic solvent.

POSITIVE FUNCTIONS

Vitamin A is essential in maintaining epithelial cells, preventing atrophy and keratinization of epithelial tissues. It is combined with a protein in visual purple and is important in night vision. Vitamin A also prevents xerophthalmia in rats and young children. Vitamin A promotes growth of new cells and aids in maintaining resistance to infection.

THE SYNDROME OF DEFICIENCY OR EXCESS

Hypovitaminosis A is characterized by poor growth, poor vision, keratinization of epithelial tissue, xerophthalmia, night blindness, haemorrhage in the anterior chamber of the eye, haemorrhage of the base of the fins, and abnormal bone formation. Nerve degeneration has been reported in pigs, chickens, rats, rabbits, and ducks, but only occasionally observed in fish after long periods of deficiency. Hypervitaminosis A has been described in fish and in other animals and involved enlargement of liver and spleen, abnormal growth, skin lesions, epithelial keratinization, hyperplasia. of head cartilage, and abnormal bone formation resulting in ankylosis and fusion of vertebrae. Hypervitaminosis A is reflected in very high liver oil vitamin A content and elevated serum alkaline phosphatase. Removal of excess vitamin A from the diet promotes rapid recovery.

SOURCES AND PROTECTION

Cod liver oil is the best known source of vitamin A although black sea bass, swordfish, or ling cod oils contain much more of the vitamin. Whale liver oil contains kitol which has little or no biological activity until heated above 200° C when one molecule of biologically active vitamin A is generated per molecule of whale kitol.

This biologically inactive kitol may be deposited in the whale to avoid hypervitaminosis A during excessive vitamin A intake. The possibility of hypervitaminosis A occurs when tuna, shark, or ling cod viscera are used in preparation of moist fish diets. Synthetic vitamin A preparations, such as vitamin A palmitate, are available and are often used to supplement rations low in fish meal, fish viscera, or carotenes. Some fish species seem able to utilize b-carotene as a vitamin A source, whereas others are unable to split the g-carotene molecule and vitamin A must be added to the diet.

CLINICAL ASSESSMENT

Vitamin A status can best be assessed by the absence of deficiency signs and by assay of liver oil for vitamin A content. Assay for the vitamin in blood or plasma has not been found to be useful.

VITAMIN D

Rickets was induced with test diets by Hopkins, and Mellanby cured the disease in dogs by adding cod liver oil to the ration. Ultraviolet light exposure was shown to have an anti-rachitic effect, and provitamin D activity had been ascribed to ergosterol. Crystalline vitamin D was isolated by Angus and activated 7-dehydrocholesterol was isolated by Windaus.

CHEMISTRY

Vitamin D_2 or ergocalciferol, is one of several biologically active forms of vitamin D. Vitamin D_3 or activated 7-dehydrocholesterol has the chemical formula C27H44O and contains a more simplified, unsaturated 8-carbon side chain. Vitamin D_3, also known as cholecalciferol, is formed in most animal tissue by the rupture of one of the ring bonds of 7-dehydrocholesterol by ultraviolet radiation. Cholecalciferol is a white, crystalline compound soluble in fat and organic solvents and is stable to heat and oxidation in mild alkali or acid solution.

POSITIVE FUNCTIONS

Vitamin D is essential for maintaining calcium and inorganic phosphate homeostasis. It is involved in alkaline phosphatase activity, promotes intestinal absorption of calcium, and influences the action of parathyroid hormone on bone. Fish may sequester calcium from water through the gill membrane; thus the major function of vitamin D for other animals may not be necessary to satisfy calcium requirements for fish.

SYNDROME OF DEFICIENCY OR EXCESS

Hypovitaminosis D for fish has not been described. Rickets and abnormal bone formation described in detail for animals has not been observed in fish fed low vitamin D diets. However little work has been done under carefully controlled conditions with young growing fish and only alteration in alkaline phosphatase activity has been reported on different vitamin D diet intake.

Hypervitaminosis D, however, has been reported. Brook trout fed large doses of vitamin D showed impaired growth, lethargy, and dark colouration. High intake of vitamin D mobilizes phosphorus and calcium from the bone and tissues and may result in fragile bones, poor growth, and poor appetite related to the nausea described in man afflicted with hyper-vitaminosis D.

Arterial and kidney lesions due to excessive intake of vitamin D reported for rats and dogs have not been described histologically for fish nor has hypercalcaemia been described. This area needs to be explored because of the potential for hypervitaminosis D in fish fed diets containing various fish viscera which might contain large amounts of the vitamin D. Tuna liver oil may contain, for example, 100 to 1 000 times as much active vitamin D as cod liver oil.

SOURCES AND PROTECTION

Vitamin D requirements of many animals can be met by ultraviolet irradiation of the skin where cholesterol derivatives are converted to calciferol. Since the vitamin is fat soluble and accumulates in lipid stores, fish liver oil is a rich source of the material.

Content varies tremendously in liver oil, however, with values of about 25 I.U./g present in shark liver oil and over 200 000 I.U./g in albacore tuna liver oil. Cod liver oil contains from 100-500 I.U./g and animal liver contains some vitamin D. One international unit (I.U.) is equal to 0.025 mg of crystalline vitamin D.

CLINICAL ASSESSMENT

Absorption maxima in the ultraviolet region can be used to detect provitamins D in the non-saponifiable fraction of oils. Concentrated preparations of vitamin D can be assayed by the Carr-Price antimony trichloride reaction when assay is necessary to determine biologically active materials in liver oil of fish on different treatments. The chick assay may not apply to fish liver storage levels because the biologically active form for fish has not been determined. Therefore, clinical assessment for hypervitaminosis A must rely on crude methods for determination of vitamin D in liver oil samples by the Carr-Price reaction or by measuring absorption in the UV spectrum.

VITAMIN E

The existence of an antisterility vitamin was postulated by Evans and Bishop. This factor was named 'vitamin E' by Sure. The active tocopherol was isolated, characterized, and synthesized by Karrer.

CHEMISTRY

Vitamin E is composed of a class of compounds known as tocopherols. One of the most important tocopherols is -tocopherol. Eight naturally occurring tocopherol derivatives have been isolated and all belong to the D series. Synthetic -tocopherol is a racemic DL- -tocoperhol mixture. The derivatives of tocol or of tocotrienol are named, alpha, beta, gamma, delta, epsilon, eta, zeta1 and zeta2-tocopherol as shown in Table. The pure tocopherols are fat-soluble oils which are capable of esterification to form crystalline compounds.

The tocopherols are stable to heat and acids in the absence of oxygen, but are rapidly oxidized in the presence of nascent oxygen, peroxides, or other oxidizing agents. The tocopherols are sensitive to ultraviolet light and are excellent antioxidants in the free form. The esters are more stable and are commonly used as dietary supplements-anticipating hydrolysis in the gut and absorption of the free alcohol to act as an active intra- and intercellular antioxidant. Ethyl derivatives on the aromatic ring are also active. Oxidation products of a-tocopherol can be reduced with hydrosulphite to -tocopherylhydroquinone or, in the presence of ascorbic acid, to -tocopherol.

Table. The Tocopherols

Tocopherol (alpha)	5,7,8-trimethyltocol
Tocopherol (beta)	5,8-dimethyltocol
Tocopherol (gamma)	7,8-dimethyltocol
Tocopherol (zeta$_2$)	5,7-dimethyltocol
Tocopherol (eta)	7-methyltocol
Tocopherol (delta)	8-methyltocol
Tocopherol (epsilon)	5,8-dimethyltocotrienol
Tocopherol (zeta$_1$)	5,7,8-trimethyltocotrienol

POSITIVE FUNCTIONS

The tocopherols act as extra and intracellular antioxidants to maintain homeostasis of labile metabolites in the cell and tissue plasma. As physiological antioxidants, these usually protect oxidizable vitamins and labile unsaturated fatty acids. Vitamin E prevents encephalomalacia in chicks, erythrocyte haemolysis in several animals and steatitis in mink, pigs, and farm animals.

The tocopherols also prevent exudative diathesis in chicks, white muscle disease in lambs and calves, and dietary liver necrosis in rats. They act with selenium and with vitamin C for normal reproductive activity and are involved in prevention of nutritional muscular dystrophy in the chick, the yellowtail and carp. The tocopherols act as free radical traps to stop the chain reaction during peroxide formation and stabilize unsaturated carbon bonds of polyunsaturated fatty acids and other long-chain labile compounds.

Vitamin E is involved in the maintenance of normal blood capillary permeability and the integrity of heart muscle. It was first shown to be involved in prevention of sterility and foetal resorption in rats. It may likewise affect embryo membrane permeability and hatch-ability of fish eggs.

SYNDROME OF DEFICIENCY OR EXCESS

Deficiency signs in fish are listed in Table. One of the first signs is erythrocyte fragility closely followed by anaemia, ascites, xerophthalmia, poor growth, poor food conversion, epicarditis, and ceroid deposits in spleen and liver. Muscle dystrophy and xerophthalmia have been described in yellowtail and carp. Non-specific forms of degenerative conditions have been described in several species of fish fed large quantities of polyunsaturated fatty acids with inadequate tocopherol in the ration. Hypervitaminosis E results in poor growth, toxic liver reaction, and death.

SOURCES AND PROTECTION

Vegetable oils are rich sources of tocopherols. Synthetic -tocopherol in esterfied acetate or phosphate form is commonly used as a diet supplement. These esters are much more stable than the free form which is rapidly lost by air oxidation or in the presence of labile compounds like the polyunsaturated fish oils. Addition of antioxidants such as BHT (butylated hydroxyanisole) and BHT (butylated hydroxytoluene) protects fats and other labile compounds in the ration from oxidation, but these antioxidants have no vitamin E activity.

CLINICAL ASSESSMENT

The erythrocyte fragility test indicates the physiological state of fish. Absence of histologically detectable ceroid in liver and spleen from representative samples in the population is a good clue on the presence of adequate amounts of physiological antioxidants in the fish. A barbituric acid test for oxidation of components has not been applied to fish tissues as a clinical tool for assessment of nutrition state except when liver oils become saturated with very labile polyunsaturated fatty acids such as when feeding squid or saury oils. These assays are good indicators of the state of oxidation of the finished ration, but the absence of deficiency signs and normal erythrocyte fragility are better clinical tests. Analysis for tocopherol is difficult and time-consuming and only applicable under critical research experiment situation.

VITAMIN K

The name vitamin K (for 'koagulation') was proposed by Dam. Dam isolated the vitamin from alfalfa and from fish meal in 1939. It was synthesized later that year.

CHEMISTRY

The K vitamins K_2 contain 6,7, or 9 isoprene units in the side chain which varies from 30-45 carbon atoms. Many isosteres of vitamin K have been identified in animal tissues, plant tissues, and micro-organisms. The structure of phthiocol and menadione (K_3) are shown below: Although fairly stable compounds, they are destroyed upon oxidation and exposure to ultraviolet radition. Menadione is very reactive and is subject in aqueous media to chemical interaction.

POSITIVE FUNCTIONS

Vitamin K is involved in the hepatic synthesis of blood clotting proteins - prothrombin and proconvertin. Substituted forms of vitamin K are strongly bacteriostatic and may serve as an alternate defence mechanism for bacterial infections. Vitamin K is structurally related to flavo compounds which function as electron carriers in the electron transport system. The primary role of vitamin K is to maintain a fast normal blood clotting rate which is so important to fish living in a water environment.

SYNDROME OF DEFICIENCY OR EXCESS

Prothrombin time in salmon fed diets devoid of vitamin K was increased three to five times and, during prolonged deficiency states, anaemia and haemorrhagic areas appeared in the gills, eyes, and vascular tissues.

Increased blood clotting time has also been reported for other fish reared on diets with low vitamin K content. Interrelationships with other vitamins have not been documented in fish where the primary deficiency signs are slow blood clotting and haemorrhage, severe anaemia, or death in wounded fish. Haemorrhagic areas often appear in fragile tissues such as the gills.

SOURCES AND PROTECTION

Vitamin K sources are green, leafy vegetables. Alfalfa leaves are among the best sources of vitamin K. Low levels are found in soybeans and animal liver. Synthetic menadione is also available. Although vitamin K in ground alfalfa is fairly stable, synthetic material should be protected from exposure to ultraviolet light and to excessive oxidizing or reducing conditions. The use of rapidly cured dehydrated alfalfa is essential to minimize formation of the physiological antagonist, dicumarol. The diet should be kept dry, prepared with minimum exposure to air oxidation, and fed as soon as practicable after manufacture to minimize vitamin K loss.

ANTIMETABOLITES AND INACTIVATION

Two vitamin K antagonists, dicumarol and warfarin are shown below:

Dicumarol was isolated from spoiled sweet clover hay and shown to prevent the normal function of vitamin K in maintenance of normal blood clotting times. Early work was with cattle and hogs but was soon extended to rats, other experimental animals, and man.

Dicumarol is an anticoagulant and has been used to prevent thrombosis in animals and man. It is not an antimetabolite competing for vitamin K sites but plays another role in preventing normal blood clotting. Vitamin K counteracts the dicumarol effect quantitatively. Another anticoagulant, warfarin, is a common rat poison which has five to ten times the anticoagulant activity of dicumarol. The effects of warfarin can also be reversed by administration of vitamin K.

CLINICAL ASSESSMENT

Vitamin K status of fish is determined by measuring blood clotting time. Chemical methods for determining menadione content of active metabolite after storage are available but have not been applied to measure vitamin K activity in the liver or tissues of young, growing fish.

3

Amino Acids

One of a class of organic compounds containing the amino (NH_2) and the carboxyl (COOH) group, occurring naturally in plant and animal tissues and forming the chief constituents of protein; many of them are necessary for human and animal growth and nutrition and hence are called Essential Amino Acids. Amino Acids are the building blocks of protein.

PRIMARY FUNCTIONS OF AMINO ACIDS

Depending upon the source of dated reference material studied, some reports state"20" Amino Acids have been identified. More modern reference materials state"22" Amino Acids have been identified. In my personal research, identifying all possibilities available to me at this time, I have listed"28" Amino Acids/ or Amino Acid Derivatives. Of my 28 listings, 22 have been identified as"Commonly Occurring" in Proteins.

In living things, Amino Acids serve as the primary unit of synthesis both of Tissue Proteins and other Non-Protein Nitrogenous Compounds.

Amino acids may undergo complex inner conversions in the body into macromolecules such as carbohydrates, nucleic acids, lipids, etc. Mineral co-factors also play an important role in that they interact with enzymes to assist with their catalytic activities. They are also capable of complexing with the individual free amino acids. For example, magnesium (MG2+) is important in the phosphorylation process or the transfer of high-energy phosphate groups into the body.

Upon entering the body any free amino acids undergo complex inner-conversions into those substances necessary to maintain the metabolic pathways in the body. For example, Phenylalanine goes to Tyrosine and Arginine goes to Citrulline and Ornithine. (The Arginine pathway is a very important cycle.) Amino acid inner-conversions are not limited to other amino acids. Many free form amino acids play important roles as precursors for metabolic intermediaries.

An example of this, Tyrosine's and therefore Phenylalanine's ability to give rise to two hormones Thyroxsine and Epinephrine. The Sulfur containing amino acids often precurse Taurine, which is an important bio-acid component. As W.C. Rose observed in 1938,"Perhaps undo emphasis was placed upon the quantity of protein ingestion and that scant consideration was given to possible difference in nutrient quality."

Upon ingestion of foodstuff all nutrients and amino acids do not immediately diffuse into the surrounding tissue of the blood stream, but must first undergo a series of steps of biochemical reactions in the digestive tract. These reactions reduce the protein into its' individual amino acids. Protein cleaving enzymes are activated in the digestive process, but in order for this cleavage to occur energy must be supplied on the part of the individuals metabolism to break the peptide bonds. By starting with a free amino acid mixture in which all amino acids are in their L conformation little digestion is necessary. The result is an energy saving and rapid absorption into the blood stream and surrounding tissues."Amino acid supplementation of foods and feeds bring about many beneficial results. Of practical importance is the attainment of a better balance of protein and therefore a better balanced diet, an extension or savings of the available protein supply, and an improvement in the efficiency of the protein and food utilization.

FUNCTIONAL IDENTITY OF AMINO ACIDS

Of all the fundamental amino acids, 10 (ten) are classified as"ESSENTIAL" and are NOT biosynthesized by the body. All commonly occurring amino acids can be subdivided into 5 fundamental groups each of which has functional significance.

- *Group* #1: This group includes all the amino acids possessing aliphatic properties (those whose carbon atoms are joined in open chains), Glycine, Alanine, Leucine, w, Serine, Threonine, and Isoleucine.
- *Group* #2: These are the Sulfur containing amino acids, Methionine, Cystine, Cysteine.
- *Group* #3: These amino acids are the acidic amino acids, (those that can donate a hydrogen ion), Aspartic Acid, Glutamic Acid, and their associated amids.
- *Group* #4: These are the diamino acids (those with additional NH_2 groups), Arginine and Lysine.
- *Group* #5: These are the cyclic and aromatic amino acids, Phenylalanine, Tyrosine, Tryptophan, Histidine, Proline, and Hydroxyproline that is synthesized from Prolamine.

ESSENTIAL AMINO ACIDS

Table. Estimated Minimum Requirements of Adults for Essential Amino Acids

Amino Acid	Women/gms/day	Men/gms/day
Cystine	0.055	1.010
Isoleucine	0.450	0.700

Leucine	0.620	1.100
Lysine	0.500	0.800
Methionine	questionable	1.100
Phenylalanine	0.220	1.010
Threonine	0.310	0.500
Tryptophan	0.160	0.250
Tyrosine	0.900	questionable
Valine	0.650	0.800

Cystine

An essential, Sulfur-containing amino acid, produced by digestion or acid hydrolysis of proteins, sometimes found in the urine and kidneys, and readily reduced to two molecules of Cysteine.

- Reported helpful in dermatological conditions.
- Promotes faster recovery of tissue after surgery.
- Part of the Insulin molecule.
- Found high in hair (sulfur bonds).

Isoleucine

An essential amino acid produced by hydrolysis of fibrin and other proteins; ESSESNTIAL for optimal INFANT growth and for nitrogen equilibrium in adults.

- Needed along with other branch chain amino acids for all rebuilding of muscle tissue.
- Role in the release of energy during muscular work. (Metabolizes along the same pathway as fat.)

Leucine

An essential amino acid, ESSENTIAL for optimal growth in INFANTS, and for nitrogen equilibrium in adults.

- Metabolized along the same pathways as fat.
- Precursors of cholesterol.
- Involved in the role of energy release during any work of the muscles.

Lysine

An essential, naturally occurring amino acid, ESSESNTIAL for optimal growth in human INFANTS, and for maintenance of nitrogen equilibrium in adults.

- Reported to inhibit growth and replication of herpes simplex and Epstein Barr viruses (EBV).
- Promotes bone growth in infants.
- Stimulates secretion of gastric juices.

- Found in abundance in muscle tissue, connective tissue and collagen.
- Low in vegetarian diets.

Methionine

An essential, naturally occurring amino acid, which is an essential component of the diet, furnishing both methyl groups and sulfur necessary for normal metabolism.

- Prevents deposits and cohesion of fats in the liver.
- Gives rise to Taurine (an important inhibitory neuro modulator in the brain).
- Involved in synthesis of Choline (Must be given with Vitamin B6 (paradoxial-5-phosphate) to inhibit synthesis of homocysteine which promotes plaque deposition in the arteries.)

Phenylalanine

An essential, naturally occurring amino acid, ESSESNTIAL for optimal growth in INFANTS, and for nitrogen equilibrium in human adults.

- May be useful in appetite control by stimulating CCK (Chaleceptokinin Enzyme) secretion.
- Shown to be useful in management of certain types of depression.
- Increases blood pressure in hypotension.
- Gives rise to Tyrosine.
- It is one of the amino acids in the dypeptide sweetener Aspartain. (Aspartic acid and Phenolalanine).

Threonine

An essential, naturally occurring amino acid, essential for human metabolism.

- Rises to three times it's normal value at pregnancy.
- Acts as a lipotropic factor.
- Recently found to increase brain Glycine content, greatly reducing ALS symptoms. (Amyotrophic Lateral Sclerosis or"Lue Gehrig's Disease.")

Tryptophan

An essential, naturally occurring amino acid, existing in proteins and essential for human metabolism.

- Reported useful in the management of depression and schizophrenia.
- Produces Serotonin, which induces sleep. Has a Serotonengenic effect.
- Precursor of the vitamin Niacin.
- Vasoconstrictor, which appears to aid in blood clotting mechanism, aids in elevating the threshold of pain.

Tyrosine

An essential, naturally occurring amino acid present in most proteins. It is a product of Phenylalanine metabolism and a precursor of thyroid hormones, catecholamines, and melanin.

- Along with Phenylalanine, it is a useful anti depressant due to increased production of Tacolamine. (Recommended to be taken with Vitamin B6 (paradoxial-5-phosphate).)
- Reported to stabilize blood pressure by lowering energy in some cases and elevating it in others.
- Involved in tissue pigmentation
- Is important in the formation of thyroid hormone

Valine

An essential, naturally occurring amino acid, essential for human metabolism.

- Needed for all muscle building.
- Required in the precursors of cholesterol.

Histidine

An amino acid obtainable from many proteins by the action of sulfuric acid and water; it is ESSENTIAL for optimal growth in INFANTS. Its decarboxylation results in formation of histamine.

- Release of histamines from body stores are required for sexual arousal.
- Reported useful in alleviating pain associated with rheumatoid arthritis. (Should be taken with vitamin C; Should be used with caution with Manic Depressives with elevated histamines; Should be used with caution in woman with severe depression or suicidal tendencies due to Premenstrual Syndrome; Is considered a mild neurotransmitter).

ESSENTIAL AND NON-ESSENTIAL AMINO ACIDS

Salmon, trout and channel catfish fed diets devoid of arginine, histidine, isoleucine, leucine, lysine, methionine, phenylalanine, threonine, tryptophan or valine failed to grow. These same fish fed diets devoid of other L-amino acids grew as well as fish receiving all 18 amino acids tested.

The nitrogen component in the test diets was made up of 18 L-amino acids in the pattern found in whole egg protein. All fish on test recovered rapidly when the missing amino acid was replaced in the diet. The slope of the growth curve of the recovery group was identical with that of fish receiving the complete amino acid test diet. Dispensable amino acids tested were alanine, aspartic acid, cystine, glutamic acid, glycine, proline, serine and tyrosine. These amino acids were found to be not essential for the growth of salmon, trout and channel catfish. Quantitative studies on the requirements of the 10 indispensable amino acids used a casein-gelatin mixture supplemented with crystalline L-amino acids. The test diet had an amino acid pattern of 40 per cent whole egg protein for the nitrogen component. Experiments conducted with carp and eel showed a similar lack of growth when an indispensable amino acid was absent from the diet.

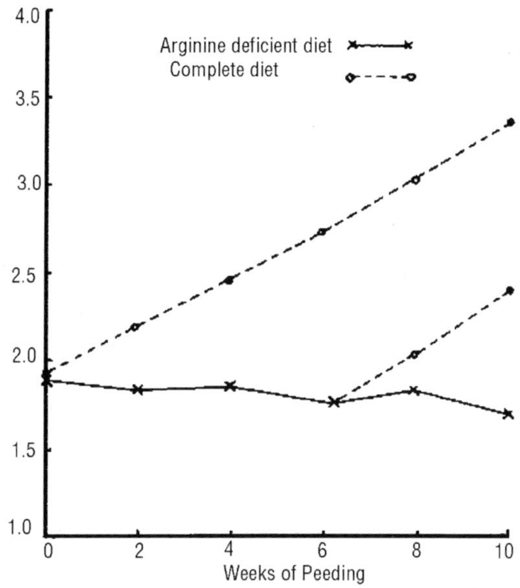

Fig. Growth of Arginine Deficient Fish. The Deficient Group was Divided After Six Weeks on the Deficient Diet and the missing Amino Acid was Replaced in One of the Two Sub-lots

Essential Amino Acids and Protein Quality

If the essential amino acid requirements of fish are known, it should be possible to meet these needs in culture systems in a number of ways from different food proteins or combinations of food proteins.

Phenylalanine is spared by tyrosine. It is not known to be chemically modified nor rendered unavailable by the harsh conditions to which feedstuff proteins are normally subjected during processing. Measurement of phenylalanine in proteins is uncomplicated so that the provision and evaluation of phenylalanine in proteins in practical diets presents little difficulty.

Lysine is a basic amino acid. In addition to the a -amino acid group normally bound in peptide linkage, it also contains a second, a -amino group. This a -amino group must be free and reactive, otherwise the lysine, although chemically measurable, will not be biologically available. During the processing of feedstuff proteins the a -amino group of lysine may react with non-protein molecules present in the feedstuff to form additional compounds that render the lysine biologically unavailable. Methionine is spared by cystine. However, measurement of the methionine content of feed proteins is not easy as the amino acid is subject to oxidation during processing. After processing, methionine may be present as such or as the sulphoxide or as the sulphone. The sulphoxide may be formed from methionine during acid hydrolysis of the feed protein prior to measurement of its any-no acid composition. Acid hydrolysis of proteins before analysis disturbs the original equilibrium between the two compounds so that the composition of the hydrolysate no longer reflects that of the protein. In determining the methionine content of pure proteins, oxidation of the amino acid to methionine sulphone is normally quantitative. In the case of feed proteins, however, this will not reveal how much methionine or methionine sulphoxide was present in the protein prior to performate oxidation and hydrolysis.

Methionine sulphoxide may have some biological value for fish which may have some capability of reconverting it to methionine and thus partially make up for some of the methionine oxidized during processing. Methods have recently been reported for measurement of methionine in proteins using an iodoplatinate reagent before and after reduction with titanium trichloride, to give values for both methionine and the sulphoxide in the original protein. A method for measuring methionine specifically by cyanogen bromide cleavage has also been described.

Both methods remain to be independently assessed. Microbiological assay of methionine in feed proteins is a valuable tool although there is the danger that oxides of methionine may differ in their activity for micro-organisms and misrepresent values.

QUANTITATIVE REQUIREMENTS OF AMINO ACID

Quantitative requirements by salmonids for the ten indispensable amino acids were determined by feeding linear increments of one amino acid at a time in a test diet containing an amino acid profile identical with whole egg protein except for the amino acid tested.

Replicate groups of fish were fed the diet treatments until gross differences appeared in the growth of test lots. An Almquist plot of growth response indicated the level of amino acids required for maximum growth under those specific test conditions.

Diets were designed to contain protein at or slightly below the optimum protein requirement for that species and test condition to assure maximum utilization of the limiting amino acid. A comparison of the requirements for the ten indispensable amino acids.

A recent innovation has been the use in test diets of proteins relatively deficient in a given essential amino acid. Thus combinations of fishmeal and zein have been used in test diets to define the requirement of rainbow trout for arginine. Diets containing different relative amounts of casein and gelatin showed that an increase in the level of protein-bound arginine from 11 to 17 g/kg resulted in a significant increase in the growth of channel catfish.

Table. Amino Acid Requirements of Seven Animals

Amino acid	Eel fingerling	Carp fry	Channel catfish	Chinook salmon fingerling	Chick	Young Pig	Rat
Arginine	3.9 (1.7/42)	4.3 (1.65/38.5)		6.0 (2.4/40)	6.1 (1.1/18)	1.5 (0.2/13)	1.0 (0.2/19)
Histidine	1.9 (0.8/42)			1.8 (0.7/40)	1.7 (0.3/18)	1.5 (0.2/13)	2.1 (0.4/19)
Isoleucine	3.6 (1.5/42)	2.6 (1.0/38.5)		2.2 (0.9/41)	4.4 (0.8/18)	4.6 (0.6/13)	3.9 (0.5/13)
Leucine	4.1 (1.7/42)	3.9 (1.5/38.5)		3.9 (1.6/41)	6.7 (1.2/18)	4.6 (0.6/13)	4.5 (0.9/19)
Lysine	4.8 (2.0/42)		5.1 (1.23/24.0)	5.0 (2.0/40)	6.1 (1.1/18)	4.7 (0.65/13)	5.4 (1.0/19)
Methionine [2/]	4.5 (2.1/42) [3/]	3.1 (1.2/38.5)	2.3 (0.56/24.0)	4.0 (1.6/40) [3/]	4.4 (0.8/18)	3.0 (0.6/20)	3.0 (0.6/20)
Phenylalanine [4/]				5.1 (2.1/41) [5/]	7.2 (1.3/18)	3.6 (0.45/13)	5.3 (0.9/17)
Threonine	3.6 (1.5/42)			2.2 (0.9/40)	3.3 (0.6/18)	3.0 (0.4/13)	3.1 (0.2/19)
Tryptophan	1.0 (0.4/42)			0.5 (0.2/40)	1.1 (0.2/18)	0.8 (0.2/25)	1.0 (0.2/19)
Valine	3.6 (1.5/42)			3.2 (1.3/40)	4.4 (0.8/18)	3.1 (0.4/13)	3.1 (0.4/13)

[1/] Expressed as per cent of dietary protein.

In parentheses, the numerators are requirements as per cent of dry diet and the denominators are per cent total protein in the diet

[2/] In the absence of cystine

[3/] Methionine plus cystine

[4/] In the absence of tyro sine

[5/] Phenylalanine plus tyrosine

Arginine requirement of rainbow trout has been determined from a conventional dose/response (growth) curve and also by measuring the tissue (blood and muscle) levels of free arginine in groups of trout given increasing amounts of dietary arginine. After the dietary requirement of the trout for arginine has been met, any further increase in arginine intake led to an increase in the concentration of free arginine in blood and muscle.

Good agreement was obtained between the two methods. The data shown in Table suggest that real differences exist between fish species in their requirement for certain amino acids. This leads to difficulties in formulating the protein component of practical diets for those species whose amino acid requirements are not yet known. A possible solution is to use, for each amino acid, the highest level required by any of those species for which data is available.

The need for further quantitative data on the amino acid requirements of fish, especially those actually or potentially useful as farm animals, is obvious.

SUPPLEMENTING DIETS WITH AMINO ACIDS

One solution to the use of proteins that are relatively deficient in one or more amino acids is to supplement the protein with appropriate amounts of the amino acid needed in practical diets. Fish appear to utilize free amino acids at various degrees of efficiency.

Young carp, Cyprinus carpio, were shown to be unable to grow on diets in which the protein component (casein, gelatin) was replaced by a mixture of amino acids similar in overall composition. A trypsin hydrolyzate of casein was equally ineffective. However, if a diet containing free amino acids as the protein component is carefully neutralized with NaOH to pH 6.5-6.7 then some growth of young carp does occur. This growth was markedly inferior to that occurring on a comparable casein diet under the same conditions.

Channel catfish are also unable to utilize free amino acids given as supplements to deficient proteins. When soybean meal was substituted isonitrogenously for menhaden meal, growth and feed efficiency of channel catfish were substantially reduced. Addition of free methionine, cystine or lysine, the most limiting amino acids, to these soy-substituted diets did not enhance weight gain. Raising the arginine level of catfish diets from 11 to 17 g/kg by isonitrogenous substitution of gelatin for casein enhanced weight gain significantly but the addition of free arginine, cystine, tryptophan or methionine to casein had little effect on growth or food conversion.

Salmonids are able to utilize free amino acids for growth. A zein-gelatin diet supplemented with lysine and trytophan was shown to be markedly superior to an unsupplemented zein-gelatin diet for rainbow trout when weight gain and protein utilization were used as criteria.

Several investigators have demonstrated the potential of supplementing amino acid deficient proteins with limiting amino acids in diets for salmonids. Casein supplemented with six amino acids produced feed conversion ratios with Atlantic salmon similar to those obtained when an isolated fish protein was used as the dietary protein source.

Soybean meal supplemented with five or more amino acids (including methionine and lysine) was a superior protein source to soybean meal alone for rainbow trout. Single additions of methionine and lysine did not, however, improve the value of soybean meal. These results suggest that the amino acid spectrum of the isolated fish protein they used may possibly approximate the amino acid requirement of rainbow trout. The nutritional value of a soy protein isolate could be enhanced by supplementing it with the first limiting amino acid; i.e., methionine.

Diets containing, as protein component, fishmeal, meat and bone meal and yeast and soybean meal could be improved by supplementing with cystine (10 g/kg) and tryptophan (5 g/kg) together.

Fishmeal can be entirely replaced without a reduction in food conversion rate in diets for rainbow trout by a mixture of poultry by-product meal and feather meal together with 17 g lysine HCL/kg, 4.8 g DL-methionine/kg and 1.44 g DL-tryptophan/kg.

TYPES OF AMINO ACIDS

Aminoacetic Acid

A nonessential amino acid, derived from many proteins, used as a gastric antacid and dietary supplement, and in the treatment of various myopathies and peripheral vascular insufficiencies. A solution is used as an irrigation fluid. Its hydrochloride salt is use as a source of hydrochloric acid in the treatment of achlorhydria.

Alanine

A natural amino acid occurring in two forms, alpha-Alanine, and beta-Alanine.

- Hypoglycemia (Prolongs the stabilization of blood glucose over long periods of time.)
- Alanine may be used for the production of glucose in gluconeogenesis.

Arginine

An amino acid occurring in proteins; it is involved in the urea cycle, which converts ammonia to urea.

- Induces growth hormone release from the pituitary gland.
- Major component of seminal fluid.
- Helpful in burn treatment, elevated ammonia levels, and cirrhosis of the liver.
- Stimulates immune response by enhancing the production of T-Cells.
- Protective effect of toxicity of hydrocarbons and intravenous diuretics.

(Should be used with caution in schizophrenic cases.)

Intake should be kept low in persons with Herpes simplex and Epstein Barr Virus (EBV. Mechanism of interaction is believed because of possible replications of the Herpes and EB viruses.)

Aspartic Acid

A nonessential, natural dibasic amino acid, involved in transamination reactions, the ornithine cycle, and the formation of carnosine, anserine, purines, and pyrimidines.

- Has a protective function over the liver.
- Helps in detoxification of ammonia.
- Promotes mineral uptake in the intestinal tract.
- It is part of the sweetener Aspartain that is a dypeptide of the amino acids; Aspartic Acid and Phenylalanine.

Carnitine

A betaine derivative involved in the transport of fatty acids into mitochondria, where they are metabolized.

- Suggested as useful in mobilization of surface fats such as cellulite.
- Helpful in treatment of fatigue and when muscle weakness is present. Useful in the oxidation of long chain fatty acids. A major source of energy for tissues.
- Reported as useful in myocardial ischemia (low energy metabolism by the heart).
- Useful in clearing triglycerides from the blood.
- Useful in metabolic abnormalities.

Citrulline

An alpha-amino acid involved in urea production; formed from ornithine and is itself converted into Arginine in the urea cycle.

- Produces the amino acids, Arginine and Ornithine.
- Detoxifies nitrogen containing waste products such as ammonia (a cellular toxin) and is therefore part of ammonia detoxification (urea cycle).
- Stimulates growth hormone production.

Cysteine

A sulfur-containing amino acid produced by enzymatic or acid hydrolysis of proteins, readily oxidized to Cystine, sometimes found in urine.

- Assists tissues damaged by alcohol abuse, cigarette smoke and air pollution through the detoxification of Acetalaldahide.
- Helps maintain skin flexibility and texture by slowing abnormal cross linking of collagen (connective tissue protein which holds the skin together).
- Promotes red and white blood cell reproduction and tissue restoration in lung affecting diseases.
- Useful in iron deficient anemia and promotion of iron absorption.
- Helpful in prevention of peroxidized fats and free radicals.

- Converts to Cystinine in the absence of vitamin C.
- One of the amino acids found in tripeptide Glutathione and GTF.
- Hair contains between 10 to 16 per cent Cysteine.
- Heavy metal chelator. (Should be used with caution in diabetics due to possible three dimensional structure changes in the insulin cycle)

Cystine

An essential, Sulfur-containing amino acid, produced by digestion or acid hydrolysis of proteins, sometimes found in the urine and kidneys, and readily reduced to two molecules of Cysteine.

- Reported helpful in dermatological conditions.
- Promotes faster recovery of tissue after surgery.
- Part of the Insulin molecule.
- Found high in hair (sulfur bonds).

GABA - (Gamma Amino Butyric Acid)

An amino acid that is one of the principle inhibitory neurotransmitters in the central nervous system.

- Useful in schizophrenia and epilepsy, depression, high blood pressure, high stress levels, manic Behaviour, and acute agitation due to increased regulation of nerve firings and enhancement of Niacinamide binding receptors.
- May induce calmness and tranquility by inhibiting neurotransmitters, which decrease the activity in neurons.
- May be useful in reducing an enlarged prostate due to the suppression of Prolactia released by the pituitary gland.

Glutamic Acid

Glutamate is a salt of Glutamic Acid; in biochemistry, the term is often used interchangeably with Glutamic Acid. Glutamic Acid is a crystalline dibasic nonessential amino acid, widely distributed in proteins, which is thought to be a neurotransmitter, inhibiting neural excitation in the central nervous system; its hydrochloride salt is used as a gastric acidifier. The monosodium salt of L-Glutamic acid (sodium glutamate) is used in treating encephalopathies associated with hepatic diseases, and to enhance the flavour of foods and tobacco.

- Reported useful in Muscular Dystrophy.
- Does not cross blood brain barriers.
- It is an excitatory neurotransmitter in the CNS; (Precursor of GABA.)

Glutamine

The monoamide of Glutamic Acid, an amino acid occurring in proteins;

it is an important carrier of urinary ammonia and is broken down in the kidney by the enzyme glutaminase.

- Is useful in treatment of Alcoholism by reducing the desire to drink.
- May be helpful in the improvement of Autism and mental retardation in children.
- Reported improvement in mental functions such as memory and dexterity.
- Aids peptic ulcer healing due to antacid quality.
- Crosses Blood brain barriers.
- Participates in nucleonic acid synthesis.
- Converts to Glutanic acid.

Glutathione

Reduced Glutathione, a tripeptide of Glutamic Acid, Cysteine, and Glycine, which serves as a reducing agent in many biochemical reactions being converted to oxidized Glutathione (GSSG) in which the Cysteine residues of two Glutathione molecules are connected by a disulfide bridge. Reduced Glutathione is important in protecting erythrocytes (red blood cells) from oxidation and hemolysis; deficiency causes sensitivity to oxidant drugs.

- It is a tripeptide containing amino acid that contains: Cysteine, Glycine, and Glutamate.
- Inhibits peroxide formations, reducing free radical damage.
- Detoxifies aromatic hydrocarbons common in air pollution, such as chlorine. Reported to have protective function against radiation therapy.
- Transports other amino acids into the interior of the cell.

Glycine

A nonessential amino acid, occurring as a constituent of proteins and functioning as an inhibitory neurotransmitter in the central nervous system; used as a gastric antacid and dietary supplement, and in the treatment of various myopathies.

- Is the simplest and sweetest of the amino acids. Can be used as a sweetener in herbal beverages.
- Reported useful in degenerative diseases such as MD (Muscular Dystrophy)
- Detoxifies benzoic acid (a common food additive) and aromatic hydrocarbons in the liver.
- Involved in synthesis of nucleic acids and bio-acids. May be useful in conditions characterized by abnormal nerve firing such as epilepsy, inhibition of tripeptides Glutathione and GTF.
- Is one of the amino acids in the tripeptides Glutathione and GTF; low brain concentrations of Glycine have recently been found in ALS. (Amyotrophic Lateral Sclerosis or"Lue Gehrig's Disease.")

Histidine

Is not listed as essential for adults, but is very essential for infants. An amino acid obtainable from many proteins by the action of sulfuric acid and water; it is ESSENTIAL for optimal growth in INFANTS. Its decarboxylation results in formation of histamine.

- Release of histamines from body stores are required for sexual arousal.
- Reported useful in alleviating pain associated with rheumatoid arthritis. (Should be taken with vitamin C; Should be used with caution with Manic Depressives with elevated histamines; Should be used with caution in woman with severe depression or suicidal tendencies due to Premenstrual Syndrome; Is considered a mild neurotransmitter).

Hydroxyproline

An amino acid produced in the digestion of hydrolytic decomposition of proteins, especially collagens.

Isoleucine

An essential amino acid produced by hydrolysis of fibrin and other proteins; ESSESNTIAL for optimal INFANT growth and for nitrogen equilibrium in adults.

- Needed along with other branch chain amino acids for all rebuilding of muscle tissue.
- Role in the release of energy during muscular work. (Metabolizes along the same pathway as fat.)

Leucine

An essential amino acid, ESSENTIAL for optimal growth in INFANTS, and for nitrogen equilibrium in adults.

- Metabolized along the same pathways as fat.
- Precursors of cholesterol.
- Involved in the role of energy release during any work of the muscles.

Lysine

An essential, naturally occurring amino acid, ESSESNTIAL for optimal growth in human INFANTS, and for maintenance of nitrogen equilibrium in adults.

- Reported to inhibit growth and replication of herpes simplex and Epstein Barr viruses (EBV).
- Promotes bone growth in infants.
- Stimulates secretion of gastric juices.

- Found in abundance in muscle tissue, connective tissue and collagen.
- Low in vegetarian diets.

Methionine

An essential, naturally occurring amino acid, which is an essential component of the diet, furnishing both methyl groups and sulfur necessary for normal metabolism.

- Prevents deposits and cohesion of fats in the liver.
- Gives rise to Taurine (an important inhibitory neuro modulator in the brain).
- Involved in synthesis of Choline (must be given with Vitamin B6 (paradoxial-5-phosphate) to inhibit synthesis of homocysteine which promotes plaque deposition in the arteries)

Autistic Patients

"Taurine is deficient in Autistic patients; Methionine has an altered abnormally normal to high level, with anything distal to the Methionine pathway being low. Since Taurine is so important in the intelligence quotients of most species, this seems to be a significant." Alpha A - Leverton, Protein and Amino Acid Nutrition (1959) B - Rose, Journalism of Biological Chemistry 217:977, 1955.

Ornithine

An amino acid obtained from Arginine by the splitting of urea; it is an intermediate in urea biosynthesis.

- May reduce fat and increase muscle mass by promoting fat metabolism and stimulating growth hormone production.
- Helps in detoxification of ammonia in the urea cycle.
- May be useful in autoimmune disease such as arthritis.

Phenylalanine

An essential, naturally occurring amino acid, ESSESNTIAL for optimal growth in INFANTS, and for nitrogen equilibrium in human adults.

- May be useful in appetite control by stimulating CCK (Chaleceptokinin Enzyme) secretion.
- Shown to be useful in management of certain types of depression.
- Increases blood pressure in hypotension.
- Gives rise to Tyrosine
- It is one of the amino acids in the dypeptide sweetener Aspartain.

Proline

A cyclic amino acid occurring in proteins; it is a major constituent of collagen.

- Is an anti-hypotensive agent in lowering high blood pressure.

- Reported helpful in repairing muscle and tendon damage.
- Reported useful in promoting skin flexibility in relation to aging and sun exposure.
- A major amino acid found in collagen (connective tissue) in the presence of Vitamin C.

Serine

A naturally occurring Amino Acid, present in many proteins.

Taurine

A crystallized acid, ethylamine sulfonic acid, from the bile; found also in small quantities in lung and muscle tissue.

- Low levels seen in newborn infants fed low Taurine diets.
- Associated with retinal degenerations.
- The role of Taurine as a nutrient is to protect the cell membranes by attenuating such toxic compounds as oxidants, secondary bioacids and antibiotics.
- Recommended for children on long-term parenteral nutrition.
- Helpful in balancing calcium and potassium flux in heart muscle.
- Helps patients suffering from congenitive heart failure by alleviating their physical signs and symptoms.
- Increases left ventricular performance without any significant changes in atrial pressure.
- Often times considered a neuro modulator.
- Helpful in treating some types of epilepsy.

Autistic Patients:"Taurine is deficient in autistic patients; Methionine has an altered abnormally normal to high level, with anything distal to the Methionine pathway being low. Since Taurine is so important in the intelligence quotients of most species, this seems to be a significant.

Threonine

An essential, naturally occurring amino acid, essential for human metabolism.

- Rises to three times it's normal value at pregnancy.
- Acts as a lipotropic factor.
- Recently found to increase brain Glycine content, greatly reducing ALS symptoms.

Tryptophan

An essential, naturally occurring amino acid, existing in proteins and essential for human metabolism.

- Reported useful in the management of depression and schizophrenia.
- Produces Serotonin, which induces sleep. Has a Serotonengenic effect.

- Precursor of the vitamin Niacin.
- Vasoconstrictor, which appears to aid in blood clotting mechanism, aids in elevating the threshold of pain.

(Should be taken with Vitamin B6 (paradoxial-5-phosphate) and in the presence of carbohydrates such as fruit or vegetable juice to maximize uptake in the brain.)

Tyrosine

An essential, naturally occurring amino acid present in most proteins. It is a product of Phenylalanine metabolism and a precursor of thyroid hormones, catecholamines, and melanin.

- Along with Phenylalanine, it is a useful anti depressant due to increased production of Tacolamine. (Recommended to be taken with Vitamin B6 (paradoxial-5-phosphate).)
- Reported to stabilize blood pressure by lowering energy in some cases and elevating it in others.
- Involved in tissue pigmentation
- Is important in the formation of thyroid hormone (Should not be used when MAO's are prescribed or when cancer melanoma is present)

Valine

An essential, naturrally occurring amino acid, essential for human metabolism.

- Needed for all muscle building.
- Required in the precursors of cholesterol.

STRUCTURE OF AMINO ACIDS

Proteins are polymers of amino acids.. Amino acids, with rare exception, contain an á carbon that is connected to an amino (NH_3) group, a carboxyl group (COOH) and a variable side group (R). The side group gives each amino acid its distinctive properties and helps to dictate the folding of the protein.

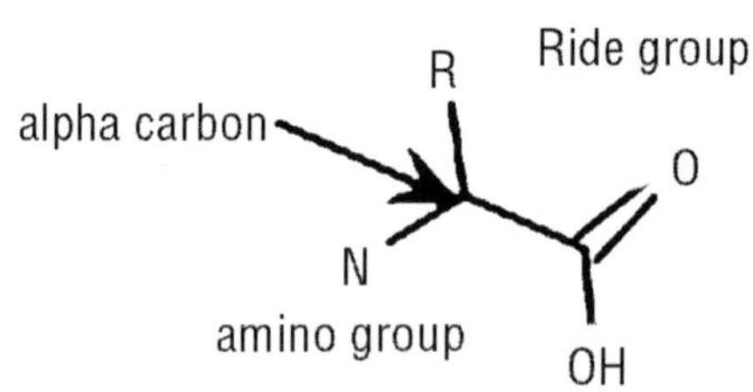

Fig. A General Amino Acid

Each amino acid contains a carboxyl group, an amino group and a variable side group (R). These all connect to a central carbon termed the á-carbon.

THE TWENTY AMINO ACIDS OF PROTEINS

The hydrophobicity index tells the relative hydrophobicity among amino acids. More positive value indicates stronger hydrophobicity. Hydrophilic amino acids have negative values.

In a protein, hydrophobic amino acids are more likely to be located in the protein interior, whereas hydrophilic amino acids are more likely to face the aqueous environment.

Phenylalanine	Phe	F	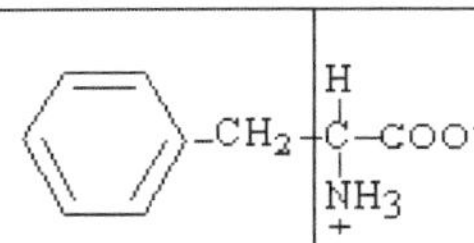	2.8
Cysteine	Cys	C	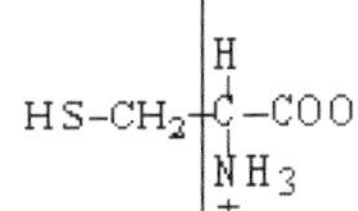	2.5
Methionine	Met	M	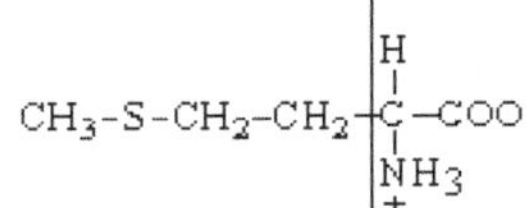	1.9
Serine	Ser	S	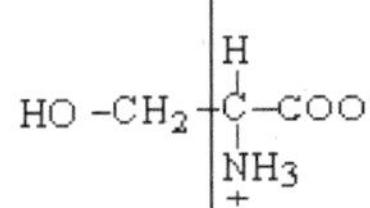	-0.8
Threonine	Thr	T	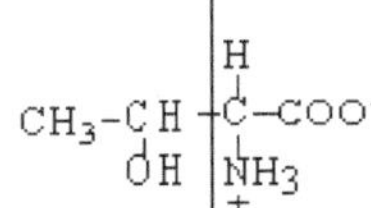	-0.7
Asparagine	Asn	N	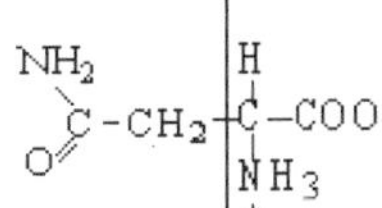	-3.5
Glutamine	Gln	Q	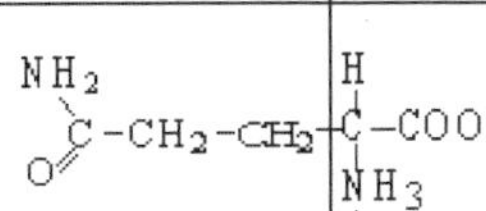	-3.5

Glycine	Gly	G	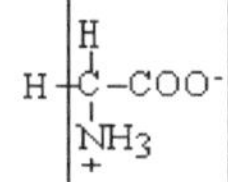	-0.4
Alanine	Ala	A	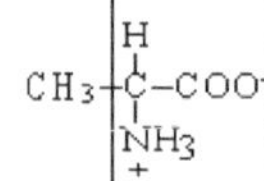	1.8
Valine	Val	V	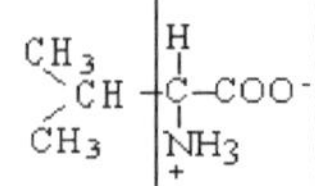	4.2
Leucine	Leu	L	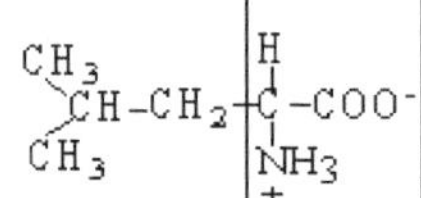	3.8
Isoleucine	Ile	I	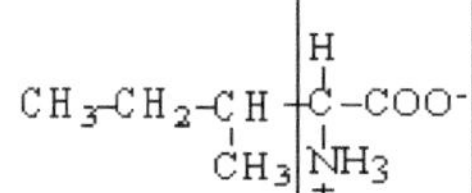	4.5
Proline	Pro	P	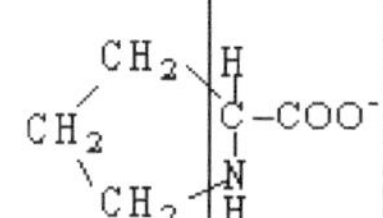	-1.6

Based on the physical and chemical properties of R groups, the 20 amino acids of proteins may be classified as follows.

- *Acidic:* Including aspartic acid (aspartate) and glutamatic acid (glutamate). In a neutral solution, the R group of an acidic amino acid may lose a proton and become negatively charged.
- *Basic:* Including lysine, arginine and histidine. In a neutral solution, the R group of a basic amino acid may gain a proton and become positively charged. Interaction between positive and negative R groups may form a salt bridge, which is an important stabilizing force in proteins.
- *Aromatic:* Including tyrosine, tryptophan and pheny-lalanine. Their R groups contain an aromatic ring.
- *Sulfur:* including cysteine and methionine. Their R groups contain a sulfur atom (S). The disulfide bond formed between two cysteine residues provides a strong force for stabilizing the globular structure. A unique feature about methionine is that the synthesis of all peptide chains starts from methionine.

- Uncharged hydrophilic: including serine, threonine, asparagine and glutamine. Their R groups are hydrophilic and capable of forming hydrogen bonds.
- Inactive hydrophobic: including glycine, alanine, valine, leucine and isoleucine. These amino acids are more likely to be buried in the protein interior. Their R groups do not form hydrogen bonds and rarely participate in chemical reactions.
- Special structure: including proline. In most amino acids, the R group and the amino group are not directly connected. Proline is the only exception among 20 amino acids found in protein. Due to this special feature, proline is often located at the turn of a peptide chain in the three-dimensional structure of a protein.

PH AND PKA

A molecule, or an atom group in a molecule, may lose or gain a proton when the molecule is placed in an aqueous solution. The exact probability that a molecule will be protonated or deprotonated depends on the pK_a of the molecule and the pH of the solution.

Let AH be an atom group in a molecule. AH could be neutral or charged. After AH loses a proton, it is denoted by A^-. The protonation/ deprotonation reaction may be written as

$$AH \rightleftarrows H^+ + A^-$$

Define its equilibrium constant,

$$K_a \equiv [H^+][A^-]/[AH]$$

where [..] represents the concentration.

Taking logarithm on both sides,

$$\log K_a = \log [H^+] + \log [A^-]/[AH]$$

Define

$$pK_a \equiv -\log K_a\,; \quad pH \equiv -\log [H^+]$$

we obtain

$$pK_a = pH - \log [A^-]/[AH]$$

or

$$\boxed{[A^-]/[AH] = 10^{(pH - pK_a)}}$$

pH	$[H^+]$ (mol/l)	
0	10^{-0}	
1	10^{-1}	↑
2	10^{-2}	
3	10^{-3}	Increasing acidity
4	10^{-4}	
5	10^{-5}	
6	10^{-6}	
7	10^{-7}	Neutral
8	10^{-8}	
9	10^{-9}	
10	10^{-10}	Increasing alkalinity
11	10^{-11}	
12	10^{-12}	
13	10^{-13}	↓
14	10^{-14}	

From the last equation, we can see that half of molecules will lose protons if they are in a solution with pH = pK_a. The higher the pH value, the more likely a molecule will lose a proton. On the other hand, from the definition of pH, we find that the proton concentration of a solution becomes lower at higher pH, thereby capable of accepting more protons. This physical concept is consistent with the result we just obtained from the equation.

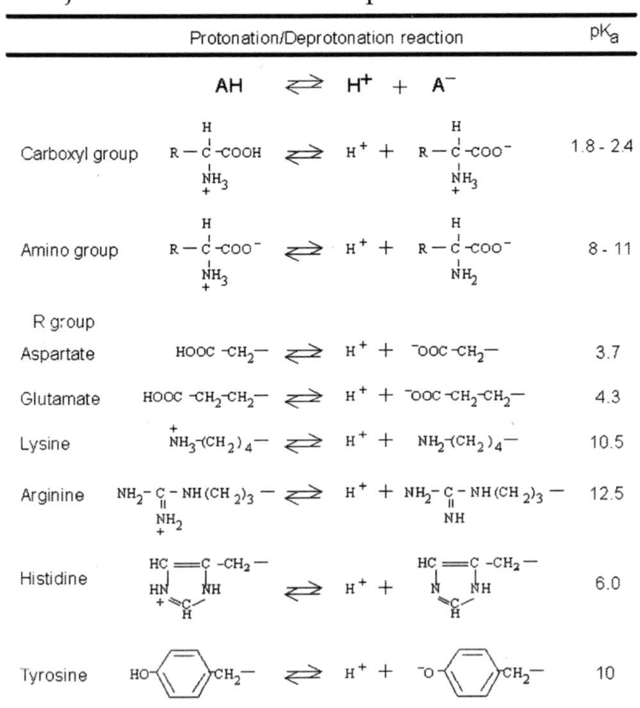

Fig. pK_a Values of Amino Group, Carboxyl Group and a Few R Groups.

This Information Tells us in a Neutral Solution:

- The carboxyl group is most likely negatively charged.
- The amino group is most likely positively charged
- The R group of aspartate and glutamate are most likely negatively charged.
- The R group of lysine and arginine are most likely positively charged.
- The R group of tyrosine at pH = 7 is most likely neutral.
- The R group of histidine has 10 per cent probability to become positively charged at pH = 7, but the probability increases to 50 per cent at pH = 6. Thus, histidine is very sensitive to pH change in the physiological range

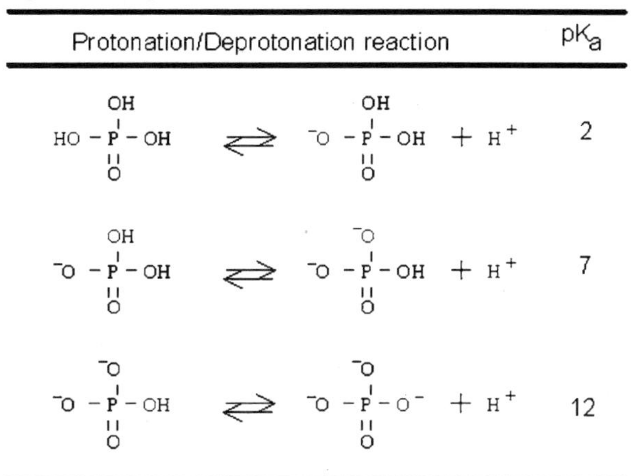

Fig. pK_a Values of Phosphate Groups.

This information tells us that in a neutral solution H_3PO_4 is very likely to lose a proton and become H_2PO4^-.

The latter has a 50 per cent chance of losing another proton, becoming HPO_4^{2-}. The chance for HPO_4^{2-} to lose an additional proton is very small. Hence, the average charge of a phosphate group is about 1.5 electron charges.

BIOSYNTHESIS OF AMINO ACIDS

NITROGEN SOURCES

In order to synthesize amino acids, a source of nitrogen is needed. In animals glutamate and glutamine play the pivotal roles. The ±"amino group of most of the amino acids comes from the transamination reaction transferring the amino group from glutamate to an ±-ketoacid acceptor. Glutamate is synthesized from ammonia and ±-ketoglutarate by the action of glutamate dehydrogenase.

We have already discussed this enzyme in the degradation of amino acids. It catalyzes the following reaction:

NH_4^+ α-ketoglutarate + NAD(P)H + H^+ ⇄ Glutamate + $NAD(P)^+$ + H_2O

This reaction occurs in two steps first a Schiff base is formed between the ammonia and the ketone of aketoglutarate. The Schiff base is reduced by hydride transfer from either NADH or NADPH to form glutamate.

The importance of the hydride transfer is that it establishes the stereochemistry at the ±-carbon. Glutamate dehyrogenase binds ±-ketoglutarate is such a way that the hydride is transferred to specific face to form only the L-isomer of glutamate.

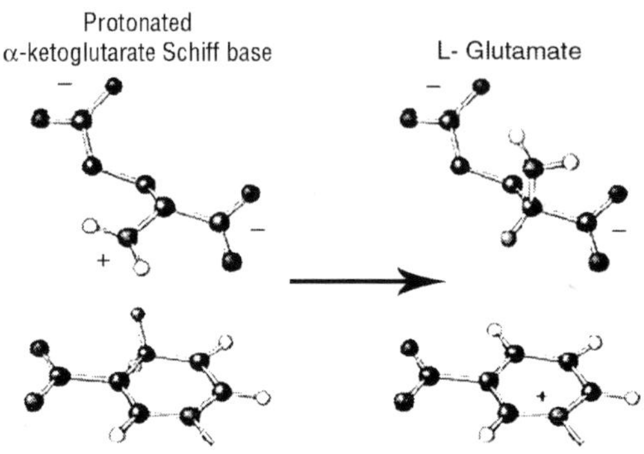

A second ammonium ion is incorporated into glutamate by the action of glutamine synthetase to form glutamine. This amidation is driven by ATP as we have previously discussed. The regulation of glutamine synthetase plays a crucial role in controlling nitrogen metabolism. The dynamic duo of glutamate dehydrogenase and glutamine synthetase are found in all living organisms.

II. AMINO ACIDS ARE SYNTHESIZED FROM METABOLITES OF OTHER PATHWAYS.

Plants and bacteria can synthesize all 20 of the amino acids. Where as we humans cannot synthesize of them. These nine amino acids must come from our diets and are called essential amino acids. The essential amino acids are: Histidine, Isoleucine, Leucine, Lysine, Methionine, Phenylalanine, Threonine, Tryptophan and Valine.

The other 11 amino acids are called nonessential amino acids they include: Alanine, Arginine, Asparagine, Aspartate, Cysteine, Glutamate, Glutamine, Glycine, Proline, Serine and Tyrosine. These nonessential amino acids are synthesize by simple pathways. Whereas the biosynthesis of the essential amino acids are quite complex. Shown above are the major metabolic precursors of the amino acids.

The amino acids that are precursors for other amino acids are shown in yellow. The nine essential amino acids are shown in boldface. The carbon skeletons come from intermediates of glycolysis, the pentose phosphate pathway and the citric acid cycle. On the basis of the starting points the 20 amino acids can be group into 6 categories depending on the precursor: oxaloacetate, PEP, ±-ketoglutarate, pyruvate, 3-phosphoglycerate, ribose-5P.

III. ASPARTATE, GLUTAMATE AND ALANINE BIOSYNTHESIS

Oxaloacetate, pyruvate and ±-ketoglutarate are all ±-ketoacids which are substrates for transamination reactions that we very familiar with by now. Asparatate aminotransferase transfers an amino group from glutamine to oxaloacetate to form aspartate and ±-ketoglutarate. There are a number of transaminase that will transfer an amino group from an amino acid such as aspartate or alanine and transfer it to ±- ketoglutarate to form glutamate. Alanine aminotransferase transfers the amino group of glutamine to pyruvate to form alanine and ±-ketoglutarate. All of these transamination reactions are catalyzed by enzymes that require pyridoxal phosphate as a cofactor. Review the pyridoxal phosphate mechanism.

The enzyme reaction begins with the formation of the external aldimine between PLP and the amino group donor displacing the active site lysine residue.

The amino group is transferred to PLP to form the pyridixoamine phosphate intermediate releasing the ±-ketoacid remnant of the donor. The second half of the reaction is the reverse of the first. The amino group acceptor is another ±-ketoacid which forms a Schiff base with the PMP to form a ketimine intermediate.

The amino group is transferred to the acceptor to form the amino acid. The important step in the transamination reaction is the protonation of the ±-carbon of the quinonoid intermiate. This protonation determines the stereochemistry at the ±-carbon. In the case of aspartate amino transferase (the postor child of transaminases) the chirality of the amino acid is determined by binding interactions with the substrate. Particularly, the interaction of the conserved arginine residue (Arg-386) with the carboxylate group of the substrate. This interaction orients the substrate so that when Lys-268 protonates the ±"carbon of the quinonoid intermediate it generates an external aldimine with the L-configuration at the ±-carbon.

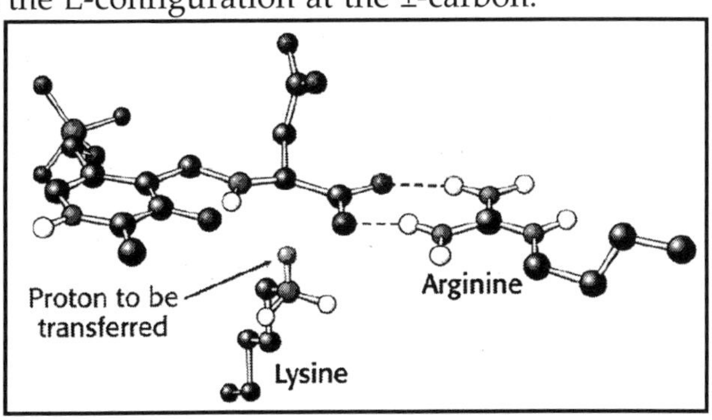

ASPARAGINE BIOSYNTHESIS

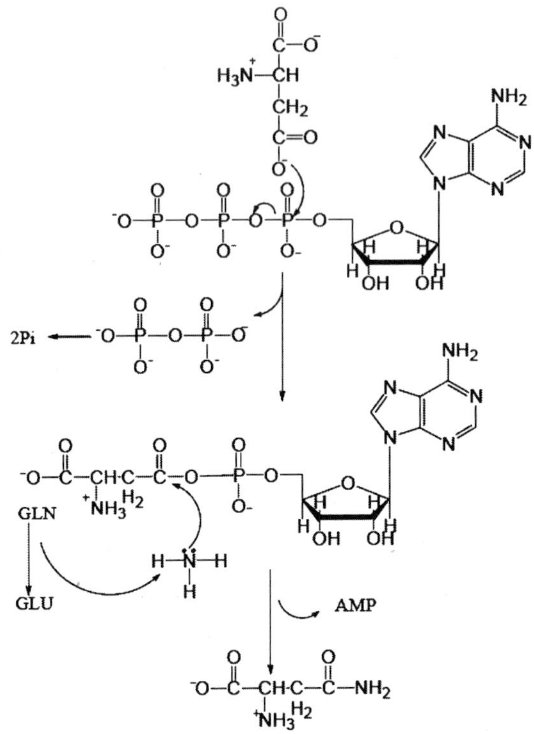

Asparagine synthetase catalyzes the formation of asparagine from aspartate. The first step is to activate the carboxylate group of aspartate by reacting with ATP to form an aspartyladenylate intermediate. The activated carbonyl of aspartyl-adenylate intermediate is attacked by an ammonia nucleophile to produce asparagines. In bacteria ammonia is the nitrogen source for this reaction. In mammals the nitrogen source for this reaction is derived from the hydolysis of glutamine into glutamate. The ammonia produced in this reaction is channeled to the aspartyl-adenylate intermediate to from asparagine. The advantage of this pathway of using glutamine as the nitrogen source is that the ammonia is produced by the enzyme and channeled directly to the intermediate. The cells are not exposed to ammonia which is toxic to animals.

PROLINE AND ARGININE BIOSYNTHESIS

Proline and arginine are both derived from glutamate. The first step is to form glutamate 3-semialdehyde. The 3-carboxylate group of glutamate is activated by phosphorylation with ATP to from a 3- glutamylphosphate intermediate. The mixed anhydride is then reduced with NADPH liberating the phosphate and producing glutamate 3-semialdehyde. Glutamate 3-

semialdehyde will spontaneously cyclize eliminating H2O and forming pyrroline 5- carboxylate which is then reduced by pyrolline-5-carboxylate reductase to form proline. Glutamate ³-semialdehyde is also a substrate for ornithine-′-aminotransferase which will transfer and amino group from glutamate to glutamate ³-semialdehyde to from ornithine and ±-ketoglutarate. Ornithine of course is a metabolite of the urea cycle which reacts with carbamoyl phosphate to form citrulline. Citrulline is condensed with aspartate to form argininosuccinate which is then cleaved to form arginine and fumarate.

SERINE, CYSTEINE AND GLYCINE BIOSYNTHESIS.

Serine is synthesized from 3-phosphoglycerate. Serine is a precursor for both cysteine and glycine. The first step is the oxidation of 3-phosphoglycerate into 3-phosphohydroxypyruvate.

The enzyme that catalyzes the oxidation is 3-phosphoglycerate dehydrogenase. This oxidation generates an ±-ketoacid which can now be transaminated by transferring an amino group from glutamate to 3-phosphoglycerate to generate 3-phosphoserine and ±-ketoglutarate. The phosphate group is removed by phosphoserine phosphatase to produce serine. Serine is converted into glycine by serine hydroxy-methyltransferase which contains both a pyridoxal phosphate cofactor and a tetrahydrofolate cofactor. It is time to take a closer look as this remarkable cofactor and this remarkable enzyme.

Tetrahydrofolate-Carrier of Activated One-Carbon Units

Tetrahydrofolate: Tetrahydrofolate is a versatile carrier of one carbon units. It consists of 3 groups. A pteridine, an amino benzoate and a chain of one or more glutamate residues. Mammals can synthesize all three components but lack the enzymes that conjugate them together. Microorganisms do contain the necessary conjugating enzymes and can produce folate.

We must obtain folate from our diet or from the microorganism that inhabit our intestinal tract. Folic acid is member of the vitamin B complexes and is found in green vegetables, fruits, yeast and liver. Folate is reduced by dihydrofolate reductase to form THF. This is a two step reduction requiring 2 hydride transfers from NADPH. One carbon units may be bound to THF at the N5 and N10 nitrogens. The carbons carried can exist in three oxidation state. The most reduced state of a THF carried carbon is a methyl group, the intermediate oxidation state carries a methylene group, the most oxidized state carries a formimino group. The next higher oxidation state is the carboxylate which is carried by the biotin cofactor rather than THF. All of these interconvertible tetrahydrofolate derivatives carry activated one carbon units which can be used for a variety of biosyntheses. THF also serves as one carbon unit acceptor in degradative reactions.

Sulfonamides such as sulfanilamide are analogs of the p-aminobenzoate substituent of folic acid. The are competitive inhibitors of the bacterial conjugating enzyme for the p-aminobenzoate binding site. They block the synthesis of folate producing a deficiency which inactivates all of the enzymes that require a THF cofactor. Since mammals do not possess the conjugating enzymes, sulfonamides are not toxic which is why sulfonamides are such widely used antibiotics.

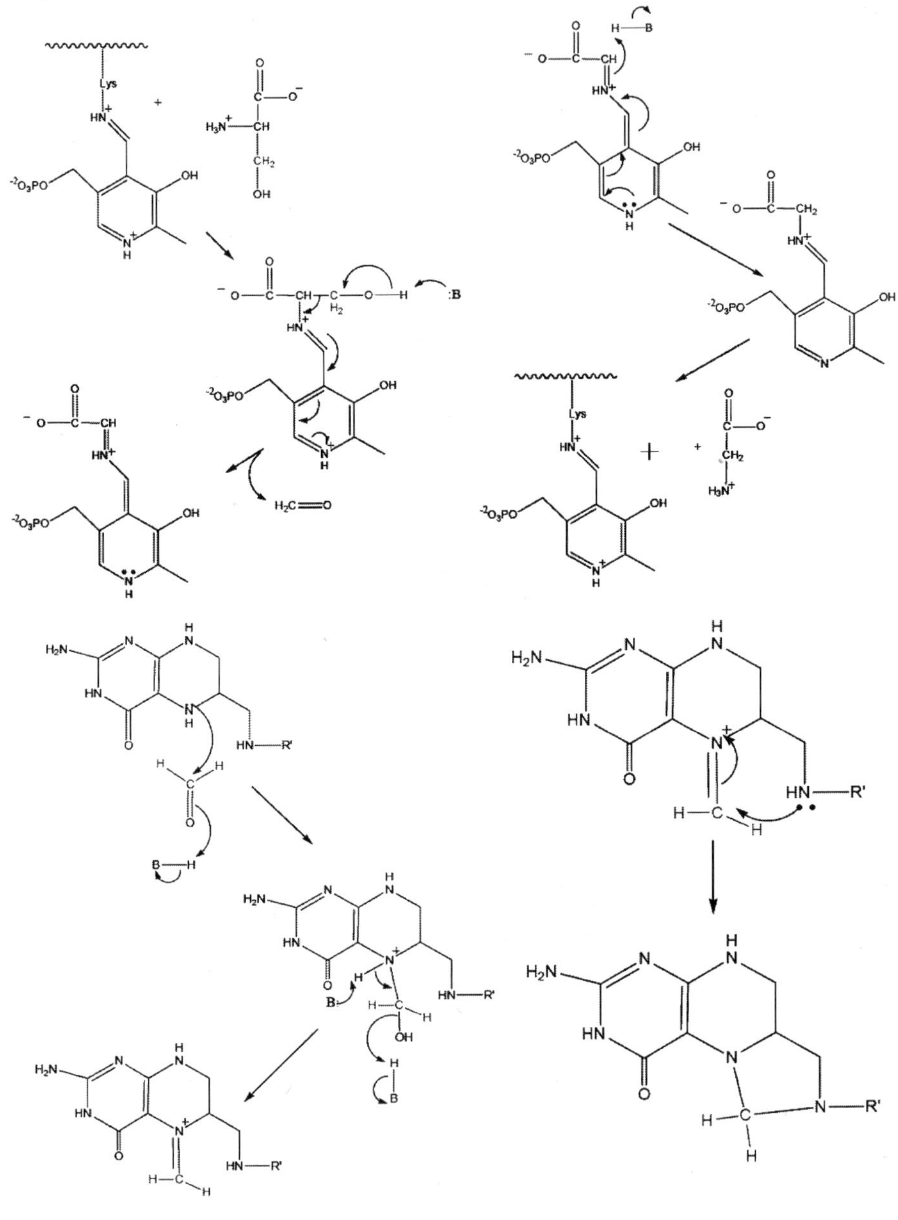

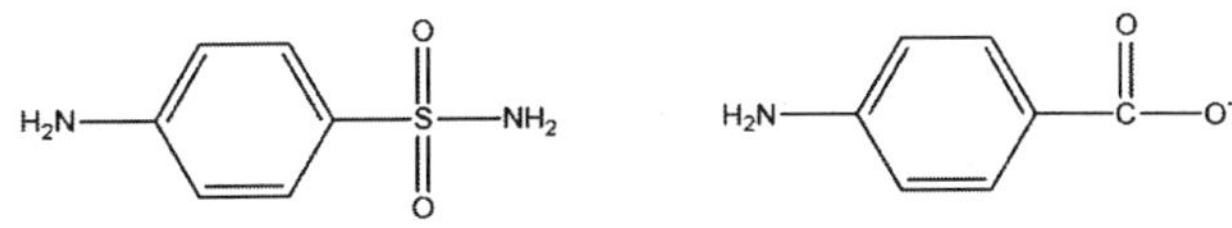

Sulfonamide p-Aminobenzoate

SAM, the Major Donor of Methyl Units

THF can carry a methyl group, but the transfer potential is not very high enough for most biochemical methylations. The activated methyl carrier of choice is S-Adenosylmethionine. SAM is produced by the transfer of an adenosyl group from ATP to methionine. The methyl group of the methionine group is activated by the positive charge on the adjacent sulfur atom. This positive charge makes the methyl transfer more energetic than from N5-methyl-THF. SAM carries the methyl groups used for the methylation reactions of DNA.

AMINO ACID DEGRADATION

The carbon skeletons of amino acids are broken down into metabolites that can either be oxidized into CO2 and H2O to generate ATP, or can be used for gluconeogenesis. The catabolism of amino acids accounts for 10 to 15 per cent of the human body's energy production. Each of the 20 amino acids has a separate catabolic pathway, yet all 20 pathways converge into 5 intermediates, all of which can enter the citric acid cycle. From the citric acid cycle the carbon skeletons can be completely oxidized into CO2 or diverted into gluconeogensis or ketogenesis.

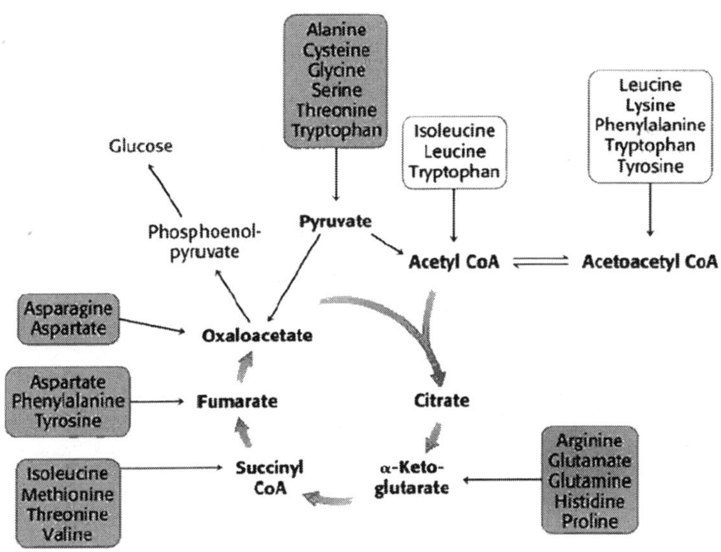

Glucogenic amino acids are broken down into one of the following metabolites: pyruvate, ±- ketoglutarate, succinyl CoA, fumarate or oxaloacetate. Ketogenic amino acids are broken down into acetoacetate or acetyl-CoA. Larger amino acids, tryptophan, phenylalanine, tyrosine, isoleucine and threonine are both glucogenic and ketogenic. Only 2 amino acids are purely ketogenic they are lysine and leucine. If 2 of the amino acids are purely ketogenic and 5 amino acids are both ketogenic and glucogenic, than that leaves 13 amino acids that are purely glucogenic: Arg, Glu, Gln, His, Pro, Val, Met, Asp, Asn, Ala, Ser, Cys and Gly.

Amino Acids that are Catabolized into Pyruvate.

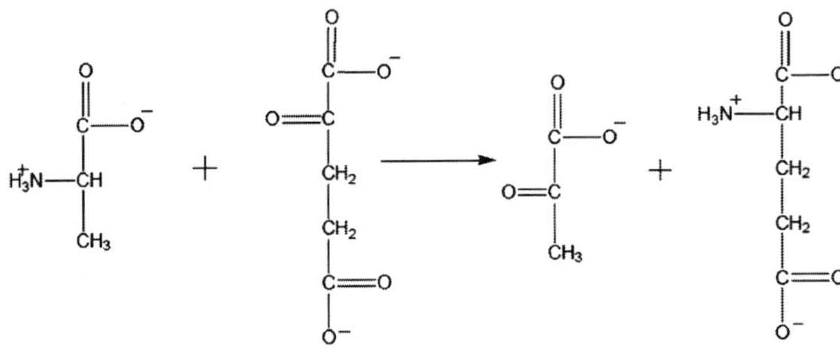

Alanine Transaminase

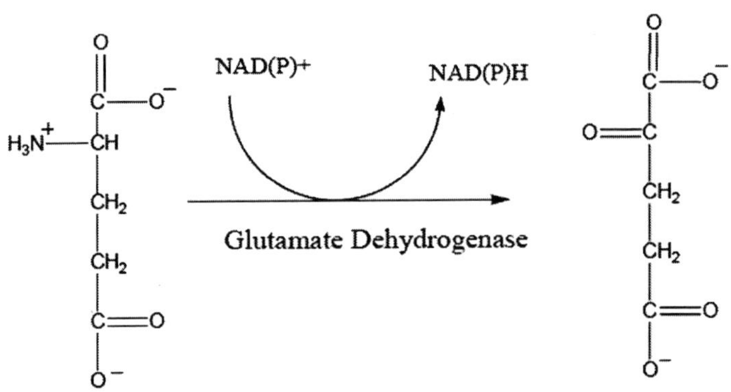

Pyruvate is the entry point for amino acids that contain 3 carbons, alanine, serine and cysteine. Alanine transaminase reversibly transfers the amino group from alanine to ±-ketoglutarate to form pyruvate and glutamate. Note that enzyme requires a pyridoxal phosphate cofactor. The ±-ketoglutarate is regenerated by glutamate dehydrogenase.

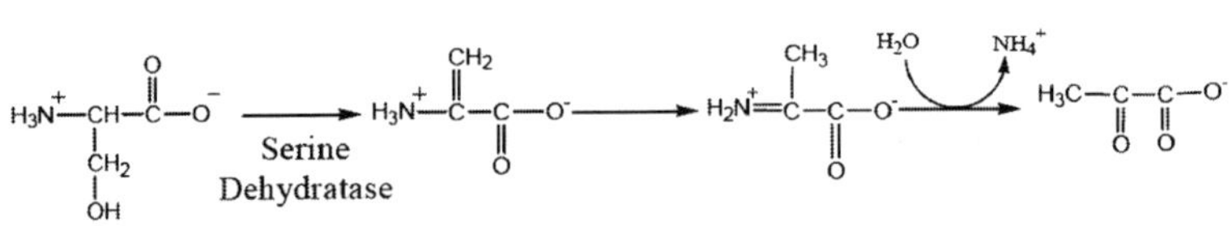

Serine dehyratase is another enzyme that requires a pyridoxal phosphate cofactor. This enzyme catalyzes the 2-elimination of the hydroxyl group of serine to form an amino acrylate intermediate which tautomerizes into the imine which is then hydrolyzed to produce ammonia and pyruvate.

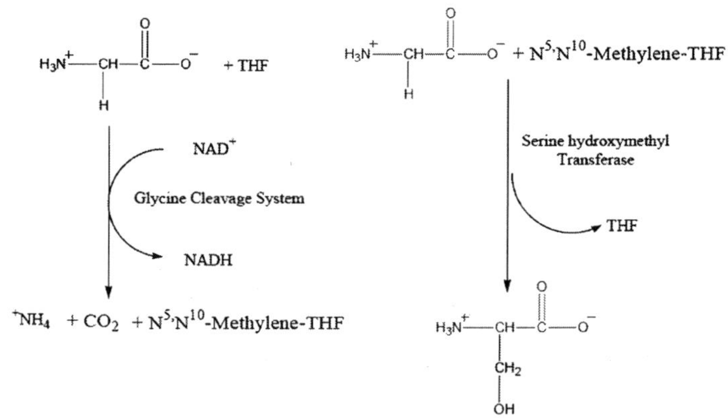

Glycine is converted into pyruvate via conversion of glycine to serine by serine hydroxymethyl transferase which is an incredibly interesting enzyme. It contains a pyridoxal phosphate cofactor and a *N*5,*N*10-methylene-tetrahydrofolate which is a cofactor we have not encountered yet. The *N*5,*N*10-methylene-tetrahydrofolate is produced by the glycine cleavage system which transfers a methylene group from glycine to THF. The THF cofactor is a one carbon acceptor and donor. We will discuss this cofactor further when we get to amino acid biosynthesis. They are several pathways by which cysteine is converted into pyruvate.

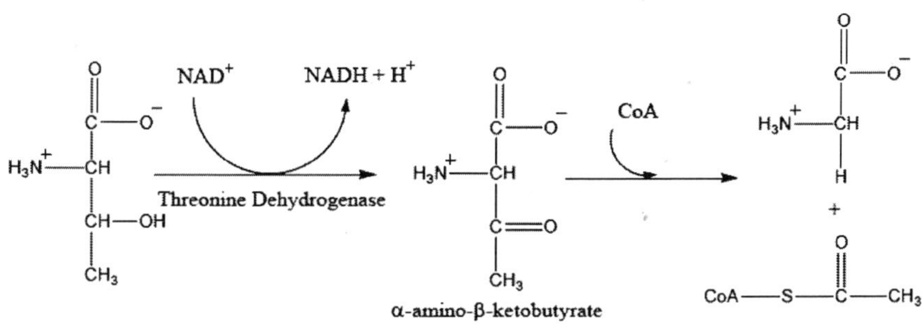

α-amino-β-ketobutyrate lyase

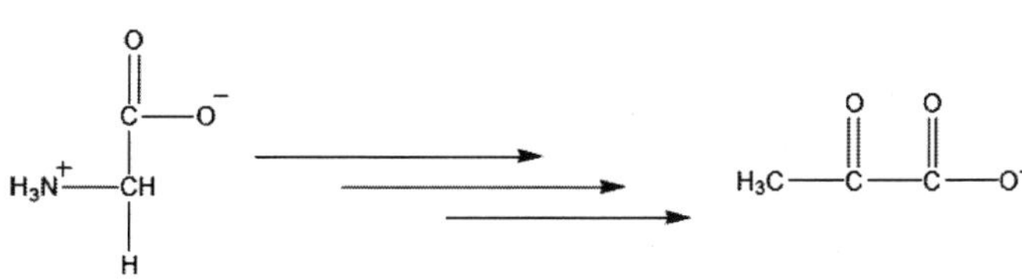

The three alkyl carbons of trypophan are converted into alanine which is then converted by alanine transaminase into pyruvate. Threonine is both glucogenic and ketogenic. There are a couple of routes for the degradation of threonine. The major route is shown below. Threonine is converted into acetyl CoA and glycine. Glycine is then converted into serine by serine hydroxymethyl transferase and serine is then converted into pyruvate by serine dehydratase.

Amino Acids Degradated to Oxaloacetate

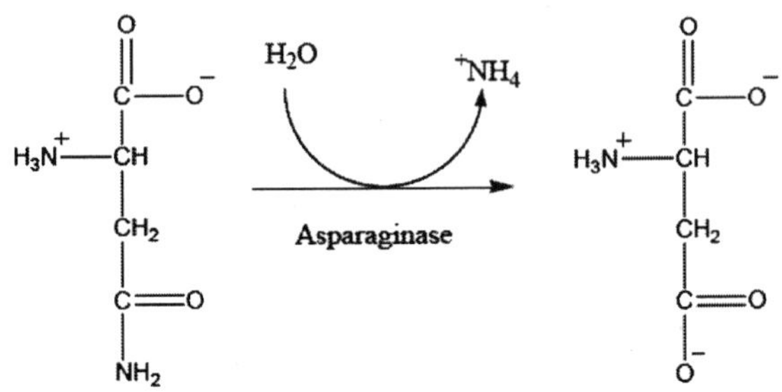

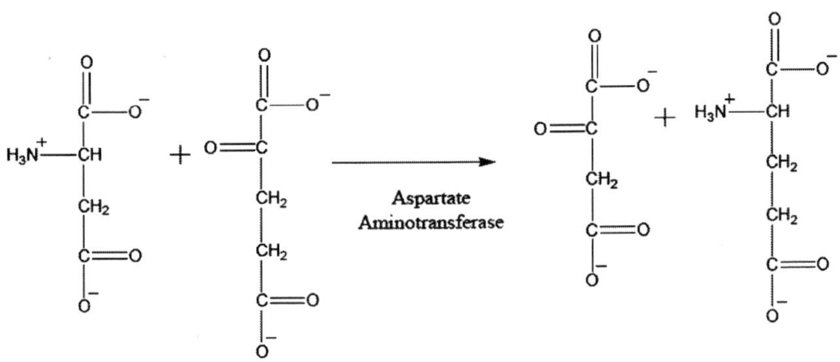

Aspartate and asparagines are both degraded into oxaloacetate. Asparagine is hydrolyzed into aspartate and ammonia by asparaginase. Aspartate is converted into oxaloacetate by aspartate amino transferase which is a PLP enzyme that transfers an amino group from aspartate to ±"ketoglutarate to form glutamate and oxaloacetate.

Amino Acids Degraded to ±-Ketoglutarate

Glutamine, proline, arginine and histidine are converted into glutamate which is then deaminated by a transaminase to form ±-ketoglutarate.

Glutamine is converted into glutamate by glutaminase. Proline is oxidized by proline oxidase to form pyrroline 5-carboxylate which spontaneously hydrolyzes to from glutamate ³-semialdehyde. From the urea cycle we know that arginase converts arginine into ornithine and urea. Ornithine ′-aminotransferase transfers the ′-amino group of ornithine to ±-ketoglutarate to form glutamate ³-semialdehyde and glutamate.

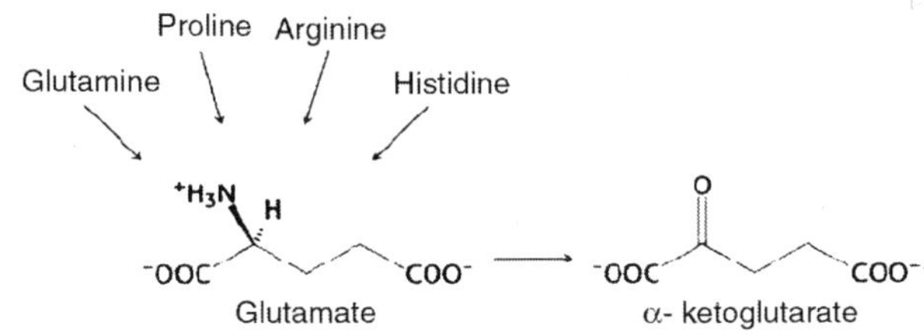

Glutamate ³-semialdehyde is oxidized to form glutamate by glutamate -5-semialdehyde dehydrogenase. Histidine is deaminated by histidine ammonia lyase which forms urocanate. Urocanate hydratase adds water to form 4-Imidazolone-5-propionate which is hydrolyzed by imidazalone propionase to form Mformiminoglutamate. Glutamate formimino transferase transfers the formimino group to tetrahydrofolate to generate glutamate and N^{5-}formimino-THF.

Amino Acids that are Broken Down in Succinyl-CoA.

Methionine, valanine and isoleucine are broken down into propoinyl CoA. By studying ²-oxidation of odd chain fatty acids we know that propionyl CoA is converted into D-methylmalonyl CoA by propionyl CoA carboxylase. D-methylmalonyl CoA is racemized into L-methylmalonyl CoA by methylmalonyl CoA racemase. Methylmalonyl mutase produces succinyl CoA.

The degradation of methionine requires 9 steps. One of which involves the synthesis of Sadenoylmethionine (SAM). The methyl group of SAM is highly reactive making it an important methylating reagent. SAM is a common methyl-group donor in the cell.

The first step is catalyzed by methionine adenosyl transferase which tranfers the adenosyl group of ATP to the Sulpher of methionine to form SAM. Sam methylase transfers the activated methyl group to an acceptor to form S-adenosylhomocysteine which is hydrolyzed by adenosylhomocysteinase to form homocysteine. Cystathionine ²-synthase is a PLP dependent enzyme that catalyzes the condensation of a serine residue with homocysteine to form cystathionine.

Cystathioniine ³-lyase cleaves cystathionine into cysteine and ±-ketobutyrate. ±"ketobutyrate is converted into propionyl CoA by ±-ketobutyrate dehydrogenase which catalyzes a reaction that is analogous to pyruvate dehydrogenase and ±-ketoglutarate dehydrogenase.

Branched Chain Amino Acids

The degradation of branched chain amino acids uses some of the enzymes we have already encountered in the citric acid cycle or ²-oxidation.

Most amino acid catabolism occurs in the liver. The branched chain amino acids are not catabolized in the liver. Branched chain amino acids are catabolized mainly in the muscle, adipose, kidney and the brain. The liver does not contain the branched amino acid aminotransferase enzyme which these other tissues contain. Branched chain ±-ketoacid dehydrogenase is a huge multienzyme complex homologous to pyruvate dehydrogenase and ±-ketoglutarate dehydrogenase.

This enzyme contains a thiamine pyrophosphate cofactor, a lipoamide cofactor, a FAD prosthetic group. The chemistry, mechanism and structure of these enzymes is very similar. Branched chain ±-ketoacid dehydrogenase is phosphorylated by a kinase which inactivates the enzyme in a similar manner that pyruvate dehydrogenase is phosphorylated and inactivated. The intake of dietary branched amino acids activates a phosphatase which activates this enzyme.

A genetic deficiency in the branched chain ±-ketoacid dehydrogenase enzyme is called maple syrup urine disease. The deficiency causes an excessive buildup of branched ±-ketoacids in the blood and the urine. The urine of these patients has the odor of maple syrup and hence the name of the disease. Maple syrup disease usually leads to mental retardation unless the patient is placed on diet that is low in valine, isoleucine and leucine early in life.

Amino Acids that are Degraded into Acetyl CoA and Acetoacetate

There are only two amino acids that are purely ketogenic, lysine and leucine. Leucine catabolism is similar to the branched amino acids valine and isoleucine. First leucine is transaminatedby branched amino acid aminotransferase to form ±-ketoisocaproate which is then oxidatively decarboxylated to form isovaleryl CoA by the branched chain ±-ketoacid dehydrogenase complex we just discussed. In the next step isovaleryl CoA is dehydrogenated to form β- methylcrotonyl CoA. The enzyme that catalyzes this dehydrogenation is isovaleryl CoA dehydrogenase. β-methylcrotonyl CoA is then carboxylated by a biotin containing enzyme called methylcrotonyl CoA carboxylase to form β-methylglutaconyl CoA.

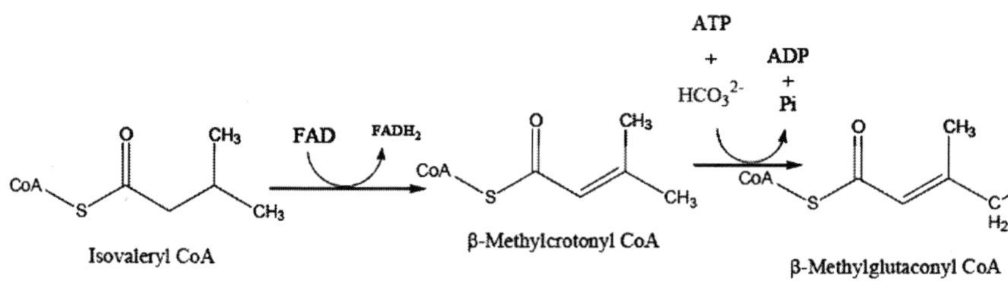

β-methylglutaconyl CoA is then hydrated by β-methylglutaconyl CoA. hydratase to form β-hydroxy-²- methylglutaryl CoA which is then cleaved into acetyl CoA and acetoacetate. The enzyme that catalyzes the last step is HMG-CoA lyase, a familiar enzyme from ketogenesis.

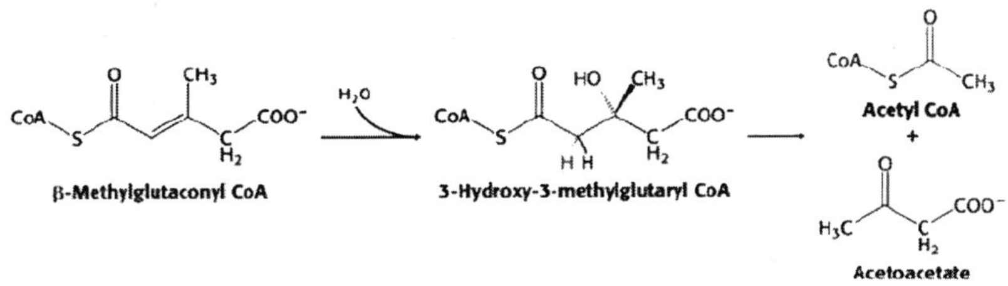

Catabolism of Aromatic Amino Acids

- The degradation of aromatic amino acids requires molecular oxygen to break down the aromatic rings.
- The degradation of phenylalanine begins with a monooxygenase, phenylalanine hydroxylase which adds a hydroxyl group to phenylalanine to from tyrosine.
- Tyrosine aminotransferase deaminates tyrosine to form phydroxyphenylpyruvate.
- p-hydroxyphenylpyruvate dioxygenase catalyzes the formation of homogentisate.
- Homogentisate 1,2-dioxygenase catalyzes the formation of maleylacetoacetate.
- Maleylacetoacetate isomerase produces fumarylacetoacetate.
- Fumarylacetoacetase produces fumarate and acetoacetate.

Many genetic defects of phenylalanine catabolism in humans has been identified. A deficiency in phenylalanine hydroxylase is responsible for the disease phenylkotonuria (PKU) which is caused by elevated concentrations of phenylalanine. Individuals with a deficiency in phenylalanine hydroxylase rely on a secondary catabolic pathway which in normal individuals is not used.

In this pathway phenylalanine is converted into phenylpyruvate by a transaminase which transfers the amino group to pyruvate to form alanine. Phenylalanine and phenylpyruvate accumulate in the blood and excreted in the urine hence the term phenylketonuria.. Some of the phenylpyruvate is decarboxylated to form phenylacetate. Some of the phenylpyruvate is reduced to phenyllactate. Phenyllactate gives the urine a distinctive odor used for diagnosis.

The high concentration of phenylalanine in the blood limits the transport of amino acids across the blood-brain barrier resulting in impairment of normal brain development, causing severe mental retardation. This can be avoided by early detection and rigid dietary control. The diet must only provide enough phenylalanine and tyrosine to meet the needs of protein synthesis.

Alkapotonuria is another inheritable disease of phenylalanine catabolism. In this case the defective enzyme is homogentisate dioxygenase. This disease is less severe than PKU. Large amounts of homogentisate are excreted in the urine. The oxidation of homogentisate turns the urine black. Individuals who have this disease develop arthritis at an early age.

4

Enzymatic Activity

The activity displayed by enzymes is affected by a variety of factors, some of which are essential to the harmony of metabolism.

1. The enzymatic rate, v = d[P]/dt, "slows down" as product accumulates and equilibrium is approached. The apparent decrease in rate is due to the conversion of P to S by the reverse reaction as [P] rises. Once [P]/[S] = Keq, no further reaction is apparent. Keq defines thermodynamic equilibrium. Enzymes have no influence on the thermodynamics of a reaction. Also, product inhibition can be a kinetically valid phenomenon: Some enzymes are actually inhibited by the products of their action.
2. The availability of substrates and cofactors will determine the enzymatic reaction rate. In general, enzymes have evolved such that their Km values approximate the prevailing in vivo concentration of their substrates. (It is also true that the concentration of some enzymes in cells is within an order of magnitude or so of the concentrations of their substrates.)
3. There are genetic controls over the amounts of enzyme synthesized (or degraded) by cells. If the gene encoding a particular enzyme protein is turned on or off, changes in the amount of enzyme activity soon follow. Induction, which is the activation of enzyme synthesis, and repression, which is the shutdown of enzyme synthesis, are important mechanisms for the regulation of metabolism. By controlling the amount of an enzyme that is present at any moment, cells can either activate or terminate various metabolic routes. Genetic controls over enzyme levels have a response time ranging from minutes in rapidly dividing bacteria to hours (or longer) in higher eukaryotes.
4. Enzymes can be regulated by covalent modification, the reversible covalent attachment of a chemical group. For example, a fully active enzyme can be converted into an inactive form simply by the covalent attachment of a functional group, such as a phosphoryl moiety. Alternatively, some enzymes exist in an inactive state unless

specifically converted into the active form through covalent addition of a functional group. Covalent modification reactions are catalyzed by special converter enzymes, which are themselves subject to metabolic regulation. Although covalent modification represents a stable alteration of the enzyme, a different converter enzyme operates to remove the modification, so that when the conditions that favoured modification of the enzyme are no longer present, the process can be reversed, restoring the enzyme to its unmodified state. Many examples of covalent modification at important metabolic junctions will be encountered in our discussions of metabolic pathways. Because covalent modification events are enzyme-catalyzed, they occur very quickly, with response times of seconds or even less for significant changes in metabolic activity. The 1992 Nobel Prize in physiology or medicine was awarded to Edmond Fischer and Edwin Krebs for their pioneering studies of reversible protein phosphorylation as an important means of cellular regulation.

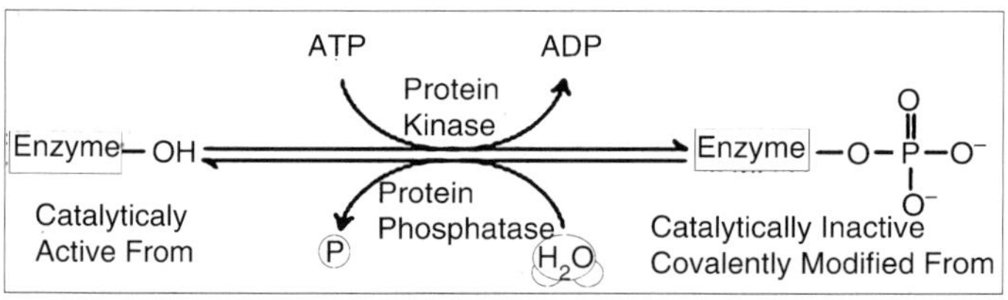

Fig. Enzymes Regulated by Covalent Modification are Called Interconvertible Enzymes.

The enzymes (protein kinase and protein phosphatase, in the example shown here) catalyzing the conversion of the interconvertible enzyme between its two forms are called converter enzymes..

Enzymatic activity can also be activated or inhibited through noncovalent interaction of the enzyme with small molecules (metabolites) other than the substrate. This form of control is termed allosteric regulation, because the activator or inhibitor binds to the enzyme at a site other than (allo means "other") the active site. Further, such allosteric regulators, or effector molecules, are often quite different sterically from the substrate. Because this form of regulation results simply from reversible binding of regulatory ligands to the enzyme, the cellular response time can be virtually instantaneous. Specialized controls: Enzyme regulation is an important matter to cells, and evolution has provided a variety of additional options, including zymogens, isozymes, and modulator proteins.

ZYMOGENS

Most proteins become fully active as their synthesis is completed and they spontaneously fold into their native, three-dimensional conformations.

Some proteins, however, are synthesized as inactive precursors, called zymogens or proenzymes, that only acquire full activity upon specific proteolytic cleavage of one or several of their peptide bonds. Unlike allosteric regulation or covalent modification, zymogen activation by specific proteolysis is an irreversible process. Activation of enzymes and other physiologically important proteins by specific proteolysis is a strategy frequently exploited by biological systems to switch on processes at the appropriate time and place.

Insulin

Some protein hormones are synthesized in the form of inactive precursor molecules, from which the active hormone is derived by proteolysis. For instance, insulin, an important metabolic regulator, is generated by proteolytic excision of a specific peptide from proinsulin.

ENZYMES

As major gene products, enzymes in the CNS are likely candidates for heritable variation. The range of enzymes active in neural tissue is wide and provides a wealth of material for pharmacological study. At a time when research on Behavioural control by various neurotransmitter systems is so predominant, it is perhaps easy to overlook other kinds of enzyme systems that could markedly influence drug response. Not the least of these are those biochemical pathways involved in energy metabolism.

Small perturbations in cerebral energy metabolism can have potentially major effects on a wide variety of Behaviours. The example provided by hypoglycemia and its effects on cognitive and affective characters is clear. The brain is a high energy utilization organ. All specialized functions of neural tissue depend on the more basic biochemistry of energy sources. Although we are not aware of genetic studies of enzymes involved in such energy metabolism and pharmacological alterations of activity, the caveat is relevant in the context of the distinction between direct and indirect drug effects on Behaviour.

A drug such as alcohol may very well modify neurotransmitter activity and some related Behaviour by actions on synthesis, degradation, receptor binding characteristics, or channel alteration. Nu merous other "side" effects might also occur. For example, ethanol consumption produces a marked alteration in general metabolism, producing among other things a general acidosis and increasing blood a-hydroxybutyrate levels. As â hydroxybutyrate is a usable energy source for brain tissue, marked changes in cerebral energy metabolism are likely to follow.

Such "indirect" effects on neural function, and perhaps Behaviour, would certainly appear subject to the same sort of heritable variation as other biochemical systems. Thus the degree of Behavioural modification by indirect as well as direct action of the pharmacological agent might also be influenced by genetic factors. While hoping not to be redundant, we believe the conclusion is obivous. That is, gene action may affect pharamcological responsiveness in a variety of ways other than via the specific neural system controlling the phenotype of interest. Keeping the foregoing caution in mind, numerous informative avenues of research on neuro-transmitter system regulation by genetic variables are available.

The most well researched of these is probably the area of catecholamine neurotransmitters. Norepinephrine and dopamine systems are subject to functional alteration by numerous pharmacological agents; many of these agents are of use in a clinical setting precisely because of their Behavioural effects. These functional alterations can occur as a result of inhibition or induction of the enzymes responsible for synthesis and degradation of these catecholamine transmitters. Early work by Ciaranello, Barchas, Kessler, and Barchas first established the fact of strain differences in brain activity of tyrosine hydroxylase (TH), the rate limiting enzyme in catecholamine biosynthesis.

The same report also demonstrated marked strain differences in adrenal phenylethanolamine-N methyl transferase (PNMT), the enzyme responsible for epinephrine production from norepinephrine. We here describe further examination of these mouse strain differences in TH activity in a context that serves as an example for pharmacogenetic study as well as neural development.

In midbrain of CBA/J mice, the activity of TH is 20per cent less than in BALB/cJ mice. A comparable difference in immunocyto chemically detectable dopamine neurons in midbrain suggests that this difference in TH activity is due to more cells with the enzyme, rather than to greater specific activity per cell. A correspondingly more dense innervation of the striatum, by what are presumably the axons of these midbrain nigral cells, occurs in the BALB/c mice.

It is likely that this difference is related in a general way to the fact that the BALB/c brain is simply a larger one. Nonetheless, this represents a clear example of genetic control over neuronal organization and neurochemical characteristics of a Behaviourally important system. The differences in density of striatal innervation apparently predispose a differential response to two agents with dopaminergic activity, amphetamine and apomorphine.

BALB/c mice are more responsive to the locomotor stimulant effects of amphetamine and less sensitive to the stereotypy producing effects of apomorphine, a dopamine receptor agonist. Interestingly, these charac teristics follow a developmental time course that lags slightly behind the appearance of the TH activity difference. Apparently both pre and postsynaptic functions of the dopamine system are different between the two strains as reflected in the actions of the two drugs.

A more detailed understanding of the genetic control over dopamine cell proliferation, growth, and effects on target tissue in this model system will answer many questions not only of pharmacogenetic interest, but also of basic importance to developmental neurobiology.

In this example of TH activity differences, the enzyme activity itself is probably not the focal point of either gene or pharmacological action. As noted, a more basic developmental issue is central. Yet the enzyme is a clear marker for a dramatic genetic influence. In other work, TH is seen to be more directly involved. Gamma-butyrolactone (GBL), through its metabolite gamma-hydroxy butyric acid (GHB), produces an inhibition of impulse transmission in dopamine neurons. A compensatory activation of TH occurs in these cells and an accumulation of dopamine occurs. In the LS and SS mice, GLB has hypnotic effects that differ in the same direction as ethanol effects. Of interest here is the fact that dopamine accumulation is 50per cent greater in the LS mice after GBL treatment.

Presumably the activation of TH was greater in LS mice. This activation occurs via kinetic rather than molecular genetic induction and apparently reflects a differential effectiveness of GHB at a subcellular site in LS and SS dopamine neurons.

Although several studies of neurotransmitter systems' enzymes exist, there is yet to be a complete characterization of all biosynthetic and degradatory enzymes, transmitter levels and turnover rate, and receptor kinetics for any single system where genetic variation is also of interest. The difficulties of such a massive undertaking are obvious, but the benefits accruing to all three fields, neuropharmacology of drug action, neurochemical organization and dynamics, and mechanisms of genetic action are equally impressive. The field must surely progress along these lines.

ALCOHOL

The genetic effects on a given Behavioural response do not necessarily imply similar genetic effects on other phenotypes. The degree to which a similar genetic pattern occurs for more than one phenotype is a measure of the genetic correlation of these phenotypes. Yet taskor situation specific genetic effects are not rare. The LS and SS mice are usually very different in their response to ethanol.

Yet in one study, across a wide range of doses, a particular Behaviour was not differentially affected. When ethanol was used as a poison (UCS) in a conditioned taste aversion experiment, the magnitude of aversion produced was similar in the two lines. Thus, whatever systems underlie the loss of righting relfex differences do not control response to ethanol in the taste aversion experiment.

A second example may be drawn from the literature on strain differences in ethanol response. Although BALB/c mice lose the righting reflex longer than C_{57}BL/6 mice, they show a greater locomootor stimulation following low dose ethanol administration. It may be of interest that, in these two strains, amphetamine shows exactly opposite effects on locomotor activity. BALB/c mice are not activated, whereas C_{57}BL/10 mice are thus a negative genetic correlation is suggested for these effects of amphetamine and ethanol.

MEASURES OF RESPONSES TO OPIATES

It is becoming increasingly apparent that strain differences are readily obtained using the standard measures of opiate induced analgesia and locomotor activity. However, it is equally apparent that both the magnitude and the direction of these differences change with the particular measures and procedures employed. For example, C_{57}BL mice have been characterized as being sensitive to morphine induced hyperactivity in open field tests and toggle boxes. However, Eidelberg, Erspamer, Kreinick, and Harris have reported that C_{57}BL mice respond to several doses of morphine with decreased activity when the measure is recorded on an activity wheel. Similarly, measures of analgesia, such as the hot plate technique, tail flick procedures, and abdominal writhing induced by agents such as acetic acid often yield very different strain distribution patterns. The exact nature of these differences among measures has been reviewed elsewhere.

For the present, the disparate results of these procedures lead us to argue for careful descriptions of the methods employed when general phenomena associated with opiate responsivity. However, we share the optimism of Crabbe and Belknap that consistent patterns among several dimensions of opiate responsiveness may yet emerge as the results of further investigations are known.

GENETIC CORRELATIONS

The basis of many inferences in pharmacogenetic research is the comparison of effects of more than one drug on the same genetic material. This is implicitly an example of genetic correlation. Both correlated drug effects and correlated effects of a single drug on several Behaviours are means for analyzing the domain of effects of that pharmacological agent, or the common physiological substrate of a variety of agents.

The quantitative genetic methods for deriving genetic correlations are readily available and are not presented in detail here. Instead, the major experimental approaches to genetic correlation methods and the general usefulness of such correlations.

One of the most utilized approaches to genetic correlations has been the examination of correlated characters in selected lines. A generalized example would find a set of lines selected for response to drug X, characterized at a variety of physiological levels, and perhaps other types of responses to the same drug being assessed as well. The usual argument is that care must be taken in interpretation of any correlated characters especially if replicated lines and unselected control lines are not available. The basis of this assertion is that when only two lines are sampled and two characters are measured on those genotypes, then the degrees of freedom for assessing the correlation are $N - 2 = 0$. Two points are worthwhile in counter. A set of once replicated lines would produce a situation with $df = 2$, hardly a powerful statistical test.

Only when the numbers of genotypes exceeds 20 will actual correlations below.40 be detectable. In addition, correlational degrees of freedom are based on the assumption that both variables generating the covariance are free to vary. This is clearly not the case after a selection has been accomplished.

The selected phenotype is at that point fixed. Only the as yet unmeasured phenotype is "free" to vary. Perhaps a more Bayesian approach is desirable.

The most rational alternative to numerous replicates is a simultaneous measurement of several phenotypes during the course of selection. As selection progresses, those characters sharing a true genetic covariance with the selected phenotype will diverge in somewhat parallel fashion. As this approach forces one to identify relevant variables even before the selection begins, it is a less than perfect one.

Obviously a more practical approach is required. We would advocate a rational approach that makes use of a priori knowledge of the pharmacological system being examined and physiological intervention if the question being asked is one of mechanism of drug action. For example, in the LS and SS mice it is not surprising that alcohols other than ethanol have a differential effect.

However, if a correlated character is, for example, an altered preference for saccharin sweetened water, then one must wonder if drift is responsible. In sum, no single approach seems completely suitable. Therefore, only careful and objective scrutiny can avoid erroneous conclusions. An additional tool in genetic correlation analysis is the diallel cross and also the triple test cross. The inclusion of non homozygous animals in the diallel cross analysis is probably an advantage, but again, unless the number of genotypes is large, statistical power is low. The use of genetic correlations to assess commonalities of drug action is a powerful approach. Only the use of cross tolerance studies can so clearly demon strate common substrates of action as can the presence of a genetic correlation. The aforementioned effects of higher order alcohols on LS and SS mice led to the conclusion that considerably different mechanisms of action underlie the effects of alcohols and other sedative hypnotics that did not differentially affect the two lines.

It is conceivable that, in a genetic system with high heritabilities, an experimenter could test a very large range of drug classes and attempt to classify them via one of a number of classification techniques on the basis of the genetic correlations.

The alternate approach examines the effects of a single drug on numerous Behaviours. Examination of the kinds of genotype patterns of Behaviours affected similarly by a given drug should indicate similar substrates for those Behaviours. It would also be of interest to examine the pattern of genetic correlations in drugged and undrugged mice separately. Changes in those correlations would suggest an ability of the drug to dissociate genetically based covariation of Behaviours. Thus even at a strictly Behavioural level of analysis, the genetic correlation could be a very useful tool.

DEVELOPMENTAL PHENOMENA

As already mentioned, the examination of drug responses in the developing organism and concomitant study of neural development is a powerful approach to the study of drug action and neural development. The addition of the genetic variable as a unique tool to tease apart neuropharmacological mechanisms is a potentially fruitful tactic. Correlated biochemical and pharmacological changes are more easily linked if the correlation holds in genotypes that show different time courses of developmental events.

An example from the literature on alcohol preference in mice is instructive. Wood reviews this literature on age related differences in response to and preference for alcohol by labouratory rodents. Two studies in particular are instructive both for purposes of the genetic variable and its interaction with an experiential variable. BALB/c mice are, as adults, avoiders of ethanol. Yet, from 4-9 weeks of age, they will voluntarily consume a fair amount of ethanol. Consumption diminished to near zero by 9-10 weeks unless a choice between alcohol and water was provided during the 4 to 8-week period. Interestingly, mice forced to drink alcohol during this period would typically not consume the alcohol as adults. Thus the fact of choice in young mice appears to predispose alcohol consumption in adults. This plasticity of phenotype or lack of phenostability makes clear the complex nature of the neural and Behavioural systems on which the genes can act only very indirectly.

A more general view of pharmacogenetic research must realise that a most critical variable in drug effectiveness is the functional organization of the complex target organ called the nervous system. It is ultimately the genetic influence on the assemblage of the brain and other physiological systems that provides the substrate for individual differences produced in drug responsivity.

A given drug acting on similar receptors in two brains that differ in organization of the neural network in which these receptors reside might produce different Behavioural outcomes. The mechanism of such gene effects on neural development may be as simple as cell proliferation and migration rates or as complex as regulation of the timing of critical and sensitive periods for Behavioural and emotional development.

The study of such biochemical characters as those already enumerated may yield important findings. Ultimately, however, a concern with developmental issues in pharmacogenetic work will yield much more information about the mechanisms of gene and drug action.

PROTEOLYTIC ENZYMES OF THE DIGESTIVE TRACT

Enzymes of the digestive tract that serve to hydrolyze dietary proteins are synthesized in the stomach and pancreas as zymogens.

Table. Pancreatic and Gastric Zymogens

Origin	Zymogen	Active Protease
Pancreas	Trypsinogen	Trypsin
Pancreas	Chymotrypsinogen	Chymotrypsin
Pancreas	Procarboxypeptidase	Carboxypeptidase
Pancreas	Proelastase	Elastase
Stomach	Pepsinogen	Pepsin

Only upon proteolytic activation are these enzymes able to form a catalytically active substrate-binding site. The activation of chymotrypsinogen is an interesting example. Chymotrypsino-gen is a 245-residue polypeptide chain cross-linked by five disulfide bonds. Chymotrypsinogen is converted to an enzymatically active form called π-chymotrypsin when trypsin cleaves the peptide bond joining Arg15 and Ile16.

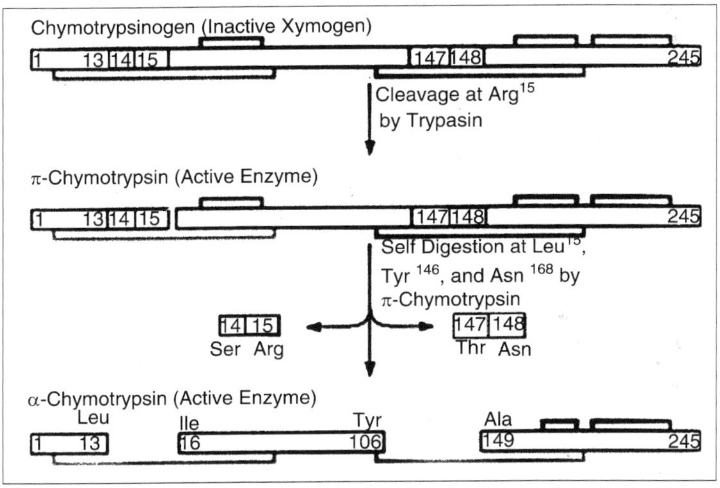

Fig. The Proteolytic Activation of Chymotrypsinogen.

The enzymatically active π-chymotrypsin acts upon other π-chymotrypsin molecules, excising two dipeptides, Ser^{14}-Arg^{15} and Thr^{147}-Asn^{148}. The end product of this processing pathway is the mature protease a -chymotrypsin, in which the three peptide chains, A (residues 1 through 13), B (residues 16 through 146), and C (residues 149 through 245), remain together because they are linked by two disulfide bonds, one from A to B, and one from B to C. It is interesting to note that the transformation of inactive chymotrypsinogen to active π-chymotrypsin requires the cleavage of just one particular peptide bond.

BLOOD CLOTTING

The formation of blood clots is the result of a series of zymogen activations. The amplification achieved by this cascade of enzymatic activations allows blood clotting to occur rapidly in response to injury. Seven of the clotting factors in their active form are serine proteases: kallikrein, XIIa, XIa, IXa, VIIa, Xa, andthrombin. Two routes to blood clot formation exist. The intrinsic pathway is instigated when the blood comes into physical contact with abnormal surfaces caused by injury; the extrinsic pathway is initiated by factors released from injured tissues. The pathways merge at Factor X and culminate in clot formation.

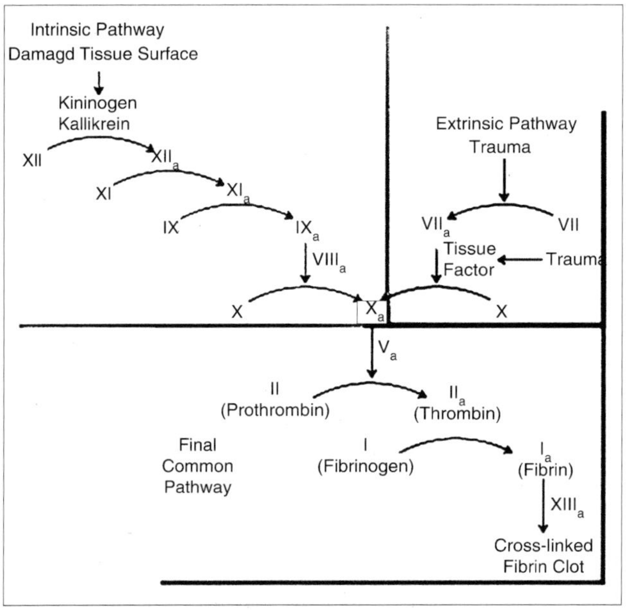

Fig. The Cascade of Activation Steps Leading to Blood Clotting.

The intrinsic and extrinsic pathways converge at Factor X, and the final common pathway involves the activation of thrombin and its conversion of fibrinogen into fibrin, which aggregates into ordered filamentous arrays that become cross-linked to form the clot. Thrombin excises peptides rich in negative charge from fibrinogen, converting it to fibrin, a molecule with a different surface charge distribution. Fibrin readily aggregates into ordered fibrous arrays that are subsequently stabilized by covalent cross-links. Thrombin specifically cleaves Arg-Gly peptide bonds and is homologous to trypsin, which is also a serine protease (recall that trypsin acts only at Arg and Lys residues).

Protein Kinases

Protein kinases are converter enzymes that catalyze the ATP-dependent phosphorylation of serine, threonine, and/or tyrosine hydroxyl groups in target proteins. Phosphorylation introduces a bulky group bearing two negative charges, causing conformational changes that alter the target protein's function. (Unlike a phosphoryl group, no amino acid side chain can provide two negative charges.) Protein kinases represent a protein superfamily whose members are widely diverse in terms of size, subunit structure, and subcellular localization. Nevertheless, all share a common catalytic mechanism based on a conserved catalytic core/kinase domain of approximately 260 amino acid residues. Protein kinases are classified as Ser/Thr- and/or Tyr-specific and are subclassified in terms of the allosteric activators they require and the consensus amino acid sequence within the target protein that is recognized by the kinase. For example, cAMP-dependent protein kinase (PKA) phosphorylates proteins having Ser or Thr residues within an R(R/K)X (S*/T*) target consensus sequence (* denotes the residue that becomes phosphorylated). That is, PKA phosphorylates Ser or Thr residues that occur in an Arg-(Arg or Lys)-(any amino acid)-(Ser or Thr) sequence segment.

Targeting of protein kinases to particular consensus sequence elements within proteins creates a means to regulate these kinases byintrasteric control. Intrasteric control occurs when a regulatory subunit (or protein domain) has a pseudosubstrate sequence that mimics the target sequence but lacks a OH-bearing side chain at the right place. For example, the cAMP-binding regulatory subunits of PKA (R subunits in Figure) possess the pseudosubstrate sequence RRGA*I and this sequence binds to the active site of PKA catalytic subunits, blocking their activity. This pseudosubstrate sequence has an alanine residue where serine occurs in the PKA target sequence; Ala is sterically similar to serine but lacks a phosphorylatable OH-group. When these PKA regulatory subunits bind cAMP, they undergo a conformational change and dissociate from the catalytic (C) subunits, and the active site of PKA is free to bind and phosphorylate its targets.

The abundance of many protein kinases in cells is an indication of the great importance of protein phosphorylation in cellular regulation. Exactly 113 protein kinase genes have been recognized in yeast, and it is estimated that the human genome encodes more than 1000 different protein kinases. Tyrosine kinases(protein kinases that phosphorylate Tyr residues) occur only in multicellular organisms (yeast has no tyrosine kinases). Tyrosine kinases are components of signaling pathways involved in cell - cell communication.

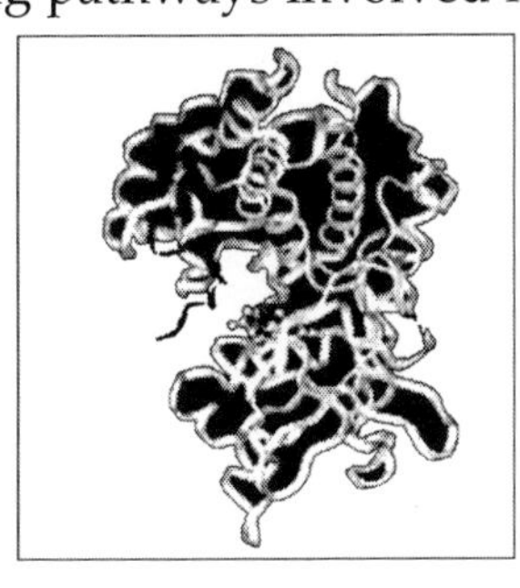

Cyclic AMP-dependent protein kinase is shown complexed with a pseudosubstrate peptide. This complex also includes ATP and two Mn2+ ions bound at the active site.

ISOZYMES

A number of enzymes exist in more than one quaternary form, differing in their relative proportions of structurally equivalent but catalytically distinct polypeptide subunits. A classic example is mammalian lactate dehydrogenase (LDH), which exists as five different isozymes, depending on the tetrameric association of two different subunits, A and B: A_4, A_3B, A_2B_2, AB_3, and B_4. The kinetic properties of the various LDH isozymes differ in terms of their relative affinities for the various substrates and their sensitivity to inhibition by product. Different tissues express different isozyme forms, as appropriate to their particular metabolic needs. By regulating the relative amounts of A and B subunits they synthesize, the cells of various tissues control which isozymic forms are likely to assemble, and, thus, which kinetic parameters prevail.

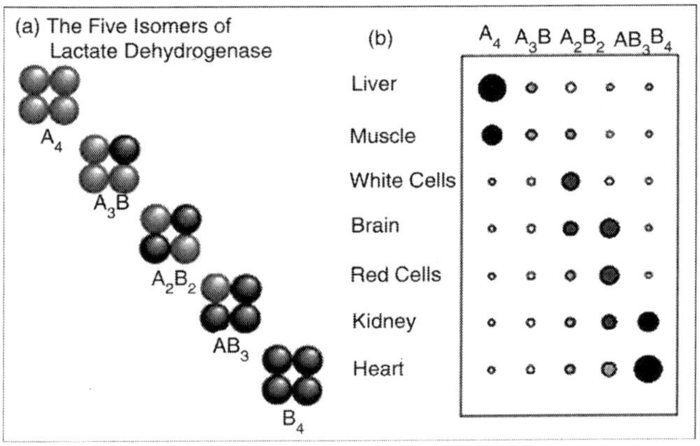

Fig. The Isozymes of Lactate Dehydrogenase (LDH).

Active muscle tissue becomes anaerobic and produces pyruvate from glucose via glycolysis. It needs LDH to regenerate NAD^+ from NADH so glycolysis can continue. The lactate produced is released into the blood. The muscle LDH isozyme (A_4) works best in the NAD^+-regenerating direction. Heart tissue is aerobic and uses lactate as a fuel, converting it to pyruvate via LDH and using the pyruvate to fuel the citric acid cycle to obtain energy. The heart LDH isozyme (B_4) is inhibited by excess pyruvate so the fuel won't be wasted.

MODULATOR PROTEINS

Modulator proteins are yet another way that cells mediate metabolic activity. Modulator proteins are proteins that bind to enzymes, and by binding, influence the activity of the enzyme. For example, some enzymes, such as cAMP-dependent protein kinase, exist as dimers of catalytic subunits and regulatory subunits. These regulatory subunits are modulator proteins that suppress the activity of the catalytic subunits. Dissociation of the regulatory subunits (modulator proteins) activates the catalytic subunits; reassociation once again suppresses activity.

Phosphoprotein phosphatase inhibitor-1 (PPI-1) is another example of a modulator protein. When PPI-1 is phosphorylated on one of its serine residues, it binds to phosphoprotein phosphatase, inhibiting its phosphatase activity. The result is an increased phosphorylation of the interconvertible enzyme targeted by the protein kinase/phosphoprotein phosphatase cycle.

We will meet other important representatives of this class as the processes of metabolism unfold in subsequent chapters. For now, let us focus our attention on the fascinating kinetics of allosteric enzymes.

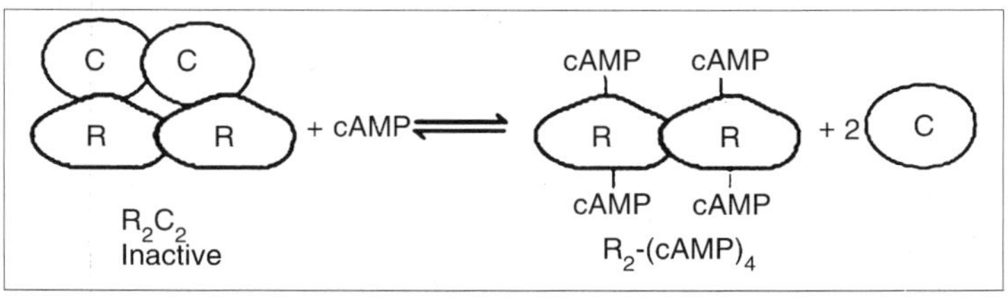

Fig. Cyclic AM π- dependent protein kinase (also known as PKA) is a 150- to 170-kD R_2C_2 tetramer in mammalian cells.

The two R (regulatory) subunits bind cAMP (KD = 3 × 10-8 M); cAMP binding releases the R subunits from the C (catalytic) subunits. C subunits are enzymatically active as monomers.

ALLOSTERIC REGULATION OF ENZYME ACTIVITY

Allosteric regulation acts to modulate enzymes situated at key steps in metabolic pathways.

Consider as an illustration the following pathway, where A is the precursor for formation of an end product, F, in a sequence of five enzyme-catalyzed reactions:

$$A \xrightarrow{\text{enz 1}} B \xrightarrow{\text{enz 2}} C \xrightarrow{\text{enz 3}} D \xrightarrow{\text{enz 4}} E \xrightarrow{\text{enz 5}} F$$

In this scheme, F symbolizes an essential metabolite, such as an amino acid or a nucleotide. In such systems, F, the essential end product, inhibits enzyme 1, the first step in the pathway. Therefore, when sufficient F is synthesized, it blocks further synthesis of itself. This phenomenon is called feedback inhibition or feedback regulation.

GENERAL PROPERTIES OF REGULATORY ENZYMES

Enzymes such as enzyme 1, which are subject to feedback regulation, represent a distinct class of enzymes, the regulatory enzymes.

As a class, these enzymes have certain exceptional properties:

1. Their kinetics do not obey the Michaelis - Menten equation. Their v versus [S] plots yield sigmoid- or S-shaped curves rather than rectangular hyperbolas. Such curves suggest a second-order (or higher) relationship between v and [S]; that is, v is proportional to [S]n, where n > 1. A qualitative description of the mechanism responsible for the S-shaped curves is that binding of one S to a protein molecule makes it easier for additional substrate molecules to bind to the same protein molecule. In the jargon of allostery, substrate binding is cooperative.

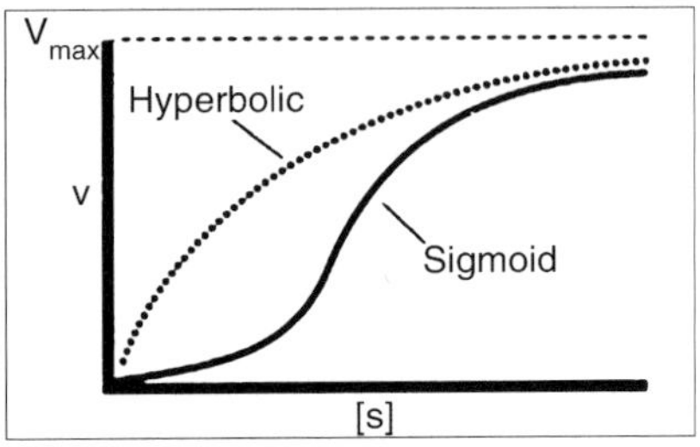

Fig. Sigmoid v Versus [S] plot.

The dotted line represents the hyperbolic plot characteristic of normal Michaelis - Menten-type enzyme kinetics.

Inhibition of a regulatory enzyme by a feedback inhibitor does not conform to any normal inhibition pattern, and the feedback inhibitor F bears little structural similarity to A, the substrate for the regulatory enzyme. F apparently acts at a binding site distinct from the substrate-binding site. The term allosteric is apt, because F is sterically dissimilar and, moreover, acts at a site other than the site for S. Its effect is called allosteric inhibition.

Regulatory or allosteric enzymes like enzyme 1 are, in some instances, regulated by activation. That is, whereas some effector molecules such as F exert negative effects on enzyme activity, other effectors show stimulatory, or positive, influences on activity.

Allosteric enzymes have an oligomeric organization. They are composed of more than one polypeptide chain (subunit) and have more than one S-binding site per enzyme molecule.

The working hypothesis is that, by some means, interaction of an allosteric enzyme with effectors alters the distribution of conformational possibilities or subunit interactions available to the enzyme. That is, the regulatory effects exerted on the enzyme's activity are achieved by conformational changes occurring in the protein when effector metabolites bind.

CITRATE SYNTHASE (CONDENSING ENZYME)

The first reaction of the cycle is condensation of the methyl carbon of acetyl-CoA with the keto carbon (C-2) of oxaloacetate (OAA). The standard free energy of the reaction, -8.0 kcal/mol, drives it strongly in the forward direction. Since the formation of OAA from its precursor is thermodynamically unfavourable, the highly exergonic nature of the citrate synthase reaction is of central importance in keeping the entire cycle going in the forward direction, since it drives oxaloacetate formation by mass action principals.

When the cellular energy charge increases the rate of flux through the TCA cycle will decline leading to a build-up of citrate. Excess citrate is used to transport acetyl-CoA carbons from the mitochondrion to the cytoplasm where they can be used for fatty acid and cholesterol biosynthesis. Additionally, the increased levels of citrate in the cytoplasm activate the key regulatory enzyme of fatty acid biosynthesis, acetyl-CoA carboxylase (ACC) and inhibit PFK-1. In non-hepatic tissues citrate is also required for ketone body synthesis.

ACONITASE

The isomerization of citrate to isocitrate by aconitase is stereospecific, with the migration of the -OH from the central carbon of citrate (formerly the keto carbon of OAA) being always to the adjacent carbon which is derived from the methylene ($-CH_2-$) of OAA. The stereospecific nature of the isomerization determines that the CO_2 lost, as isocitrate is oxidized to succinyl-CoA, is derived from the oxaloacetate used in citrate synthesis.

Aconitase is one of several mitochondrial enzymes known as non-heme-iron proteins. These proteins contain inorganic iron and sulfur, known as iron sulfur Centres, in a coordination complex with cysteine sulfurs of the protein. There are two prominent classes of non-heme-iron complexes, those containing two equivalents each of inorganic iron and sulfur Fe_2S_2, and those containing 4 equivalents of each Fe_4S_4. Aconitase is a member of the Fe_4S_4 class. Its iron sulfur Centres are often designated as $Fe_4S_4Cys_4$, indicating that 4 cystine sulfur atoms are involved in tghe complete structure of the complex. In iron sulfur compounds the iron is generally involved in oxidation-reduction events.

ISOCITRATE DEHYDROGENASE

Isocitrate is oxidatively decarboxylated to α-ketoglutarate by isocitrate dehydrogenase, (IDH). There are two different IDH enzymes. The IDH of the TCA cycle uses NAD^+ as a cofactor, whereas the other IDH uses $NADP^+$ as a cofactor. Unlike the NAD+-requiring enzyme, which is located only in the mitochondrial matrix, the $NADP^+$-requiring enzyme is found in both the mitochondrial matrix and the cytosol. IDH catalyzes the rate-limiting step, as well as the first NADH-yielding reaction of the TCA cycle. The CO_2 produced by the IDH reaction is the original C-1 carbon of the oxaloacetate used in the citrate synthase reaction.

It is generally considered that control of carbon flow through the cycle is regulated at IDH by the powerful negative allosteric effectors NADH and ATP and by the potent positive effectors; isocitrate, ADP and AMP. From the latter it is clear that cell energy charge is a key factor in regulating carbon flow through the TCA cycle.

α -KETOGLUTARATE DEHYDROGENASE COMPLEX

α-ketoglutarate is oxidatively decarboxylated to succinyl-CoA by the α-ketoglutarate dehydrogenase (α-KGDH) complex. This reaction generates the second TCA cycle equivalent of CO_2 and NADH.

This multienzyme complex is very similar to the PDH complex in the intricacy of its protein makeup, cofactors, and its mechanism of action. Also, as with the PDH complex, the reactions of the α-KGDH complex proceed with a large negative standard free energy change. Although the α-KGDH of the complex is not subject to covalent modification, allosteric regulation is quite complex, with activity being regulated by energy charge, the NAD^+/NADH ratio, and effector activity of substrates and products.

Succinyl-CoA and α-ketoglutarate are also important metabolites outside the TCA cycle. In particular, α-ketoglutarate represents a key anapleurotic metabolite linking the entry and exit of carbon atoms from the TCA cycle to pathways involved in amino acid metabolism. α-ketoglutarate is also important for driving the malate-aspartate shuttle.

The malate/aspartate shuttle is the principal mechanism for the movement of reducing equivalents (in the form of NADH) from the cytoplasm to the mitochondria. The glycolytic pathway is a primary source of NADH. Within the mitochodria the electrons of NADH can be coupled to ATP production during the process of oxidative phosphorylation.

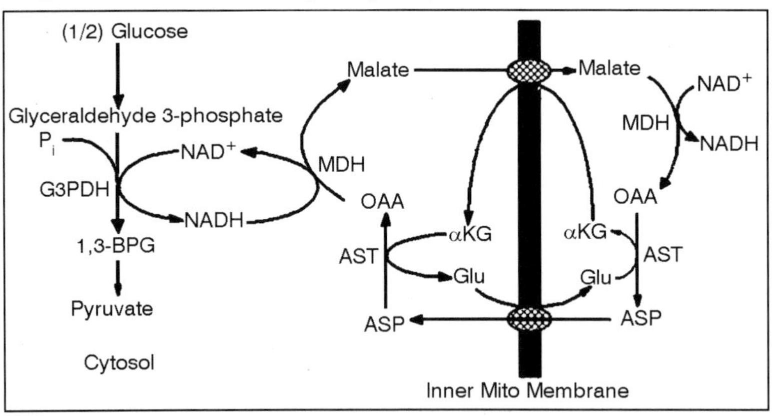

The electrons are"carried" into the mitochondria in the form of malate. Cytoplasmic malate dehydrogenase (MDH) reduces oxaloacetate (OAA) to malate while oxidizing NADH to NAD+. Malate then enters the mitochondria where the reverse reaction is carried out by mitochondrial MDH. Movement of mitochondrial OAA to the cytoplasm to maintain this cycle requires it be transaminated to aspartate (Asp, D) with the amino group being donated by glutamate (Glu, E). The Asp then leaves the mitochondria and enters the cytoplasm. The deamination of glutamate generates α-ketoglutarate (α-KG) which leaves the mitochondria for the cytoplasm. All the participants in the cycle are present in the proper cellular compartment for the shuttle to function due to concentration dependent movement. When the energy level of the cell rises the rate of mitochondrial oxidation of NADH to NAD^+ declines and therefore, the shuttle slows. G3PDH is glyceraldehyde-3-phosphate dehydrogenase.

Succinyl-CoA, along with glycine, contributes all the carbon and nitrogen atoms required for the synthesis ofprotoporphyrin heme biosynthesis and for non-hepatic tissue utilization of ketone bodies.

SUCCINYL COA SYNTHETASE

The conversion of succinyl-CoA to succinate by succinyl CoA synthetase involves use of the high-energy thioester of succinyl-CoA to drive synthesis of a high-energy nucleotide phosphate, by a process known assubstrate-level phosphorylation. In this process a high energy enzyme--phosphate intermediate is formed, with the phosphate subsequently being transferred to GDP. Mitochondrial GTP is used in a trans-phosphorylation reaction catalyzed by the mitochondrial enzyme nucleoside diphospho kinase to phosphorylate ADP, producing ATP and regenerating GDP for the continued operation of succinyl CoA synthetase.

SUCCINATE DEHYDROGENASE (SDH)

Succinate dehydrogenase catalyzes the oxidation of succinate to fumarate with the sequential reduction of enzyme-bound FAD and non-heme-iron. In mammalian cells the final electron acceptor is coenzyme Q10 (CoQ10), a mobile carrier of reducing equivalents that is restricted by its lipophilic nature to the lipid phase of the mitochondrial membrane.

FUMARASE (FUMARATE HYDRATASE)

The fumarase-catalyzed reactions specific for the trans form of fumarate. The result is that the hydration of fumarate proceeds stereospecifically with the production of L-malate.

MALATE DEHYDROGENASE (MDH)

L-malate is the specific substrate for MDH, the final enzyme of the TCA cycle. The forward reaction of the cycle, the oxidation of malate yields oxaloacetate (OAA). In the forward direction the reaction has a standard free energy of about +7 kcal/mol, indicating the very unfavourable nature of the forward direction.

As noted earlier, the citrate synthase reaction that condenses oxaloacetate with acetyl-CoA has a standard free energy of about -8 kcal/mol and is responsible for pulling the MDH reaction in the forward direction. The overall change in standard free energy change is about -1 kcal/mol for the conversion of malate to oxaloacetate and on to succinate.

The overall stoichiometry of the TCA cycle is:

acetyl-CoA + 3NAD$^+$ + FAD + GDP + P_i + 2H_2O → 2CO_2 + 3NADH + $FADH_2$ + GTP + 2H$^+$ + HSCoA

REGULATION OF THE TCA CYCLE

Regulation of the TCA cycle. like that of glycolysis, occurs at both the level of entry of substrates into the cycle as well as at the key reactions of the cycle.

Fuel enters the TCA cycle primarily as acetyl-CoA. The generation of acetyl-CoA from carbohydrates is, therefore, a major control point of the cycle. This is the reaction catalyzed by the PDH complex.

By way of review, the PDH complex is inhibited by acetyl-CoA and NADH and activated by non-acetylated CoA (CoASH) and NAD$^+$. The pyruvate dehydrogenase activities of the PDH complex are regulated by their state of phosphorylation.

This modification is carried out by a specific kinase (PDH kinase) and the phosphates are removed by a specific phosphatase (PDH phosphatase). The phosphorylation of PDH inhibits its activity and, therefore, leads to decreased oxidation of pyruvate. PDH kinase is activated by NADH and acetyl-CoA and inhibited by pyruvate, ADP, CoASH, Ca^{2+} and Mg^{2+}. The PDH phosphatase, in contrast, is activated by Mg^{2+} and Ca^{2+}.

Since three reactions of the TCA cycle as well as PDH utilize NAD+ as co-factor it is not difficult to understand why the cellular ratio of NAD+/NADH has a major impact on the flux of carbon through the TCA cycle.

Substrate availability can also regulate TCA flux. This occurs at the citrate synthase reaction as a result of reduced availability of oxaloacetate. Product inhibition also controls the TCA flux, e.g. citrate inhibits citrate synthase, α-KGDH is inhibited by NADH and succinyl-CoA. The key enzymes of the TCA cycle are also regulated allosterically by Ca^{2+}, ATP and ADP.

ENZYME CASCADES REGULATE GLYCOGEN PHOSPHORYLASE

The phosphorylation reaction that activates glycogen phosphorylase is mediated by an enzyme cascade.

The first part of the cascade leads to hormonal stimulation (described in the next section) of adenylyl cyclase, a membrane-bound enzyme that converts ATP to adenosine-3',5' -cyclic monophosphate, denoted as cyclic AMP or simply cAMP.

This regulatory molecule is found in all eukaryotic cells and acts as an intracellular messenger molecule, controlling a wide variety of processes. Cyclic AMP is known as a second messenger because it is the intracellular agent of a hormone (the "first messenger").

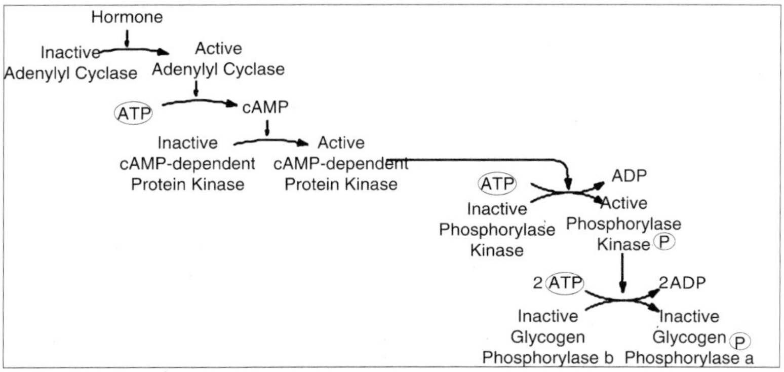

Fig. The Hormone-activated Enzymatic Cascade that Leads to Activation of Glycogen Phosphorylase.

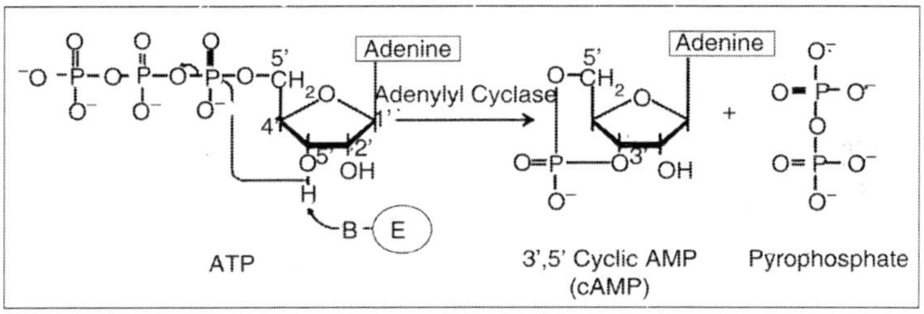

Fig. The Adenylyl Cyclase Reaction Yields 3',5' -cyclic AMP and Pyrophosphate.

The reaction is driven forward by subsequent hydrolysis of pyrophosphate by the enzyme inorganic pyrophosphatase. The hormonal stimulation of adenylyl cyclase is effected by a transmembrane signaling pathway consisting of three components, all membrane-associated. Binding of hormone to the external surface of a hormone receptor causes a conformational change in this transmembrane protein, which in turn stimulates a GTP-binding protein (abbreviated G protein). G proteins are heterotrimeric proteins consisting of a-(45-47 kD), β-(35 kD), and g-(7-9 kD) subunits. The a -subunit binds GDP or GTP and has an intrinsic, slow GTPase activity. In the inactive state, the $Ga_{\beta g}$ complex has GDP at the nucleotide site. When a G protein is stimulated by a hormone-receptor complex, GDP dissociates and GTP binds to G a, causing it to dissociate from $G_{\beta g}$ and to associate with adenylyl cyclase. Binding of G_a(GTP) activates adenylyl cyclase to form cAMP from ATP. However, the intrinsic GTPase activity of G_a eventually hydrolyzes GTP to GDP, leading to dissociation of G_a (GDP) from adenylyl cyclase and reassociation with $G_{a\beta g}$ to form the inactive $G_{a\beta g}$ complex.

This cascade amplifies the hormonal signal because a single hormone-receptor complex can activate many G proteins before the hormone dissociates from the receptor, and because the G_a-activated adenylyl cyclase can synthesize many cAMP molecules before bound GTP is hydrolyzed by G_a. More than 100 different G protein - coupled receptors and at least 21 distinct G aproteins are known.

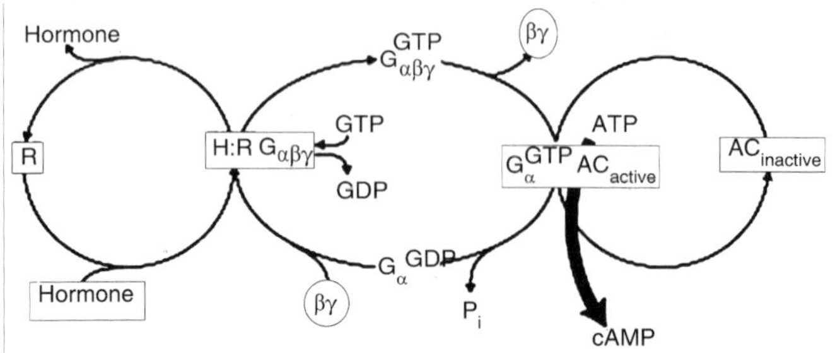

Fig. Hormone (H) Binding to its Receptor (R) Creates a Hormone; Receptor Complex (H:R) that Catalyzes GDP-GTP Exchange on the a subunit of the Heterotrimer G Protein ($G_{a\beta g}$), replacing GDP with GTP.

The G_a-subunit with GTP bound dissociates from the β g-subunits and binds to adenylyl cyclase (AC). AC becomes active upon association with G a:GTP and catalyzes the formation of cAMP from ATP. With time, the intrinsic GTPase activity of the G_a -subunit hydrolyzes the bound GTP, forming GDP; this leads to dissociation of G a:GDP from AC, reassociation of G a with the βg subunits, and cessation of AC activity. AC and the hormone receptor H are integral plasma membrane proteins; G a and $G_{\beta g}$ are membrane-anchored proteins.

Cyclic AMP is an essential activator of cAMP-dependent protein kinase (PKA). This enzyme is normally inactive because its two catalytic subunits (C) are strongly associated with a pair of regulatory subunits (R), which serve to block activity. Binding of cyclic AMP to the regulatory subunits induces a conformation change that causes the dissociation of the C monomers from the R dimer. The free C subunits are active and can phosphorylate other proteins. One of the many proteins phosphorylated by PKA is phosphorylase kinase. Phosphorylase kinase is inactive in the unphosphorylated state and active in the phosphorylated form. As its name implies, phosphorylase kinase functions to phosphorylate (and activate) glycogen phosphorylase. Thus, stimulation of adenylyl cyclase leads to activation of glycogen breakdown.

HEMOGLOBIN AND MYOGLOBIN

Ancient life forms evolved in the absence of oxygen and were capable only of anaerobic metabolism. As the earth's atmosphere changed over time, so too did living things. Indeed, the production of O_2 by photosynthesis was a major factor in altering the atmosphere. Evolution to an oxygen-based metabolism was highly beneficial. Aerobic metabolism of sugars, for example, yields far more energy than corresponding anaerobic processes. Two important oxygen-binding proteins appeared in the course of evolution so that aerobic metabolic processes were no longer limited by the solubility of O_2 in water. These proteins are represented in animals as hemoglobin (Hb) in blood and myoglobin (Mb) in muscle. Because hemoglobin and myoglobin are two of the most-studied proteins in nature, they have become paradigms of protein structure and function. Moreover, hemoglobin is a model for protein quaternary structure and allosteric function. The binding of O_2 by hemoglobin, and its modulation by effectors such as protons, CO_2, and 2,3 - bisphosphoglycerate, depend on interactions between subunits in the Hb tetramer. Subunit-subunit interactions in Hb reveal much about the functional significance of quaternary associations and allosteric regulation.

THE COMPARATIVE BIOCHEMISTRY OF MYOGLOBIN AND HEMOGLOBIN

A comparison of the properties of hemoglobin and myoglobin offers insights into allosteric phenomena, even though these proteins are notenzymes. Hemoglobin displays sigmoid-shaped O2-binding curves. The unusual shape of these curves was once a great enigma in biochemistry. Such curves closely resemble allosteric enzyme:substrate saturation graphs. In contrast, myoglobin's interaction with oxygen obeys classical Michaelis-Menten-type substrate saturation behaviour.

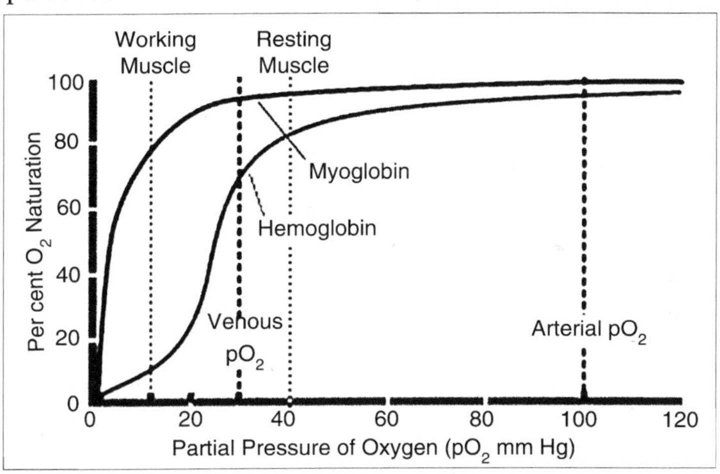

Fig. O_2-binding Curves for Hemoglobin and Myoglobin.

Before examining myoglobin and hemoglobin in detail, let us first encapsulate the lesson: Myoglobin is a compact globular protein composed of a single polypeptide chain 153 amino acids in length; its molecular mass is 17.2 kD. It contains heme, a porphyrin ring system complexing an iron ion, as its prosthetic group. Oxygen binds to Mb via its heme. Hemoglobin (Hb) is also a compact globular protein, but Hb is a tetramer. It consists of four polypeptide chains, each of which is very similar structurally to the myoglobin polypeptide chain, and each bears a heme group. Thus, a hemoglobin molecule can bind four O_2 molecules. In adult human Hb, there are two identical chains of 141 amino acids, the α-chains, and two identical β-chains, each of 146 residues. The human Hb molecule is an $\alpha_2\beta_2$-type tetramer of molecular mass 64.45 kD. The tetrameric nature of Hb is crucial to its biological function: When a molecule of O_2binds to a heme in Hb, the heme Fe ion is drawn into the plane of the porphyrin ring. This slight movement sets off a chain of conformational events that are transmitted to adjacent subunits, dramatically enhancing the affinity of their heme groups for O_2. That is, the binding of O_2 to one heme of Hb makes it easier for the Hb molecule to bind additional equivalents of O_2. Hemoglobin is a marvelously constructed molecular machine. Let us dissect its mechanism, beginning with its monomeric counterpart, the myoglobin molecule.

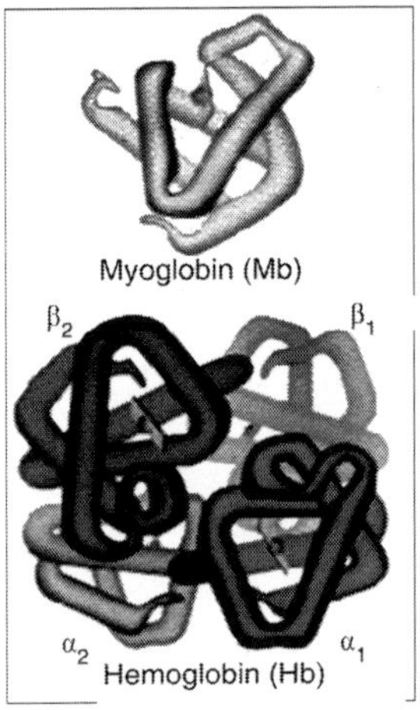

Fig. The Myoglobin and Hemoglobin Molecules.

Myoglobin (sperm whale): one polypeptide chain of 153 aa residues (mass 5 17.2 kD) has one heme (mass = 652 D) and binds one O_2. Hemoglobin (human): four polypeptide chains, two of 141 aa residues (a) and two of 146 residues (β); mass = 64.45 kD. Each polypeptide has a heme; the Hb tetramer binds four O_2. (Irving Geis)

MYOGLOBIN

Myoglobin is the oxygen-storage protein of muscle. The muscles of diving mammals such as seals and whales are especially rich in this protein, which serves as a store for O2 during the animal's prolonged periods underwater. Myoglobin is abundant in skeletal and cardiac muscle of nondiving animals as well. Myoglobin is the cause of the characteristic red colour of muscle.

THE MB POLYPEPTIDE CRADLES THE HEME GROUP

The myoglobin polypeptide chain is folded to form a cradle (4.4 × 4.4 × 2.5 nm) that nestles the heme prosthetic group.

O_2 binding depends on the heme's oxidation state. The iron ion in the heme of myoglobin is in the +2 oxidation state, that is, the ferrous form.

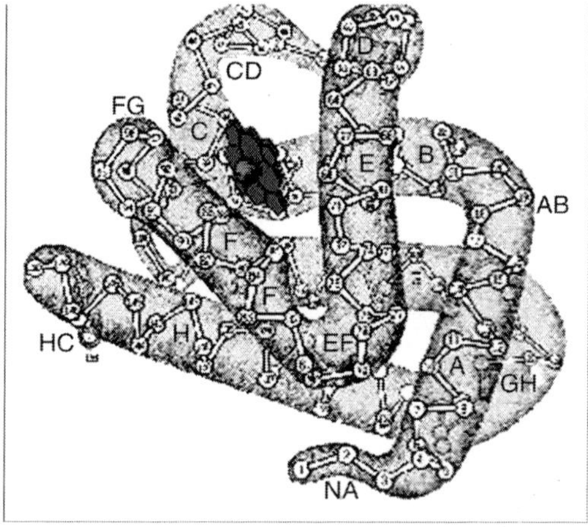

Fig. Detailed Structure of the Myoglobin Molecule.

This is the form that binds O_2. Oxidation of the ferrous form to the +3 ferric form yieldsmetmyoglobin, which will not bind O_2. It is interesting to note that free heme in solution will readily interact with O_2 also, but the oxygen quickly oxidizes the iron atom to the ferric state. Fe3+:protoporphyrin IX is referred to as hematin. Thus, the polypeptide of myoglobin may be viewed as serving three critical functions: it cradles the heme group, it protects the heme iron atom from oxidation, and it provides a pocket into which the O_2 can fit.

The myoglobin polypeptide chain consists of eight helical segments, designated by the letters A through H, counting from the N-terminus. These helices, ranging in length from 7 to 26 residues, are linked by short, unordered regions that are named for the helices they connect, as in the AB region or the EF region.

The individual amino acids in the polypeptide are indicated according to their position within the various segments, as in His F8, the eighth residue in helix F, or Phe CD1, the first amino acid in the interhelical CD region. Occasionally, amino acids are specified in the conventional way, that is, by the relative position in the chain, as in Gly153. The heme group is cradled within the folded polypeptide chain. (Irving Geis)

O_2 BINDING TO MB

Iron ions, whether ferrous or ferric, prefer to interact with six ligands, four of which share a common plane. The fifth and sixth ligands lie above and below this plane. In heme, four of the ligands are provided by the nitrogen atoms of the four pyrroles. A fifth ligand is donated by the imidazole side chain of amino acid residue His F8. When myoglobin binds O_2 to become oxymyoglobin, the O_2 molecule adds to the heme iron ion as the sixth ligand. O_2 adds end on to the heme iron, but it is not oriented perpendicular to the plane of the heme. Rather, it is tilted about 60° with respect to the perpendicular.

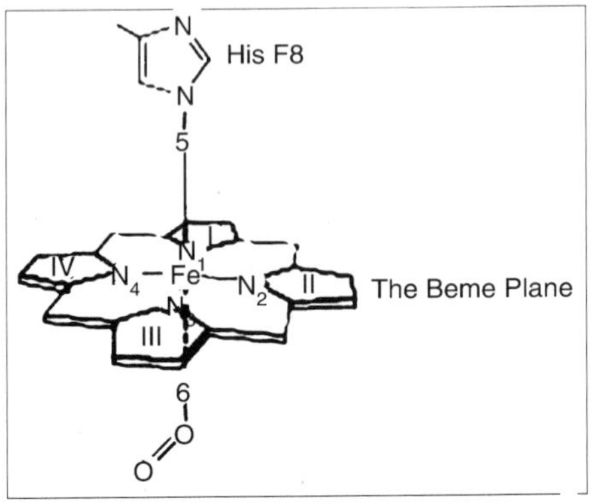

Fig. The Six Liganding Positions of An Iron Ion.

Four ligands lie in the same plane; the remaining two are, respectively, above and below this plane. In myoglobin, His F8 is the fifth ligand; in oxymyoglobin, O_2 becomes the sixth.

In deoxymyoglobin, the sixth ligand position is vacant, and in metmyoglobin, a water molecule fills the O_2 site and becomes the sixth ligand for the ferric atom. On the oxygen-binding side of the heme lies another histidine residue, His E7. While its imidazole function lies too far away to interact with the Fe atom, it is close enough to contact the O_2. Therefore, the O_2-binding site is a sterically hindered region. Biologically important properties stem from this hindrance.

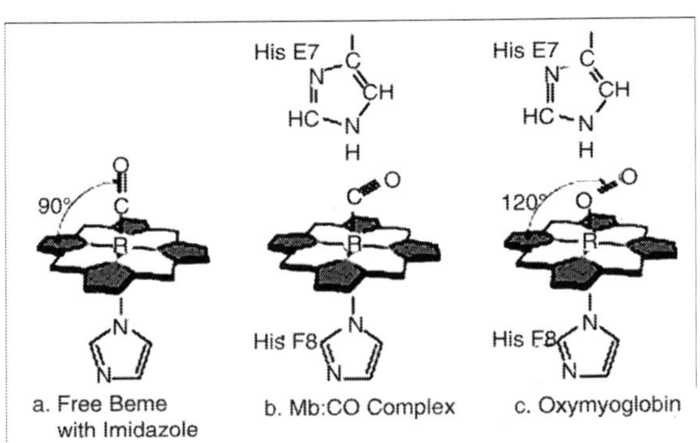

Fig. Oxygen and Carbon Monoxide Binding to the Heme Group of Myoglobin.

For example, the affinity of free heme in solution for carbon monoxide (CO) is 25,000 times greater than its affinity for O_2. But CO only binds 250 times more tightly than O_2 to the heme of myoglobin, because His E7 forces the CO molecule to tilt away from a preferred perpendicular alignment with the plane of the heme. This diminished affinity of myoglobin for CO guards against the possibility that traces of CO produced during metabolism might occupy all of the heme sites, effectively preventing O_2 from binding. Nevertheless, CO is a potent poison and can cause death by asphyxiation.

O_2 BINDING ALTERS MB CONFORMATION

What happens when the heme group of myoglobin binds oxygen? X-ray crystallography has revealed that a crucial change occurs in the position of the iron atom relative to the plane of the heme. In deoxymyoglobin, the ferrous ion has but five ligands, and it lies 0.055 nm above the plane of the heme, in the direction of His F8. The iron:porphyrin complex is therefore dome-shaped.

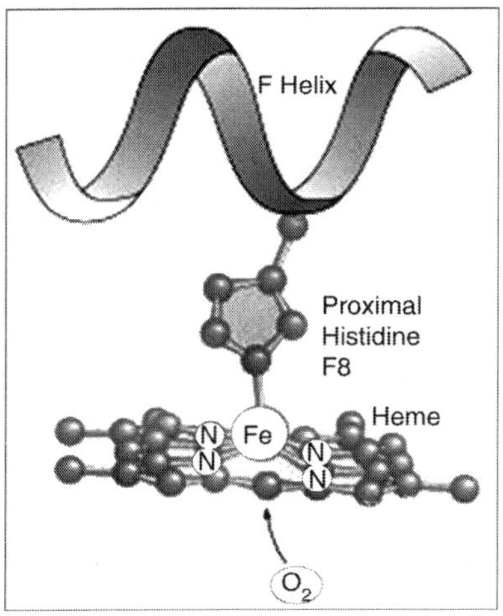

Fig. The Displacement of the Fe ion of the Heme of Deoxymyoglobin from the Plane of the Porphyrin Ring System by the Pull of His F8. In Oxymyoglobin, the Bound O^2 Counteracts this Effect.

When O_2 binds, the iron atom is pulled back towards the porphyrin plane and is now displaced from it by only 0.026 nm. The consequences of this small motion are trivial as far as the biological role of myoglobin is concerned. However, as we shall soon see, this slight movement profoundly affects the properties of hemoglobin. Its action on His F8 is magnified through changes in polypeptide conformation that alter subunit interactions in the Hb tetramer. These changes in subunit relationships are the fundamental cause of the allosteric properties of hemoglobin.

THE PHYSIOLOGICAL SIGNIFICANCE OF COOPERATIVE

The relative oxygen affinities of hemoglobin and myoglobin reflect their respective physiological roles. Myoglobin, as an oxygen storage protein, has a greater affinity for O2 than hemoglobin at all oxygen pressures. Hemoglobin, as the oxygen carrier, becomes saturated with O_2 in the lungs, where the partial pressure of O_2 (pO_2) is about 100 torr.1 In the capillaries of tissues, pO_2 is typically 40 torr, and oxygen is released from Hb. In muscle, some of it can be bound by myoglobin, to be stored for use in times of severe oxygen deprivation, such as during strenuous exercise.

THE STRUCTURE OF THE HEMOGLOBIN MOLECULE

As noted, hemoglobin is an $a_2\beta_2$ tetramer. Each of the four subunits has a conformation virtually identical to that of myoglobin. Two different types of subunits, a and β, are necessary to achieve cooperative O_2-binding by Hb.

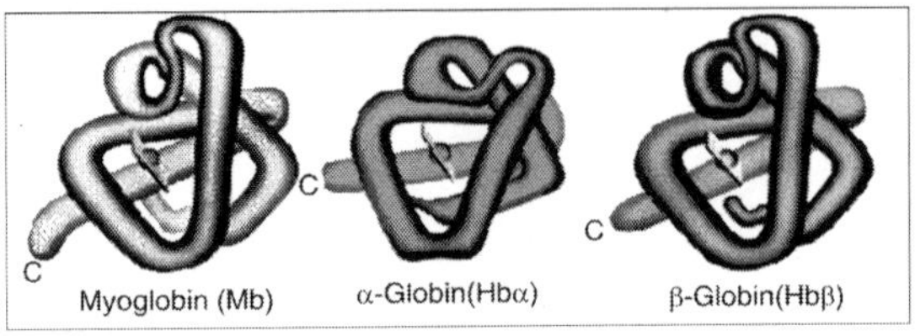

Fig. Conformational Drawings of the a - and β-chains of Hb and the Myoglobin Chain. (Irving Geis)

The β-chain at 146 amino acid residues is shorter than the myoglobin chain (153 residues), mainly because its final helical segment (the H helix) is shorter. The a -chain (141 residues) also has a shortened H helix and lacks the D helix as well.

Max Perutz, who has devoted his life to elucidating the atomic structure of Hb, noted very early in his studies that the molecule was highly symmetrical. The actual arrangement of the four subunits with respect to one another is shown in Figure for horse methemoglobin. All vertebrate hemoglobins show a three-dimensional structure essentially the same as this. The subunits pack in a tetrahedral array, creating a roughly spherical molecule 6.4 × 5.5 × 5.0 nm.

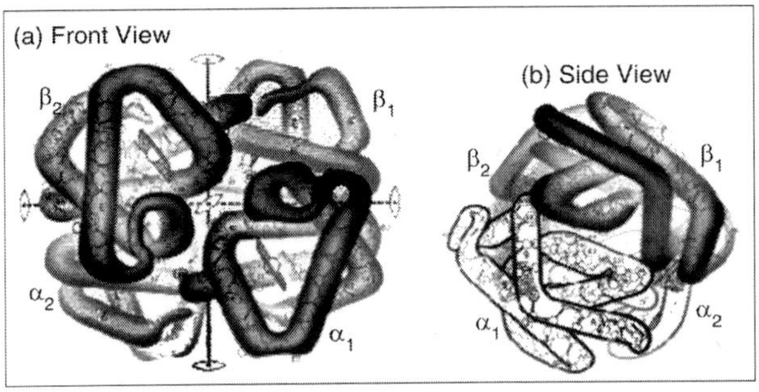

Fig. The Arrangement of Subunits in Horse Methemoglobin, the First Hemoglobin whose Structure was Determined by X-ray Diffraction. The Iron Atoms on metHb are in the Oxidized, Ferric (Fe^{3+}) State. (Irving Geis)

The four heme groups, nestled within the easily recognizable cleft formed between the E and F helices of each polypeptide, are exposed at the surface of the molecule. The heme groups are quite far apart; 2.5 nm separates the closest iron ions, those of hemes a_1 and β_2, and those of hemes a_2 and β_1. The subunit interactions are mostly between dissimilar chains: each of the a -chains is in contact with both β-chains, but there are few a-a or α-α interactions.

The Physiological Significance of the Hb:O_2 Interaction

We can determine quantitatively the physiological significance of the sigmoid nature of the hemoglobin oxygen-binding curve, or, in other words, the biological importance of cooperativity.

$$\frac{Y}{(1-Y)}=\left(\frac{pO_2}{p_{50}}\right)^n$$

The equation describes the relationship between pO_2, the affinity of hemoglobin for O_2 (defined as P_{50}, the partial pressure of O_2 giving half-maximal saturation of Hb with O_2), and the fraction of hemoglobin with O_2 bound, Y, versus the fraction of Hb with no O_2 bound, (1 – Y). The coefficient n is the Hill coefficient, an index of the cooperativity (sigmoidicity) of the hemoglobin oxygen-binding curve. Taking p O_2 in the lungs as 100 torr, P_{50} as 26 torr, and n as 2.8, Y, the fractional saturation of the hemoglobin heme groups with O2, is 0.98.

If p O_2 were to fall to 10 torr within the capillaries of an exercising muscle, Y would drop to 0.06. The oxygen delivered under these conditions would be proportional to the difference, Ylungs-Ymuscle, which is 0.92. That is, virtually all the oxygen carried by Hb would be released. Suppose instead that hemoglobin binding of O_2 were not cooperative; in that case, the hemoglobin oxygen-binding curve would be hyperbolic, and n=1.0. Then Y in the lungs would be 0.79 and Y in the capillaries, 0.28, and the difference in Y values would be 0.51. Thus, under these conditions, the cooperativity of oxygen binding by Hb means that 0.92/0.51 or 1.8 times as much O2can be delivered.

5

Carbohydrate Metabolism

CARBOHYDRATE METABOLISM DISORDERS

GLYCOGEN STORAGE DISEASES

Glycogen storage diseases are caused by deficiencies of enzymes involved in glycogen synthesis or breakdown; the deficiencies may occur in the liver or muscles and cause hypoglycemia or deposition of abnormal amounts or types of glycogen (or its intermediate metabolites) in tissues.

Inheritance for glycogen storage diseases (GSDs) is autosomal recessive except for GSD type VIII/IX, which is X-linked. Incidence is estimated at about 1/25,000 births, which may be an underestimate because milder subclinical forms may be undiagnosed. For a more complete listing of glycogen storage diseases, see Table Glycogen Storage Diseases and Disorders of Gluconeogenesis. Age of onset, clinical manifestations, and severity vary by type, but symptoms and signs are most commonly those of hypoglycemia and myopathy. Diagnosis is suspected by history, examination, and detection of glycogen and intermediate metabolites in tissues by MRI or biopsy.

Defects in glycolysis (rare) may cause syndromes similar to GSDs. Deficiencies of phosphoglycerate kinase, phosphoglycerate mutase, and lactate dehydrogenase mimic the myopathies of GSD types V and VII; deficiencies of glucose transport protein 2 (Fanconi-Bickel syndrome) mimic the hepatopathy of other GSD types (eg, I, III, IV, VI).

GALACTOSEMIA

Galactosemia is caused by inherited deficiencies in enzymes that convert galactose to glucose. Symptoms and signs include hepatic and renal dysfunction, cognitive deficits, cataracts, and premature ovarian failure. Diagnosis is by enzyme analysis of RBCs. Treatment is dietary elimination of galactose. Physical prognosis is good with treatment, but cognitive and performance parameters are often subnormal. Galactose is found in dairy products, fruits, and vegetables; autosomal recessive enzyme deficiencies cause 3 clinical syndromes.

Galactose-1-phosphate Uridyl Transferase Deficiency

This deficiency causes classic galactosemia. Incidence is 1/62,000 births; carrier frequency is 1/125. Infants become anorectic and jaundiced within a few days or weeks of consuming breast milk or lactose-containing formula. Vomiting, hepatomegaly, poor growth, lethargy, diarrhea, and septicemia (usually Escherichia coli) develop, as does renal dysfunction (eg, proteinuria, aminoaciduria, Fanconi syndrome), leading to metabolic acidosis and edema. Hemolytic anemia may also occur. Without treatment, children remain short and develop cognitive, speech, gait, and balance deficits in their teenage years; many also have cataracts, osteomalacia (caused by hypercalciuria), and premature ovarian failure. Patients with the Duarte variant have a much milder phenotype.

Galactokinase Deficiency

Patients develop cataracts from production of galactitol, which osmotically damages lens Fibres; idiopathic intracranial hypertension (pseudotumor cerebri) is rare. Incidence is 1/40,000 births.

Uridine Diphosphate Galactose 4-epimerase Deficiency

There are benign and severe phenotypes. Incidence of the benign form is 1/23,000 births in Japan; no incidence data are available for the more severe form.

The benign form is restricted to RBCs and WBCs and causes no clinical abnormalities. The severe form causes a syndrome indistinguishable from classic galactosemia, although sometimes with hearing loss.

DIAGNOSIS AND TREATMENT

Diagnosis is suggested clinically and supported by elevated galactose levels and the presence of reducing substances other than glucose (eg, galactose, galactose 1-phosphate) in the urine; it is confirmed by enzyme analysis of RBCs, hepatic tissue, or both. Most states require that neonates be screened for galactose-1-phosphate uridyl transferase deficiency. Treatment is elimination of all sources of galactose in the diet, most notably lactose, which is a source of galactose present in all dairy products, including milk-based infant formulas and a sweetener used in many foods.

A lactose-free diet prevents acute toxicity and reverses some manifestations (eg, cataracts) but may not prevent neurocognitive deficits. Many patients require supplemental Ca and vitamins.

For patients with epimerase deficiency, some galactose intake is critical to ensure a supply of uridine-5?-diphosphate-galactose (UDP-galactose) for various metabolic processes.

DISORDERS OF FRUCTOSE METABOLISM

Deficiency of enzymes that metabolize fructose may be asymptomatic or cause hypoglycemia.

Fructose is a monosaccharide that is present in high concentrations in fruit and honey and is a constituent of sucrose and sorbitol.

Fructose 1-phosphate Aldolase Deficiency

This deficiency causes the clinical syndrome of hereditary fructose intolerance. Inheritance is autosomal recessive; incidence is estimated at 1/20,000 births. Infants are healthy until they ingest fructose; fructose 1-phosphate then accumulates, causing hypoglycemia, nausea and vomiting, abdominal pain, sweating, tremors, confusion, lethargy, seizures, and coma. Prolonged ingestion may cause cirrhosis, mental deterioration, and proximal renal tubular acidosis with urinary loss of phosphate and glucose. Diagnosis is suggested by symptoms in relation to recent fructose intake and is confirmed by enzyme analysis of liver biopsy tissue or by induction of hypoglycemia by fructose infusion, 200 mg/kg IV. Diagnosis and identification of heterozygous carriers of the mutated gene can also be made by direct DNA analysis. Short-term treatment is glucose for hypoglycemia; long-term treatment is exclusion of dietary fructose, sucrose, and sorbitol. Many patients develop a natural aversion to fructose-containing food. Prognosis is excellent with treatment.

Fructokinase Deficiency

This deficiency causes benign elevation of blood and urine fructose levels (benign fructosuria). Inheritance is autosomal recessive; incidence is about 1/130,000 births. The condition is asymptomatic and diagnosed accidentally when a non-glucose reducing substance is detected in urine.

Deficiency of Fructose-1,6-bisphosphatase

This deficiency compromises gluconeogenesis and results in fasting hypoglycemia, ketosis, and acidosis. This deficiency can be fatal in neonates. Inheritance is autosomal recessive; incidence is unknown. Febrile illness can trigger episodes. Acute treatment is oral or IV glucose. Tolerance to fasting generally improves with age.

DISORDERS OF PYRUVATE METABOLISM

Inability to metabolize pyruvate causes lactic acidosis and a variety of CNS abnormalities. Pyruvate is an important substrate in carbohydrate metabolism.

Pyruvate Dehydrogenase Deficiency

Pyruvate dehydrogenase is a multi-enzyme complex responsible for the generation of acetyl CoA from pyruvate for the Krebs cycle. Deficiency results in elevation of pyruvate and thus elevation of lactic acid levels. Inheritance is X-linked or autosomal recessive.

Clinical manifestations vary in severity but include lactic acidosis and CNS malformations and other postnatal changes, including cystic lesions of the cerebral cortex, brain stem, and basal ganglia; ataxia; and psychomotor retardation. Diagnosis is confirmed by enzyme analysis of skin fibroblasts, DNA testing, or both. There is no clearly effective treatment, although a low-carbohydrate or ketogenic diet and dietary thiamin supplementation have been beneficial for some patients.

Pyruvate Carboxylase Deficiency

Pyruvate carboxylase is an enzyme important for gluconeogenesis from pyruvate and alanine generated in muscle. Deficiency may be primary, or secondary to deficiency of holocarboxylase synthetase, biotin, or biotinidase; inheritance for both is autosomal recessive and both result in lactic acidosis.

Primary deficiency incidence is < 1/250,000 births but may be higher in certain American Indian populations. Psychomotor retardation with seizures and spasticity are the major clinical manifestations. Labouratory abnormalities include hyperammonemia; lactic acidosis; ketoacidosis; elevated levels of plasma lysine, citrulline, alanine, and proline; and increased excretion of ?-ketoglutarate. Diagnosis is confirmed by enzyme analysis of cultured skin fibroblasts. Secondary deficiency is clinically similar, with failure to thrive, seizures, and other organic aciduria. There is no effective treatment, but selected patients with primary deficiency and all those with secondary deficiencies should be given biotin supplementation 5 to 20 mg po once/day.

OTHER DISORDERS

Phosphoenolpyruvate carboxykinase deficiency impairs gluconeogenesis and results in symptoms and signs similar to the hepatic forms of GSD but without hepatic glycogen accumulation.

Other deficiencies include those of glycolytic enzymes or enzymes in the pentose phosphate pathway. Common examples are pyruvate kinase deficiency (Anemias Caused by Hemolysis: Embden-Meyerhof Pathway Defects) and glucose-6-phosphate dehydrogenase (G6PD) deficiency both of which may result in hemolytic anemia. Wernicke-Korsakoff syndrome (Vitamin Deficiency, Dependency, and Toxicity: Symptoms and Signs) is caused by a partial deficiency of transketolase, which is an enzyme for the pentose phosphate pathway that requires thiamin as a cofactor.

CLASSIFICATION OF CARBOHYDRATES

The activities on which the life of the body depends involve a continuous expenditure of energy and a constant exchange of material. Ultimately the body is dependent upon food for the fuel materials which supply energy and for both the substances which are transformed in, and eliminated from, the body, and those whose presence regulates and controls these transformations.

The materials leaving the body are to be regarded not merely as wastes but as end products of an orderly and coordinated series of chemical reactions which occur in the body and by virtue of which its functions are performed.

Thus the chief functions of food are:

- To yield energy.
- To build tissue.
- To regulate body processes.

These functions involve reactions which are dependent upon the chemical composition and constitution of the food. Any food constituent which takes part in any of these functions may be regarded as having nutritive value.

Most of the nutrient material contained in food requires more or less change to bring it into the exact forms most useful in nutrition. These changes as a rule take place in the digestive tract and together constitute the process of digestion.

The changes which take place in the foodstuffs, after they have been absorbed from the digestive tract, are included under the general term"metabolism." Although the chemical changes and the energy transformations are of course inseparable, it has become customary to speak of the metabolism of matter and the metabolism of energy, and to regard the extent of the metabolism of any material substance as measured by the amount of its end products eliminated, and the extent of the energy metabolism as measured by the amount of heat, or of heat and external muscular work, which the body gives off.

The metabolism of matter and the metabolism of energy are normally supported by the food; but if no food is taken, they continue at the expense of the body substance. The expenditure of energy can never cease in the living body because it includes the work involved in carrying on the internal processes which are essential to life itself; and the expenditure of matter cannot cease because the energy for this necessary work is obtained by the breaking down of the organic compounds of the food or of the body substance into simpler compounds, many of which are of no further use to the body and must be eliminated. When the food supplies sufficient energy, the body substance is protected; when the food is insufficient, body substance is burned as fuel.

In order, then, to consider intelligently the nutritive requirements of the body as regards the substances of which it is composed, it is necessary first to know whether the fuel requirements (the requirements of the energy metabolism) have been fully met. The carbohydrates, fats, and proteins of the food all serve as fuel to yield the energy required for the activities of the body, and the proteins serve also as material for the maintenance or growth of body tissue. But of the fifteen chemical elements which are essential to the structure and functions of the body, simple proteins furnish only five. The remaining ten elements are largely constituents of the ash of the food and are known as ash constituents, inorganic foodstuffs, mineral matter, or salts.

Recent investigations have developed the fact that food of sufficient energy value and containing ample amounts of each of the chemical elements known to be essential to the body is not necessarily adequate to meet all the requirements of nutrition. Thus it appears that certain substances occurring in natural foods but not yet chemically identified are also to be included among the nutritive requirements of the body and therefore among the factors which determine the nutritive values of foods. At present these unidentified substances are referred to as"vitamines" or as"fat soluble A" and"water soluble B." *The essentials of a chemically adequate food supply may therefore be summarized as follows*:

- Sufficient of the organic nutrients in digestible forms to yield the needed energy.
- Protein, sufficient in amount and appropriate in kind.
- Adequate amounts and proper proportions of the various ash constituents or inorganic foodstuffs.
- Sufficient of each of the two unidentified"vitamine" factors, the"fat soluble A" and the"water soluble B."

In attempting to give in the following pages a general view of the chemistry of food and nutrition it has seemed best to discuss first the chemical nature and nutritive functions of the substances which serve as sources of energy in nutrition, then the nutritive requirements in terms of energy, protein, the more prominent"inorganic" elements, and the"vitamines," and finally the bearing of these various factors offood value upon problems connected with the economic use of food.

Of the constituents of the ordinary mixed food of man the carbohydrates are usually the most abundant and the most economical sources of energy. They are also considered to be the first of the three great groups of foodstuffs to be formed by synthesis from simple inorganic substances in plants;"in the long run, all the energy of living matter comes from them." The synthesis of carbohydrates in nature is therefore a logical starting point for the study of the organic foodstuffs.

In the chlorophyll cells of the leaves of green plants the energy of the sun's rays brings about reaction between carbon dioxide and water which results in the liberation of oxygen and the formation of organic compounds. There is still doubt as to the exact mechanism of the process and no certainty that it is the same in all cases. It has, however, been quite generally found that the volume of oxygen liberated is equal to that of carbon dioxide consumed. The simplest possible representation of the reaction would be $CO_2 + H_20 \rightarrow CH_20 + O_2$ according to which the first product of the synthesis would be formaldehyde. There is considerable (though not conclusive) evidence that formaldehyde is thus formed and that it is rapidly built up into less reactive compounds. Whatever the steps in the process* there is normally an early production of carbohydrate. Usually the first product which can be demonstrated as accumulating in the plant as the result of the photosynthesis is a sugar (glucose or sucrose) or starch. Assuming glucose as a typical product and neglecting the intermediate stages, the photosynthesis of carbohydrate may be represented thus:

$$6\ CO_2 + 6\ H_{20} \text{ ® } C_6H_{12}O_6 + 6O_2$$

Glucose is the most familiar representative of a group of simple sugars (monosaccharides or monosaccharoses) which are in composition direct polymers of formaldehyde (CH20) and which are classified, according to the number of carbon atoms in the monosaccharide molecule, as trioses, pentoses, hexoses, etc.

CLASSIFICATION

Definitions of the term"simple sugar" vary somewhat, depending chiefly upon the views of different authors as to how simple a compound may properly be called a sugar. According to Browne, a simple sugar or monosaccharide may be defined as an aldehyde alcohol or ketone alcohol of the aliphatic series, the molecule of which contains one carbonyl and one or more alcohol groups, one of the latter being always adjacent to the carbonyl group. All simple sugars contain, therefore, as a characteristic group upon the presence of which the chief chemical properties of the sugars depend.

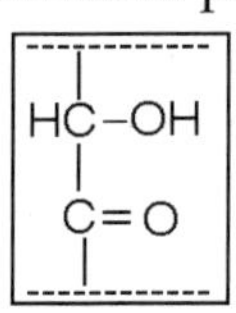

The simplest possible sugar according to this definition is glycolaldehyde, CH2OH-CHO, which (in analogy with the nomenclature of the familiar sugars) may also be called glyco-lose. The structural formulae of glucose and fructose, the most familiar representatives of the aldehyde-alcohol (aldose) and ketone-alcohol (ketose) sugars, respectively, are as follows:

Glucose

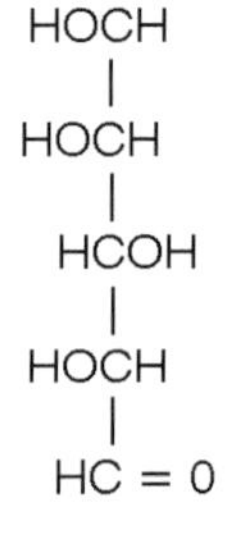

Fructose

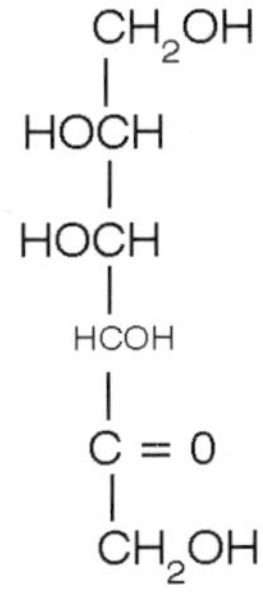

Since glucose gives aldehyde reactions but not so readily as the above structural formula would lead one to expect, it is believed that in ordinary solutions of glucose the substance exists partly in the condition indicated by the aldehyde formula and partly in a tautomeric form represented by the lactone or"oxygen bridge" formula.

Following are the aldehyde and lactone formulae written without reference to the spatial relationships of the hydrogen and hydroxyl groups:

Aldehyde form

$$CH_2OH\text{-}CHOH\text{-}CHOH\text{-}CHOH\text{-}CHOH\text{-}CHO$$

Lactone form

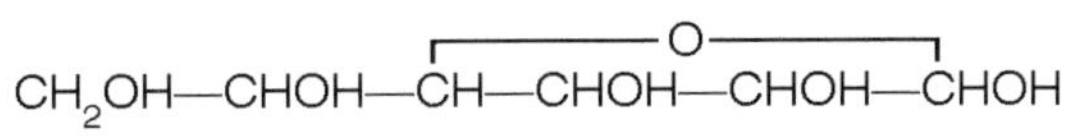

The name monosaccharide ("single sugar") implies that the monosaccharide molecule contains only one sugar radicle - that it cannot be split by hydrolysis into sugars of lower molecular weight. A substance like cane sugar which on hydrolysis splits to two molecules of simple sugar is called adisaccharide or disaccha-rose ("double sugar"). Trisaccharides and tetrasaccharides are also known.

Substances which like starch are of high molecular weight and on complete hydrolysis yield many molecules of simple sugar are called polysaccharides * or polysaccharoses. The term"carbohydrates" covers all the simple sugars and all substances which can be converted into simple sugars by hydrolysis.

The term"glucosides" is applied to substances which consist of combinations of carbohydrate radicles with radicles of other kinds and which therefore yield on hydrolysis both a simple sugar and one or more products of other than carbohydrate nature.

DIAGNOSIS AND TREATMENT OF INHERITED DISORDERS

Most of the inherited disorders of carbohydrate metabolism fall into a few broad clinical categories.

These are as follows:

- Episodic lactic acidosis in infancy, failure to thrive, and hypotonia with or without features suggesting specific disorders.
- Infantile mental retardation, hypotonia, failure to thrive, and other features
- Mental retardation/developmental delay with features that suggest storage disorders.
- Episodic vomiting in infants or young children.
- Episodic acidosis in childhood or early adult life, often with features that suggest specific disorders.
- Intermittent or episodic ataxia in childhood.
- Cramps and weakness with anaerobic exercise, often with some episodes of muscle breakdown and myoglobinuria, during or after heavy exercise.
- Symmetrical neuromuscular disease with weakness and wasting of proximal muscles.

Episodic lactic acidosis in infancy, failure to thrive, and hypotonia with or without features suggesting specific disorders.

In this category, an infant presents with signs of acidosis, tachypnea, marked irritability, and tachycardia. As the acidosis worsens, the infant becomes obtunded or comatose. Muscle tone is very poor; the infant is flaccid and has poor head control. If the infant is more than a few hours or days old, signs of failure to thrive may be seen: weight loss or at least poor gain in weight, microcephaly, and poor linear growth. Prognosis is poor even with current vigourous metabolic management. Without such management, the infant usually dies in a short time. In the absence of other clinical features, this may be the presentation of one of several defects of the PDH multienzyme complex, particularly of its E components.

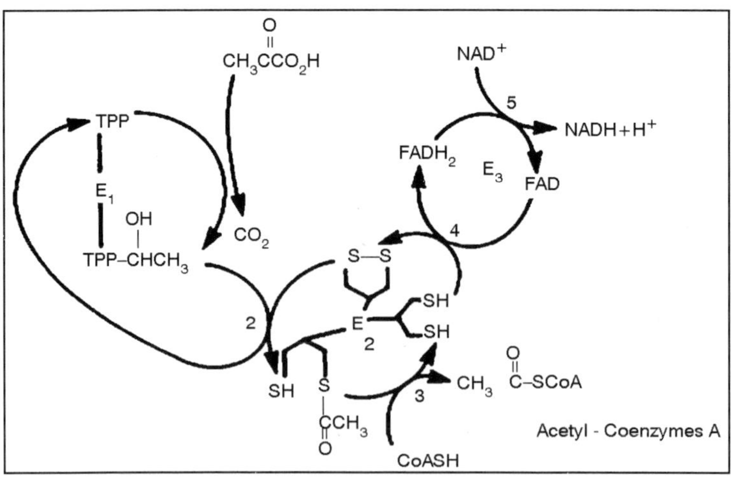

Fig. Pyruvate Dehydrogenase Complex

Scheme of the major reactions of the pyruvate dehydrogenase complex. Initial oxidative decarboxylation of pyruvate is shown as reaction

- Transfer of the 2-carbon moiety to the lipoyl side chain of lipoyl acetyltransferase (E_2) as reaction.
- Formation of acetyl coenzyme A as reaction.
- Oxidation of the lipoyl moiety as reaction.
- Reduction of NAD as reaction.
- The regulatory enzymes acting on pyruvate decarboxylase (E_1) are not shown.

The components are called E1 (ie, pyruvate decarboxylase [and the phosphorylase and dephosphorylase that modify the activity of E_1]), E_2 (ie, lipoyl acetyl transferase), and E_3 (ie, lipoamide, also known as dehydrogenase; dihydrolipoamide: NADH oxidoreductase).

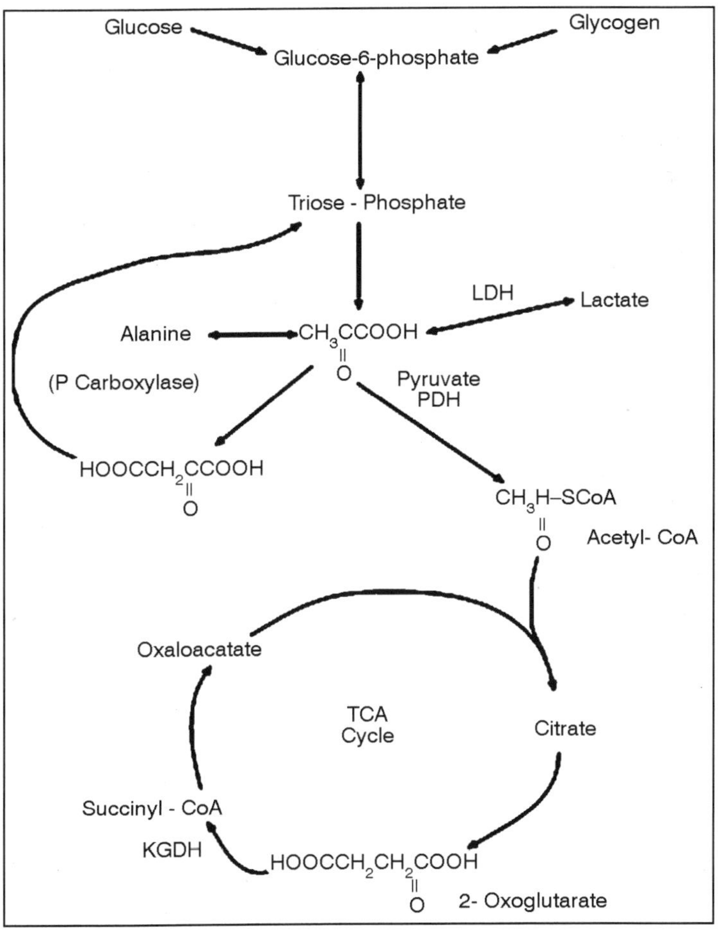

Several cases of Leigh syndrome appear to have been caused by failure of activation of pyruvate decarboxylase, presumably because of defects in phosphorylation. In theory, a profound congenital deficiency of thiamine ought to produce episodic lactic acidosis in infants with failure to thrive and hypotonia. Leigh syndrome (ie, subacute necrotizing encephalomyelopathy) usually begins in the first or second year of life. Motor and cognitive regression occur along with ataxia, chorea, and tonic spasms. Depending on the age of the patient, history and physical findings include loss of muscle tone and head control, decreased appetite, nausea, vomiting (often with lactic acidosis), irritability to the point of continuous crying, myoclonus, and primarily generalized seizures. Episodes of hyperventilation, apnea and gasps, sobbing without tears, external ophthalmoplegia with pendular or rolling nystagmoid movements of the eyeballs, loss of swallowing, muscle wasting, and loss of reflexes are observed.

In cases of postnatal lactic acidosis with elevations in the levels of ammonia, citrulline, and lysine, consider defects of the B-complex of pyruvate carboxylase, the first enzyme in the pathway that generates glucose from fatty acids and amino acids through gluconeogenesis. Death occurs by about 3 months of age.

Major metabolic relations of pyruvate. Many of the reactions, such as those between glucose-6-phosphate and pyruvate, pyruvate and alanine, and pyruvate and lactate, are of course reversible. Abbreviations: LDH - lactic dehydrogenase; PDH - pyruvate dehydrogenase complex; p carboxylase - pyruvate carboxylase; TCA cycle - Krebs tricarboxylic acid cycle; KGDH-a-ketoglutarate dehydrogenase complex.

Mental Retardation/Developmental Delay with Features that Suggest Storage Disorders

The features that suggest storage disorders are hepatomegaly, a large forehead with frontal bossing, and macrocephaly. A large number of diseases, including storage disorders caused by inborn errors of metabolism, can cause hepatomegaly in infancy without acidosis or other specific features discussed. Of the disorders of carbohydrate metabolism, these features can be seen with defects of production, storage, or catabolism of glycogen, especially deficiency of the glycogen debrancher enzyme, which leads to hepatomegaly in infancy and early death; defects of 3 of the 4 proteins that make up phosphorylase kinase; and deficiency of glucose-6-phosphatase (ie, von Gierke disease), which is accompanied by gout, lactic acidosis, and elevated lipids.

Diagnosis

Episodic vomiting in infants or young children has a large number of causes, most of which are more common than disorders of carbohydrate metabolism. Nonmetabolic causes include gastroenteritis from viruses, bacteria, and parasites; gastroesophageal reflux disorder; irritable bowel syndrome; small or large bowel obstructions; neurological conditions, such as premigrainous syndromes and brain masses; and abdominal epilepsy. In this clinical setting, inborn errors of the urea cycle and organic acid metabolism need to be considered first, before considering lactose intolerance from inherited disorders of galactose and fructose metabolism.

Workup and Treatment

Hereditary fructose intolerance and defects of fructose 1, 6-diphosphate aldolase are discussed under Episodic lactic acidosis in infancy, failure to thrive, and hypotonia with or without features suggesting specific disorders earlier in this article. Disorders of galactose metabolism are discussed in the subsection Infants with mental retardation, hypotonia, failure to thrive, and other features.

Episodic Acidosis in Childhood or Early Adult Life

This often presents with features of a specific disorder, including many of the differential diagnoses mentioned under Episodic lactic acidosis in infancy, failure to thrive, and hypotonia with or without features suggesting specific disorders. Specific features may enable rapid diagnosis.

Intermittent or Episodic Ataxia

This category is characterized by the sudden, almost abrupt, onset of ataxia, extrapyramidal symptoms in the upper extremities, and dysarthria soon after or in conjunction with an infectious illness or severe emotional upset. In one patient monitored by one of the authors, death of a grandparent and critical school examinations were sufficient stimuli to initiate attacks of ataxia. Physical findings in this patient also included a high arched palate, minimal kyphoscoliosis, minimal pes cavus, and extra rows of lashes on the upper eyelid. When the boy was 8 or 10 years old, a mild attack was precipitated by an environmental challenge with flashing red lights from scintillation counters and Geiger counters for about 30 minutes in a dimly lit room. For this patient and others, the episodes of ataxia usually last about 2 weeks. Although the neurologic symptoms usually are accompanied by increased serum lactic acid, frank acidosis is not always present.

The underlying defect in this patient was shown to be that of the first enzyme of the PDH complex, E_1. Eventually, he was placed on a diet high in thiamine and relatively low in carbohydrates; the number of attacks subsided. While this patient graduated from college, others with similar clinical presentations have had at least a mild degree of mental retardation.

Work in the 1970s and early 1980s raised the question whether familial cases of a childhood- or adolescent-onset ataxia (one with slower progression than is seen in classical Friedreich ataxia) was associated with moderate defects of the activity of PDH, E_3. The association has not been confirmed. Classic Friedreich ataxia appears to be caused by a single genetic defect and does not appear to be associated with measurable abnormalities in the E3 pathway

Differential Diagnosis

Intermittent or episodic ataxia in childhood and adolescence is a rare condition. Porphyria might induce this, and multiple sclerosis, though rare in young people, certainly can present with ataxia; however, multiple sclerosis almost always manifests in other ways. Postinfectious cerebellar dysfunction usually presents as a single episode that clears in weeks or months and does not recur. Griggs and colleagues have described patients with a familial form of intermittent ataxia, unrelated to pyruvate defects, that responds to treatment with acetazolamide.

Workup

E1 is the first enzyme of the PDH multienzyme complex. Phosphorylation inactivates this pyruvate decarboxylase, which activates by removal of the phosphate. The decarboxylase requires thiamine pyrophosphate as a substrate. Patients deficient in this enzyme have elevations of metabolic by-products related to pyruvate degradation during clinical attacks of ataxia (ie, excess lactic acid and alanine in blood or urine, or both). The primary enzyme defect can be demonstrated in cultured fibroblasts and probably can be demonstrated in WBCs or muscle tissue. The substrate, pyruvate, is so labile and the enzyme activity of this and other components of the PDH complex so sensitive to environmental contaminants that carefully controlled assays need to be performed in a research Labouratory.

Treatment

The young man discussed earlier in detail was treated with a low-carbohydrate diet and high doses of thiamine. Some patients with autosomal recessive ataxia with large-Fibre-posterior column sensory loss, areflexia, cardiomyopathy, and bony stigmata have been treated with some success with a modified (titrated) ketogenic diet. Dietary changes were introduced gradually until moderate to high levels of ketones appeared in the urine without producing frank metabolic acidosis.

Black McArdle first described these symptoms of the disease at Guy's Hospital in London. (He was so named to distinguish him from his brother"Red," also a distinguished physician there; the pseudonyms referred to the Colour of their hair.) Type V glycogenosis due to muscle phosphorylase deficiency is the clinically mildest of the glycogen storage diseases, and the symptoms may not occur until adolescence or young adulthood.

The symptoms tend to be induced during exercise when muscle depends on glycolysis rather than oxidative phosphorylation as a source of fuel. Exercise that includes sprinting or vigourous running most likely will induce symptoms. Often, onset of symptoms comes during what normally would represent the subjective"second wind." Over time, patients may learn to slow down or stop at the first hint of symptoms, and then to begin again, with what has been called a second wind but which is probably a pause in which oxidative metabolism can generate sufficient ATP in muscle so that anaerobic activity can begin again.

If this speculation is true, the"second wind" that is peculiar to myopathies from glycolytic/glycogenolytic defects differs in its nature from the second wind experienced by healthy athletes. The patients therefore can carry on anaerobic exercise in spurts over a long time, presumably with aerobic exercise in between. Affected individuals often can tailor their activities to take a break at the first sign of cramping, and then resume activities shortly thereafter.

Without this pause and resumption in athletes with glycolytic/ glycogenolytic defects, muscles suddenly become so weak that the athlete becomes virtually immobile from painful cramps, and the muscles that were exercised are very sore and tender. This disease and the finding of elevated lactic acid levels in the blood after exercise led to the erroneous notion that muscle cramps in general are caused by lactic acidosis. The muscles in the individuals affected with glycolytic defects continue to be very sore even after the cramps resolve.

If the urine is not frankly red, it must be examined to see if it still contains heme groups. In myopathy due to disorders of glycolysis, the heme pigment proves to be myoglobin, not hemoglobin, when detailed analyses are carried out. With McArdle disease, muscle enzymes and creatine levels usually are elevated in the blood after a symptomatic episode. In this condition, muscle histology demonstrates marked increase in glycogen stores in the myofibrils. Histochemical or immunoassays for specific enzymes can augment Labouratory diagnosis by identifying the precise enzyme defect. The original enzyme defect described by McArdle9 and then by Schmid and Mahler10 was localized to the glycolysis pathway. The deficiency ultimately was identified in muscle phosphorylase. Through the past decade, MR spectroscopy of forearm muscles in vivo has led to the diagnosis. Deficiencies of several other glycolytic enzymes (eg, muscle phosphofructokinase, phosphoglycerate kinase, phosphoglycerate mutase, lactate dehydrogenase) produce similar clinical pictures and are indistinguishable from muscle phosphorylase deficiency, unless enzymatic tests are performed.

Muscle breakdown occurs in disorders of glycolysis and glycogenolysis during muscle activity that is largely dependent on glycolysis or anaerobic metabolism. This is in striking contrast to the circumstances of muscle breakdown in disorders related to lipid or ketone metabolism. In these disorders, muscle breaks down, with similar symptoms, under circumstances in which muscles are relying on oxidative metabolism, especially of ketone bodies, as a fuel source. Activity most likely to produce symptoms under these circumstances is prolonged, less vigourous exercise over several hours, which is exacerbated by starvation or high-fat, high-protein meals and occurs more often in a cold or cool environment. Histologically, myofibrils may demonstrate increased lipid droplets, but in contrast to the glycogenoses, glycogen stores are either normal or even slightly depleted.

Differential Diagnosis

The differential diagnosis includes all the acute and symptomatic myopathies, including various forms of polymyositis. Many of these appear to be viral or postinfectious; still others appear to be paraneoplastic or associated with motor neuron degenerations or systemic collagen-vascular diseases but not with problems of intermediary metabolism.

The circumstances under which cramps and weakness occur may provide insight into whether a metabolic disorder is present. If evidence of metabolic etiology appears, differentiating between disorders affecting white muscle (ie, the carbohydrate errors discussed here) and disorders affecting red muscle (ie, errors of lipid and mitochondrial metabolism) is essential, because these disorders can present similar pictures but require very different metabolic treatment strategies. Making the correct diagnosis clearly is important since, for each of these groups of disorders of glycolysis/glycogenolysis, mitochondrial transport, and oxidative phosphorylation, treatment approaches are quite different.

Workup

Prior to the availability of direct enzyme assays, defects of muscle phosphorylase had to be inferred from the clinical features of muscle cramping with vigourous exercise in conjunction with accumulation of lactic acid in the blood and myoglobin in the urine. The forearm ischemia test further substantiated the diagnosis. For this test, a blood pressure cuff is inflated to systolic pressure. With the cuff on, ampules of venous blood are drawn from the antecubital fossa via an indwelling catheter attached to a stopcock before, during, and after 1 minute of forearm exercise (eg, having the patient squeeze a partially inflated blood-pressure cuff repeatedly). In reality, many patients with phosphorylase deficiency cannot continue to exercise for a full minute. Blood samples are assayed for glucose, lactic acid, and sometimes pyruvate. Affected individuals have lower glucose levels and higher lactate and pyruvate levels compared to reference values in blood samples taken 1-2 minutes after exercise.

In the last 25 years, the preferred diagnostic test for McArdle disease has been a phosphorylase assay directly on muscle tissue. Today, histochemical or biochemical techniques can be used to diagnose deficiencies in even the rare enzymes that clinically simulate McArdle disease. Again, these enzymes include muscle phosphofructokinase, phosphoglycerate kinase, phosphoglycerate mutase, and lactate dehydrogenase. In the past 2 decades, diagnosis also has been made by analysis of metabolites in vivo on human forearm muscle by means of MR spectroscopy.

Treatment

DiMauro et al gave a thorough review of strategies to treat disorders of muscle phosphorylase.11 Pyridoxine, at 50 mg/day, seems to be effective. (Note that at 100 mg/d, this vitamin can produce a painful distal sensory neuropathy.) High-protein diets also were effective, at least for several weeks, but determining the long-term success of dietary treatment will be possible only after adequate clinical trials.

Current therapeutic options include a carbohydrate-rich diet and carbohydrate ingestion shortly before strenuous exercise in combination with medically supervised aerobic training of low to moderate intensity.

Only rarely are cases of progressive neuromuscular diseases caused by defects of glycogen metabolism. When they are, children or young adults generally develop, over months or years, progressive weakness of the limb-girdle muscles and the proximal muscles of the limbs. Patients have no pain under ordinary circumstances, but heavy exercise may induce fatigue and cramps. Symptoms depend on the nature of the underlying metabolic defect. Muscles may or may not be atrophied at presentation but eventually waste. Patients do not complain of muscle tenderness unless cramping occurs. Some patients with debrancher deficiency have evidence of cirrhosis; others become asymptomatic in adult life.

Differential Diagnosis

This condition is essentially indistinguishable from Limb-Girdle Muscular Dystrophy. The symptoms can be demonstrated in a number of acquired diseases, including various forms of polymyositis or inflammatory myopathy, various neuropathic diseases (especially proximal forms of motor neuron disease and the limb-girdle pattern of painless diabetic multiple neuropathy or diabetic plexitis), inherited mitochondrial myopathies (often associated with ragged red Fibres), disorders of lipid metabolism, and/or defects of lipid as carnitine transport at the level of mitochondria. For lipids to enter the mitochondria, they need to be complexed with carnitine derivatives. Primary or secondary carnitine deficiencies as well as deficiencies of enzymes involved in the metabolic pathway can interfere with this process, resulting in a myopathic picture.

Workup

Muscle enzyme levels may be elevated in the serum, but myoglobinuria is an unusual finding. Histochemical analysis of biopsied muscle shows excessive glycogen stores. Partial acid maltase deficiency may result in the adult form of Pompe disease and may be clinically indistinguishable from muscle phosphofructokinase deficiency (reported in at least one case) and glycogen debrancher enzyme deficiency. However, patients with glycogen debrancher enzyme deficiency eventually develop physical, electrical, or radiographic evidence of cardiac ventricle dilatation. Therefore, clinicians are advised to monitor for these changes by ECGs and examinations on an annual basis.

Treatment

Defects of acid maltase in their milder forms are discussed in Mental retardation/developmental delay with features that suggest storage disorders and in the article Lysosomal Storage Disease (ie, Pompe disease). The FDA has recently designated alglucosidase alfa (Myozyme) an orphan drug for use in infants with these disorders.

The glycogen debrancher enzyme is discussed above in Mental retardation/developmental delay with features that suggest storage disorders, and muscle phosphofructokinase is discussed above in Cramps and weakness, often with some episodes of muscle breakdown and myoglobinuria, during or after heavy exercise.

FUTURE IN TREATMENT

In many instances, inborn errors of metabolism are difficult or impossible to treat. Novel therapeutic modalities, including gene therapy, need to be explored. Among available gene delivery systems, recombinant adeno-associated virus (AAV) shows promise for the treatment of metabolic disease. Despite the relatively low immunogenicity of AAV vectors, immune responses are also emerging as a factor requiring special attention as efforts accelerate toward human clinical translation. Trials targeting lipoprotein lipase deficiency are showing early evidence of efficacy.

FERTILITY CONCERNS

Increasing numbers of individuals with inherited metabolic disorders are surviving into adulthood and considering their reproductive options. Lee discusses a practical approach to supporting such individuals, focusing on issues concerning fertility, the impact of pregnancy on metabolism and the metabolic disorder itself on the pregnancy, as well as highlighting the need to pay special attention during the postpartum period. He further opines that apart from pregnancies in women with phenylketonuria, data are lacking in this area. Collection of information within registries is needed to aid our understanding of potential problems and counseling of women and their partners.

GLYCOLYSIS

We have already used the COG approach to demonstrate the complementarity of the phyletic patterns of the three forms of phosphoglycerate mutase. Figure shows the COGs that are known or predicted to include glycolytic enzymes and shows their phyletic patterns. This superposition of COGs and metabolic pathways provides a convenient framework for a detailed analysis of the phylogenetic distribution of each of the glycolytic enzymes and the general principles of evolution of carbohydrate metabolism. This figure shows, for example, that R. prowazekii, an obligate intracellular parasite and a relative of the mitochondria, does not encode a single glycolytic enzyme. In contrast, all other organisms with completely sequenced genomes encode enzymes of the lower (tri-carbon) part of the pathway. This supports the notion that glycolysis is the central pathway of carbohydrate metabolism and makes comparative analysis of variants of this pathway all the more interesting.

Glucokinase

Fermentation of glucose starts with its phosphorylation, which is catalyzed by glucokinase. Although many bacteria bypass the glucokinase step by phosphorylating glucose concomitantly with its uptake by the PEP-dependent phosphotransferase system, some of them, including E. coli, encode a glucokinase (COG0837) that shares little sequence similarity with yeast and human enzymes. There is also another bacterial form, found inS. coelicolour, Bacillus megaterium, and other bacteria.

Recently, P. furiosus has been reported to encode an ADP-dependent glucokinase. This enzyme has no detectable sequence similarity to any other glucokinase but shows significant structural similarity to enzymes of the ribokinase family. In retrospect, several conserved motifs were detected in this new glucokinase and the ribokinase family proteins, which indicates a homologous relationship. Thus, a clear-cut case of non-orthologous gene displacement is observed: a ribokinase family enzyme has been recruited to replace the typical glucokinase. So far, the ADP-dependent glucokinase has been found only in M. jannaschii and in pyrococci. The existence of at least three distinct forms of glucokinase is remarkable, especially given that this is apparently not an essential component of glycolysis. Moving down the glycolytic pathway, we find similar examples of non-orthologous gene displacement for several other, essential enzymes.

Glucose-6-phosphate Isomerase

Bacteria and eukaryotes encode several distinct but homologous forms of glucose-6-phosphate isomerase (phosphoglucomutase). The classical (E. coli) form of the enzyme is found in Gram-negative bacteria and in the cytoplasm of the eukaryotic cell. A divergent version of this enzyme is found in Gram-positive bacteria including B. subtilis, in T. maritima, and some archaea, such as M. jannaschii and Halobacterium sp. The most divergent members of this family of glucose-6-phosphate isomerases were detected in A. aeolicus and another subset of archaea, including M. thermoautotrophicum, A. pernix, and Thermoplasma spp. No enzyme of this family seems to be encoded in the genomes of A. fulgidus or pyrococci. Instead, P. furiosushas been shown to encode a novel glucose-6-phosphate isomerase, which has highly conserved orthologs in P. horikoshii and in A. fulgidus, but so far not in any other organism.

Thus, two non-orthologous (in fact, apparently unrelated) versions of this enzyme together account for the phosphoglucomutase activity in all known microbial genomes, with the exception of R. prowazekii and U. urealyticum. As indicated above, the former does not encode any glycolytic enzymes, whereas the latter apparently obtains fructose-6-phosphate by importing fructose concomitantly with its phosphorylation through the fructose-specific phosphotr-ansferase system, thus bypassing the phosphoglucomutase stage.

Phosphofructokinase

The next glycolytic enzyme, phosphofructokinase, offers an even more interesting example of non-orthologous gene displacement. It is also an example of an enzyme where several "missing" enzyme forms have been discovered just in the past year. The most common version of this enzyme, PfkA, is an ATP-dependent kinase of unique structure found in bacteria and many eukaryotes. Plants have a homologous enzyme, which, however, uses pyrophosphate as the phosphate donor. Altogether, homologs of PfkA are found in nearly all bacteria and eukaryotes but are conspicuously missing in H. pylori and in all archaeal genomes sequenced so far. In addition, E. coli encodes a second phosphofructokinase, PfkB, which is unrelated to PfkA and instead belongs to the ribokinase family of carbohydrate kinases.

A unique ADP-dependent phosphofructokinase has been described in P. furiosus. However, this enzyme appears to have a limited phyletic distribution: so far, it was found only in M. jannaschii and in pyrococci. This left the phosphofructokinase activity in other archaea unaccounted for and suggested that additional forms of this enzyme might exist. Very recently, a new ATP-dependent phosphofr-uctokinase, which is a member of the ribokinase family but is not specifically related to PfkB, has been identified in A. pernix. Close homologs of this protein (APE0012) were found in Halobacterium sp., A. fulgidus, M. thermoautotrophicum, and several other archaea. Therefore, it seems likely that these ribokinase family enzymes function as phosphofructokinases in all these archaea. Finally, Thermoplasma does not encode orthologs of any of the four forms of phosphofructokinase described above. This leaves two possibilities: either thermoplasmas lack phosphofructokinase altogether (along with fructose-1,6-bisphosphate aldolase; see below), or they might have yet another, fifth variant of this enzyme.

Fructose-1,6-bisphosphate aldolase

For more than 50 years now, it has been known that fructose-1,6-

bisphosphate aldolase exists in two distinct forms, a metal-independent one (class I) in multicellular eukaryotes and a metal-dependent one (class II) in bacteria and yeast. Certain organisms, such as Euglena, seem to have enzymes of both classes. Although these two enzyme forms have similar structures, they do not share any detectable sequence similarity.

Sequence analysis of archaeal genomes and those of chlamydia showed that they encode neither a typical class I enzyme, nor a typical class II enzyme. Instead, chlamydia and all archaea, with the exception of thermoplasmas, encode orthologs of the recently described class I aldolase DhnA (FbaB) of E. coli, which is only distantly related to the regular class I enzymes and may be considered a third class of aldolases.

Recently, fructose-1,6-bisphosphate aldolase activity was demonstrated in the P. furiosus homolog of DhnA; this enzyme has been referred to as a class IA aldolase. The phyletic patterns of the bacterial-type class II aldolase (COG0191) and the DhnA-type aldolase (COG1830) are almost complementary, except that both types of aldolases are present in E. coli and A. aeolicus, and none of them is detectable in X. fastidiosa. X. fastidiosa, a plant pathogen, encodes a eukaryotic class I aldolase, which is specifically similar to the plant class I aldolase and probably has been acquired from the plant host via HGT.

However, typical eukaryotic (class I) fructose-1,6-bisphosphate aldolase is also encoded in several other bacteria, in which cases the underlying evolutionary scenario is less clear.

Distribution of glycolysis (Embden-Meyerhoff-Parnas pathway) enzymes in organisms with completely sequenced genomes. Each rounded rectangle shows a glycolytic enzyme, denoted by its gene name and the COG number.

Although most genomes encode only one type of fructose-1,6-bisphosphate aldolase, different forms of this enzyme do coexist in several organisms. In particular, the relatively large genome of the plant symbiont M. lotiencodes fructose-1,6-bisphosphate aldolases of all three classes.

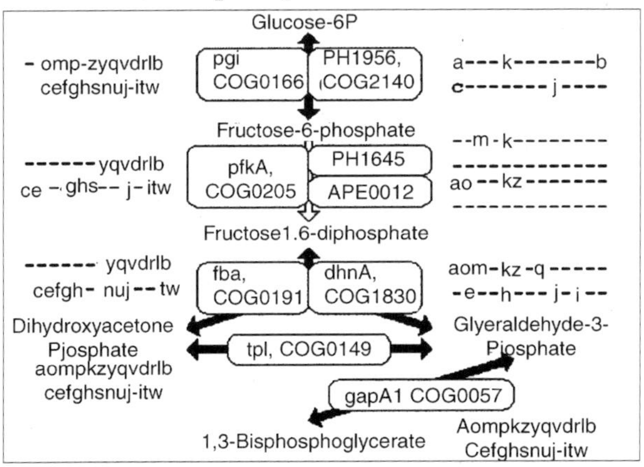

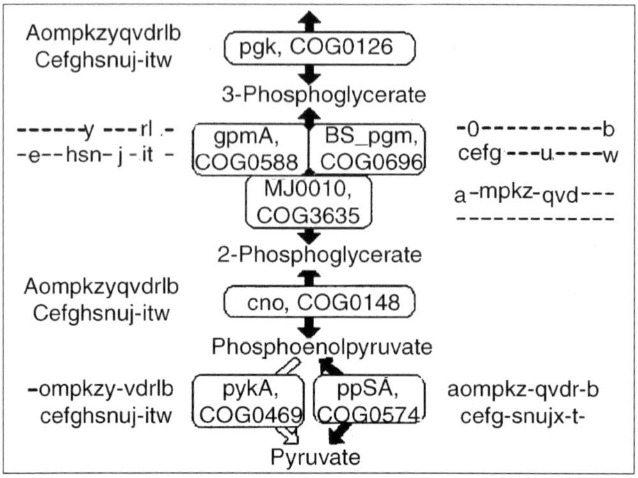

The nature of the aldolase, if any, in thermoplasmas remains unclear. The apparent absence in these archaea of both phosphofructokinase and fructose-1,6-bisphosphate aldolase might indicate that these organisms split hexoses into trioses exclusively via the Entner-Doudoroff pathway. Indeed, thermoplasmas encode close homologs of the recently described fructose-6-phosphate aldolase. Finally, given that chlamydiae are important human pathogens and that the unusual class IA fructose-1,6-bisphosphate aldolase is the only aldolase encoded in their genomes, this presumably essential enzyme might be a promising target for anti-chlamydial drug therapy.

Triose Phosphate Isomerase

Triose phosphate isomerase is conserved in all organisms, with the exception of Rickettsia. Bacterial-eukaryotic and archaeal isomerases form two clearly separated clusters. This gave rise to the notion that eukaryotic triose phosphate isomerases originated from the promitochondrial endosymbiont whose genes have been transferred into the nucleus of the eukaryotic host.

Glyceraldehyde-3-phosphate Dehydrogenase

Like triosephosphate isomerases, archaeal glyceraldehyde-3-phosphate dehydrogenases are homologous to those from bacteria and eukaryotes but form a well-defined cluster, suggesting the mitochondrial origin of this enzyme in eukaryotes.

In pyrococci and, probably, in several other archaea, the main glycolytic flow goes through a different enzyme, glyceraldehyde-3-phosphate:ferredoxin oxidore-ductase, whereas glyceraldehyde-3-phosphate dehydrogenase appears to be confined to gluconeogenesis.

In U. urealyticum, the typical NADH-dependent glyceraldeldehyde-3-phosphate dehydrogenase is missing, and this reaction is apparently catalyzed by a non-phosphorylating, NADP-dependent enzyme, similar to the well-characterized enzymes from plants and Streptococcus mutans. These enzymes belong to a large superfamily of NADP-dependent aldehyde dehydrogenases and are unrelated to the phosphorylating glyceraldeldehyde-3-phosphate dehydro-genase.

Remarkably, an archaeal member of the non-phosphorylating glyceraldeldehyde-3-phosphate dehydrog-enase family uses NAD instead of NADP.

Phosphoglycerate Kinase

Like triose phosphate isomerases and glyceraldehyde-3-phosphate dehydrogenases, phosphoglycerate kinase is conserved in all organisms that have glycolysis, and the sequences from bacteria and eukaryotes are closer to each other than they are to their archaeal counterparts, suggesting the mitochondrial origin of the eukaryotic enzyme.

Phosphoglycerate Mutase

The diversity of phosphoglycerate mutases was discussed earlier. We would only like to reiterate that there are two unrelated forms of this enzyme, 2,3-bisphosphoglycerate-dependent (animal-type) and 2,3-bisphosphoglycerate-independent (plant-type), either one of which (or both) can be found in various bacteria.

Although E. coli pgm mutants devoid of its principal (cofactor-dependent) form of phosphoglycerate mutase clearly exhibit a mutant phenotype, a recent study of the second (cofactor-independent) form of this enzyme showed that it accounts for as much as 10 per cent of the total phosphoglycerate mutase activity in E. coli.

Remarkably, neither form of phosphoglycerate mutase is encoded in any archaeal genome available to date, with the sole exception of Halobacterium spp., which has a typical cofactor-independent enzyme, similar to the one in B. subtilis. Sequence analysis of archaeal genomes showed that they encode enzymes of the alkaline phosphatase superfamily that are distantly related to the cofactor-independent phosphoglycerate mutase and contain all the principal active-site residues.

These enzymes were predicted to have a phosphog-lycerate mutase activity. This prediction was supported by the structural analysis of the cofactor-independent phosphoglycerate mutase and has been recently confirmed by direct experimental data. Thus, like phosphofructokinase and fructose-1,6-bisphosphate aldolase, phosphoglycerate mutase is found in three different (unrelated or distantly related) variants.

Enolase

Enolases encoded in bacterial, archaeal, and eukaryotic genomes are highly conserved; phylogenetic trees for enolases show a "star topology", which precludes any definitive conclusions on the evolutionary scenario for this enzyme. Pyrococci and M. jannaschii encode additional, divergent paralogs of enolase whose function(s) remains unknown.

Pyruvate Kinase

Pyruvate kinase, the terminal glycolytic enzyme, is not encoded in some bacterial (A. aeolicus, T. pallidum) and archaeal (A. fulgidus, M. thermoautotrophicum) genomes. In these organisms, the pyruvate kinase function is probably taken over by phosphoenolpyruvate synthase, which is capable of catalyzing pyruvate formation by reversing its typical reaction.

Pyruvate kinase, like phosphofructokinase, is also missing in H. pylori. Although a ribokinase-like phosphofructokinase and phosphoenolpyruvate synthase could be considered as possible bypasses for these enzymes, it seems more likely that glycolysis is not functional in H. pylori. In contrast, this bacterium encodes the complete set of enzymes involved in gluconeogenesis. Such organization of metabolism seems to make perfect sense for H. pylori, given the challenge of maintaining near-neutral intracellular pH in the highly acidic gastric environment. Sugar fermentation, resulting in intracellular production of acid, would place an additional burden on the pH maintenance mechanism, whereas gluconeogenesis converts organic acids into sugars and thus removes H from the cytoplasm. For the purposes of energy production, H. pylori apparently depends on fermentation of amino acids and oligopeptides that are produced by gastric proteolysis and are transported into the bacterial cells by ABC-type transporters. Amino acid fermentation results in alkalinization of the cytoplasm and could relieve part of the burden of pH maintenance in H. pylori. This simple example shows that, even when seemingly plausible candidates for missing steps in a pathway can be suggested, this should be done with caution, and the resulting predicted pathways should be assessed against the biological background of the respective organism.

After a string of recent publications, it appears that most glycolytic enzymes have now been accounted for. While there are no clear candidates for phosphofructokinase and fructose-1,6-bisphosphate aldolase in Thermoplasma spp., the chances of discovering new enzyme variants in this pathway appear very slim.

GLYCOPROTEIN STRUCTURE AND STEROID HORMONE FUNCTION

Glycoproteins play an important part in hormone function. The action of hormones depends on the initial binding of the hormone to a protein receptor molecule. In many cases this molecule is a glycoprotein. Many hormones bind to receptors in the cell membrane; these hormones never actually enter the cell.

Steroids, on the other hand, bind to an intracellular protein receptor. There is still controversy over whether the steroid hormone receptor is found in the nucleus or the cytosol. However, it is clear that after the steroid binds, the hormone-receptor complex moves to the nucleus. The hormone binding domain of the receptor is found at the C-terminus. The amino acid sequence of this region is highly diverse; it is not conserved from one protein to the next.

Binding of the hormone stimulates a conformational change in the hormone receptor. This change allows the hormone-receptor complex to bind to the DNA. The DNA binding domain is highly conserved and is found within the central core of the protein. The central core contains a very basic amino acid sequence. In the cellular environment, these bases will tend to pick up a hydrogen and become positively charged. The positive charge will attract the hormone-receptor complex to the negatively charged DNA. Binding of the complex to the DNA stimulates transcription (Evans 890-891 and Mathews 808-809).

SIMILARITIES BETWEEN GLYCOPROTEINS, CYTOCHROMES AND METALLOENZYMES

Cytochromes have features that are quite similar to transferrins, a particular class of glycoproteins. First of all, cytochromes make up some of the integral proteins found in membranes, as do glycoproteins. Second of all, the main function of cytochromes is the binding of ion cations. Via oxidoreduction with these cations of 2+state, cytochromes bind iron to a porphyrin prosthetic group, forming a heme group (Mathews). The function is to transport cations in and out of the cell. Transferrin is an iron binding protein as well.

It interacts with iron cations of the 2+state which also bind to a porphyrin group, Protoporphyrin IX, to form a heme group. Transferrins carry ions to many different locations in the body including in and out of cells (Gottschalk). Another group of molecules that serve a function related to glycoproteins are the metalloenzymes. These enzymes, as their name implies, bind to metal ions via the same type of heme groups just discussed (Gottschalk). One class of glycoprotein enzymes, the oxidoreductases, contain an enzyme called chloroperoxidase.

Like transferrin, chloroperoxidase contains protophyrin IX and forms the heme group with iron 3+ (Mathews). Eicosanoids, are also related to glycoproteins because they recognize the receptors on cells which are glycoproteins.

INHIBITION OF ASPARAGINE LINKED GLYCOPROTEINS BY ANTIBIOTICS

Synthesis of N-linked glycoproteins can be inhibited with the use of antibiotics (function of antibiotics). Examples are monensin, mevinolin and tunicamycin. Monensin, a monocovalent ionophore, inhibits the transport of proteins through the endoplasmic reticulum- Golgi complex (general virology 855). Mevinolin on the other hand, is an inhibitor of mevalonate synthesis. Mevalonate is the precursor of dolichol and other isoprenoid lipids (cellular physiology 284).

Finally, as mentioned above, asparagine linked glycosylation involves the first step of addition of GlcNAc to dolichol diphosphate. This reaction is catalyzed by the enzyme UDP-GlcNAc:dolichol phosphate N-acetylglucosamine-1-phosphate transferase or GPT. Tunicamycin is a specific inhibitor of GPT, therefore it can inhibit the synthesis of all N-linked glycoproteins (biological chemistry 8895). To give just an example of the effectiveness of tunicamycin, when it was injected to rats' hearts, it killed 12, 24, 50 and 60 h after the injection (histochemistry 607). Another connection could be seen with the function of the ribosomes, since the polypeptide chain of all glycoproteins.

GLUCONEOGENESIS

With the exception of reactions catalyzed by phosphofructokinase and pyruvate kinase, glycolytic reactions are reversible and function also in gluconeogenesis.

The reversal of the latter reaction, *i.e.* conversion of pyruvate into phosphoenolpyruvate, can be catalyzed by two closely related enzymes, phosphoenolpyruvate synthase and pyruvate, phosphate dikinase. The only other reaction that is specific for gluconeogenesis is the dephosphorylation of fructose-1,6-bisphosphate.

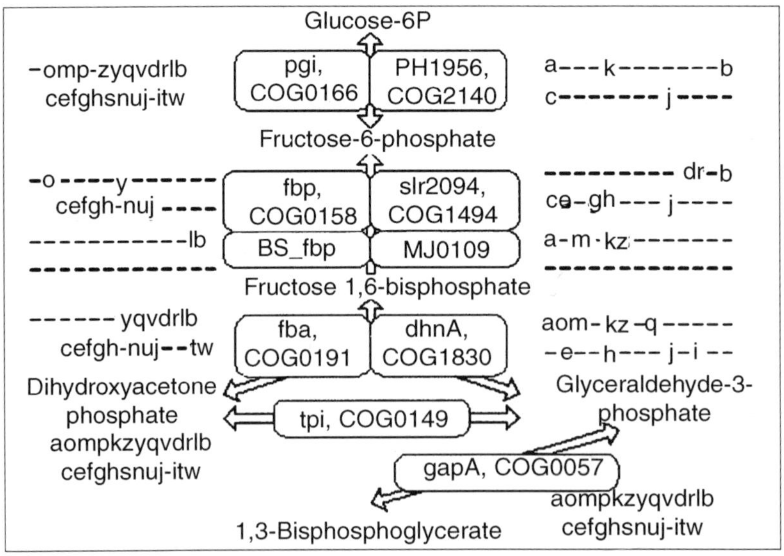

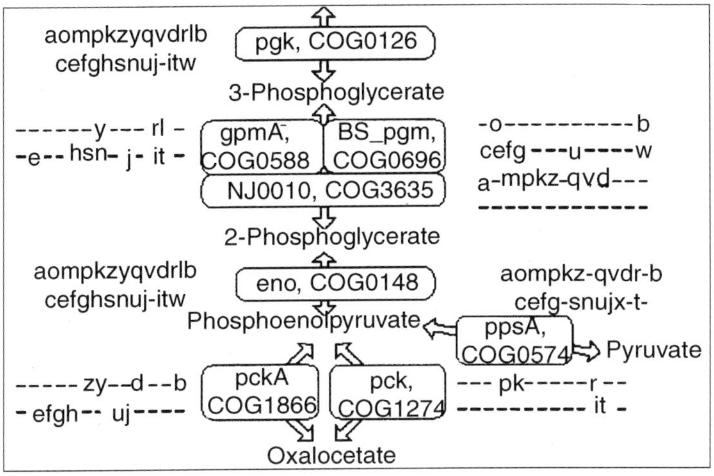

Distribution of gluconeogenesis enzymes in organisms with completely sequenced genomes. All details are as in Figure.

Phosphoenolpyruvate Synthase

Phosphoenolpyruvate synthase (pyruvate, water dikinase, EC 2.7.9.2) and pyruvate, phosphate dikinase (EC 2.7.9.1) catalyze two similar reactions of phosphoenolpyruvate biosynthesis

$$\text{Pyruvate} + \text{ATP} + H_2O = \text{Phosphoenolpyruvate} + \text{AMP} + P_i$$

$$\text{Pyruvate} + \text{ATP} + P_i = \text{Phosphoenolpyruvate} + \text{AMP} + pP_i$$

and have highly similar sequences. This enzyme is widely present in bacteria, archaea, protists, and plants but is missing in animals, where PEP is synthesized from oxaloacetates in a PEP carboxykinase-catalyzed reaction.

Phosphoenolpyruvate Carboxykinase

Phosphoenolpyruvate carboxykinase exists in two unrelated forms, which catalyze ATP-dependent (EC 4.1.1.49) or GTP-dependent (EC 4.1.1.32) decarboxylation of oxaloacetate:

$$\text{Oxaloacetate} + \text{ATP} = \text{Phosphoenolpyruvate} + \text{ADP} + CO_2$$

$$\text{Oxaloacetate} + \text{GTP} = \text{Phosphoenolpyruvate} + \text{GDP} + CO_2$$

These forms show remarkably complex phyletic distributions. The GTP-dependent form is found in animals and in a limited number of bacteria, such as Chlamydia spp., Mycobacterium spp., T. pallidum, and the green sulfur bacterium Chlorobium limicola. Among archaea, it is encoded only in the genomes of pyrococci, thermoplasmas, and Sulfolobus. In contrast, the ATP-dependent form of phosphoenolpyruvate carboxykinase is found in plants, yeast, and many bacteria. The only complete archaeal genome that has been found to encode the ATP-dependent form is that of A. pernix.

Since the typical bacterial ATP-dependent phospho-enolpyruvate carboxykinase appears to be unrelated to the GTP-dependent form found in humans, this key enzyme of central metabolism might be an interesting drug target for such pathogenic bacteria as H. influenzae and C. jejuni.

There have also been reports of a third, pyrophosphate-dependent, form of phosphoenolpyruvate carboxykinase, but they remain unconfirmed and no sequence so far has been identified with this form. Absent this third form, phosphoenolpyruvate carboxykinase appears to be missing in a large number of microorganisms, leaving room for discovery of a new enzyme.

Fructose-1,6-Bisphosphatase

The best-studied form of fructose-1,6-bisphosphatase, found in E. coli, yeast, and human (COG0158), has a limited phyletic distribution: it is not encoded in the genomes of chlamydia, spirochetes, Gram-positive bacteria, A. aeolicus, or T. maritima. Among archaea, it is present only in Halobacterium sp. A second form of this enzyme (COG1494), originally described in cyanobacteria, has been reported to function both as a fructose-1,6-bisphosphatase and as a sedoheptulose-1,7-bisphosphatase.

This form also has a limited phyletic distribution, being found in a relatively small number of bacteria Although a member of this second family is encoded in B. subtilis, this organism also has a distinct form of fructose-1,6-bisphosphatase that is unrelated to the first two and is found only in several other low-GC, Gram-positive bacteria. Finally, archaea encode yet another, fourth form of this enzyme that belongs to the inositol monophosphatase family and only recently has been shown to possess fructose-1,6-bisphosphatase activity.

Like B. subtilis, several bacterial genomes encode members of more than one protein family, which include known or potential fructose-1,6-bisphosphatases; this makes it hard to predict which of them actually has this function in gluconeogenesis. In contrast, there is no clear candidate for this function in A. aeolicus, T. maritima, X. fastidiosa, Chlamydia spp., mycoplasmas, spirochetes, and thermoplasmas. While the first three of these organisms and B. burgdorferi encode enzymes of the inositol monophosphatase family, they are not closely related to the archaeal fructose-1,6-bisphosphatase (typified by the MJ0109 protein from M. jannaschii) and might represent an independent case of enzyme recruitment. Proteins that function as fructose-1,6-bisphosphatase in Chlamydia spp., Thermoplasma spp., mycoplasmas, and T. pallidum, if any, remain to be identified.

CARBOHYDRATE METABOLISM IN FISH

Much of the carbohydrates that enter the diets of animals, including fish, is of plant origin. Carnivorous fish like the Atlantic salmon and the Japanese yellowtail, therefore, deal with little carbohydrate,. Indeed, experiments have shown that these species are ill-equipped to handle significant quantities of raw carbohydrate, in their diets. On the other hand, omnivores such as the common carp and the channel catfish are able to digest fair amounts of carbohydrates in their diets. The grass carp, a herbivore, subsists primarily on a vegetarian diet.

DIGESTION, ABSORPTION AND STORAGE

The ability of animals to assimilate starch depends on their ability to elaborate amylase. All species of fish have been shown to secrete a -amylase. It has also been demonstrated that activity of this enzyme was greatest in herbivores. In carnivores such as the rainbow trout and sea perch, amylase is primarily of pancreatic origin whereas in herbivores the enzyme is widespread throughout the entire digestive tract. In Tilapia mossambica the pancreas has been shown to be the site of greatest amylase activity followed by the upper intestine. Although the digestion of starch and dextrin by the carnivorous rainbow trout was shown to decrease progressively as levels of the carbohydrates were increased beyond the 20 per cent level, the fish could effectively utilize up to 60 per cent glucose, sucrose or lactose in the diet. This demonstrates that, contrary to earlier belief, carnivorous fish are capable of efficiently utilizing simple carbohydrate as a primary energy source.

The crystalline structure of starch appears also to influence its attack by amylase as evidenced by the two-fold increase in metabolizable energy content of fully cooked (gelatinized) maize in feeding trials with channel catfish. Rainbow trout have also been shown to have a higher tolerance for carbohydrate (present as wheat starch) in the diet when it was cooked. The process of gelatinization involves both heat and water. If an aqueous suspension of starch is heated, the granules do not change in appearance until a certain critical temperature is reached. At this point some of the starch granules swell and simultaneously lose their crystallinity. The critical temperature is that at which hydrogen bonds of the starch molecule loosen to permit complete hydration, leading to a phenomenon known as "swellingΔ.

Alpha-amylase, promotes a more or less random fragmentation of the starch molecule by hydrolyzing at the α -D-(1$\rightarrow$4) glucosidic bonds in the inner and outer chains of the compound. The result of complete hydrolysis of the amylose component are maltose and D-glucose, while the amylopectin component is reduced to maltose, D-glucose and branched limit dextrins. As a consequence of these action patterns by a -amylase on starch, other enzymes are needed for complete hydrolysis of starch to D-glucose in fish. In this regard, it has been demonstrated that even the carnivorous sea bream possess the ability to digest maltose.

On the other hand, cellulase and a -galactosidase have not been shown to be secreted by fish although cellulase of bacterial origin is present in the gut of most species of carps. The lack of a -galactosidase may partly explain the poor response by fish to dietary soybean meal which contains significant levels of the galactosidic oligosaccharides raffinose and stachyose. As has been pointed out earlier, these oligosaccharides do undergo enzymatic hydrolysis during the germination process to yield galactose and sucrose. It would, therefore, appear that the nutritive value of soybean meal will be enhanced if the bulk of this indigestible starch is first transformed. This can be achieved by soaking the beans for 48 hours prior to processing for meal production. It should also be pointed out that the nutritive value of pulses and other legume seeds can likewise be improved for fish since oligosaccharides constitute a large portion of the carbohydrates in legume seeds.

Data on glucose absorption by fish are scanty. Work with goldfish has shown that active transport of glucose is coupled with Na^+ transport as in most mammals. It is generally believed that absorption takes place on the mucosal surface of intestinal cells. The mono-saccharides which result from carbohydrate digestion consist primarily of glucose, fructose, galactose, mannose, xylose and arabinose. Although the rates of absorption of these sugars have been determined for many land mammals, similar information for fish is not available.

Glucose does not appear to be a superior energy source for fish over protein or fat although digestible carbohydrates do spare protein for tissue building. Also, unlike in mammals, glycogen is not a significant storage depot of energy despite evidence of an active and reversible Emden-Meyerhoff pathway in fish. The more efficient metabolism of amino acids over glucose for energy could be due to the ability of fish to excrete nitrogenous waste as ammonia from their gills without the high cost of energy in converting the waste to urea.

OTHER FACTORS AFFECTING METABOLISM

Apart from genetic adaptation, climatic factors also play an important role in carbohydrate metabolism in fish. Acclimation in fish, in essence, reflects enzyme acclimation, since the animal's ability to survive depends largely upon its ability to carry out normal metabolic functions. Some enzymes for metabolic acclimation show good compensation while others do not. The enzymes associated with energy liberation (enzymes of glycolysis, pentose shunt, tricarboxylic acid cycle, electron transport and fatty acid oxidation) exhibit temperature compensation whereas, those enzymes dealing largely with the degradation of metabolic products show poor or reverse compensation.

Table. Enzymes Subject to Metabolic Acclimation

Enzymes exhibiting compensation	Enzymes exhibiting reverse or no compen-sation
Phosphofructokinase	Catalase
Aldolase	Peroxidase
Lactic dehydrogenase	Acid phosphatase
6-phosphogluconate dehydrogenase	D-amino acid oxidase

Succinic dehydrogenase	Mg-ATP ase
Malic dehydrogenase	Choline acetyl transferase
Cytochrome oxidase	Acetylcholine esterase
Succinate-cytochrome C reductase	Alkaline phosphatase
NAD-cytochrome C reductase	Allantoinase
Aminoacyl transferase	Uricase
Na-K-ATPase	Amylase
	Lipase
protease	Malic enzyme
	Glucose-6-phosphate dehydrogenase

/Adapted from: Comparative Animal Physiology, edited by C.L. Prosser, 1973

It is interesting to note that two key enzymes involved in carbohydrate metabolism, amylase and glucose-6-phosphate dehydrogenase, together with an enzyme involved in fat digestion, lipase, show no temperature compensation. It is not certain if this is in any way connected with the cessation of feeding by fish at low temperatures. The molecular mechanism of thermal acclimation are not well understood and may consist of changes in synthesis or amounts of a given enzyme. Differences in kinetics, changes in the proportion of isoenzymes suitable for particular temperatures and changes in co-factors such as lipids, co-enzymes, or other factors such as pH and ions may be important in the animal's adjustment to temperature changes.

ENERGY TRANSFORMATION

Despite species differences in the tolerance of dietary carbohydrates it is generally believed that the principal end-product of carbohydrate digestion, glucose, is metabolized in a manner prevailing in all cells, i.e., via the reversible Emden-Meyerhoff pathway. In this pathway, glucose has only one principal fate: phosphorylation to glucose-6-phosphate. The major metabolic transformations are depicted as follows: All transformations proceed with a loss of free energy. Thus, the formation of two moles of lactate from glucose-6-phosphate occurs with free energy change of D G^o = -22000 cal/mole. The net result is the formation of four molecules of ATP. A functional reversal of this transformation can only occur via a different sequence requiring the input of six ATP molecules per mole of glucose-6-phosphate recovered.

Cells do not store glucose or glucose-6-phosphate. The readily available storage form is glycogen which is made from glucose-1-phosphate by one pathway and returned by another. Although in mammalian cells glucose-6-phosphate is transformed into fatty acids, such transformation does not appear to take place in fish. Studies with the common carp indicate that the precursor for lipogenesis is citrate formed when amino acids are actively metabolized through the tricarboxylic acid cycle. The major form of utilizable energy in all cells is ATP. In most cells this energy currency is generated by the oxidation of NADH by the mitochondrial electron-transport systems. The reductants of NAD^+ for this process are intermediates derived from the TCA cycle and fatty acids.

TCA Cycle

The TCA cycle showing enzymes, substrates and products. The abbreviated enzymes are: IDH = *isocitrate dehydrogenase* and α-KGDH = *α-ketoglutarate dehydrogenase*. The GTP generated during the *succinate thiokinase* (succinyl-CoA synthetase) reaction is equivalent to a mole of ATP by virtue of the presence of *nucleoside diphosphokinase*. The 3 moles of NADH and 1 mole of $FADH_2$ generated during each round of the cycle feed into the oxidative phosphorylation pathway. Each mole of NADH leads to 3 moles of ATP and each mole of $FADH_2$ leads to 2 moles of ATP. Therefore, for each mole of pyruvate which enters the TCA cycle, 12 moles of ATP can be generated.

CITRATE SYNTHASE (CONDENSING ENZYME)

The first reaction of the cycle is condensation of the methyl carbon of acetyl-CoA with the keto carbon (C-2) of oxaloacetate (OAA) to form citrate. The standard free energy of the reaction, -8.0 kcal/mol, drives it strongly in the forward direction.

Since the formation of OAA from its precursor is thermodynamically unfavorable, the highly exergonic nature of the *citrate synthase* reaction is of central importance in keeping the entire cycle going in the forward direction, since it *drives* oxaloacetate formation by mass action principals.

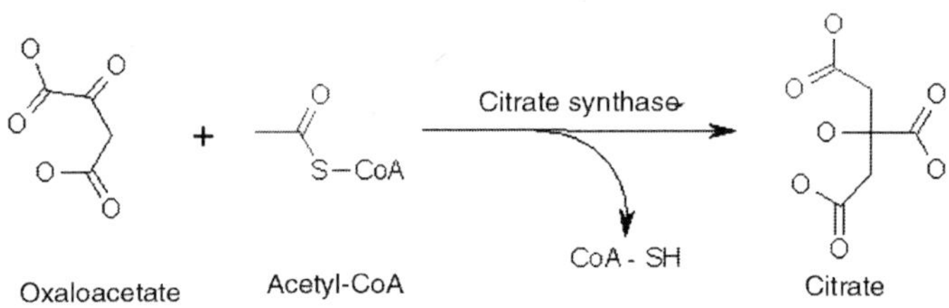

When the cellular energy charge increases the rate of flux through the TCA cycle will decline leading to a build-up of citrate. Excess citrate is used to transport acetyl-CoA carbons from the mitochondrion to the cytoplasm where they can be used for fatty acid and cholesterol biosynthesis. Additionally, the increased levels of citrate in the cytoplasm activate the key regulatory enzyme of fatty acid biosynthesis, acetyl-CoA carboxylase (ACC) and inhibit PFK-1. In non-hepatic tissues citrate is also required for ketone body synthesis.

ACONITASE

The isomerization of citrate to isocitrate by ***aconitase*** is stereospecific, with the migration of the -OH from the central carbon of citrate (formerly the keto carbon of OAA) being always to the adjacent carbon which is derived from the methylene (-CH_2-) of OAA.During the reaction the intermediate linked to the enzyme *cis*-aconitate is produced. The stereospecific nature of the isomerization determines that the CO_2 lost, as isocitrate is oxidized to succinyl-CoA, is derived from the oxaloacetate used in citrate synthesis.

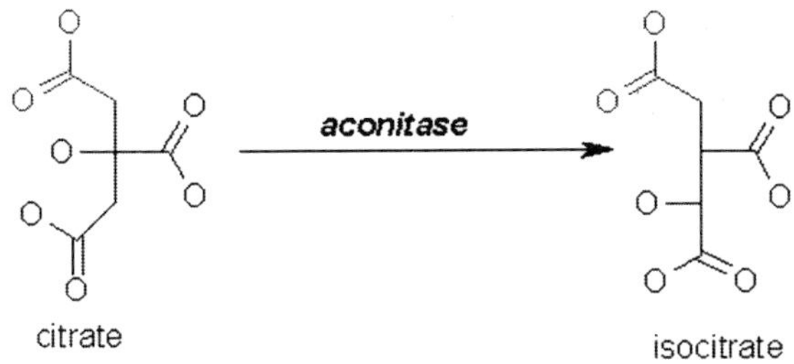

Aconitase is one of several mitochondrial enzymes known as non-heme-iron proteins. These proteins contain inorganic iron and sulfur, known as iron sulfur centers, in a coordination complex with cysteine sulfurs of the protein. There are two prominent classes of non-heme-iron complexes, those containing two equivalents each of inorganic iron and sulfur Fe_2S_2 and those containing 4 equivalents of each Fe_4S_4. Aconitase is a member of the Fe_4S_4 class. Its iron sulfur centers are often designated as $Fe_4S_4Cys_4$, indicating that 4 cystine sulfur atoms are involved in the complete structure of the complex. In iron sulfur compounds the iron is generally involved in oxidation-reduction events.

ISOCITRATE DEHYDROGENASE

Isocitrate is oxidatively decarboxylated to a-ketoglutarate by isocitrate dehydrogenase, (IDH).

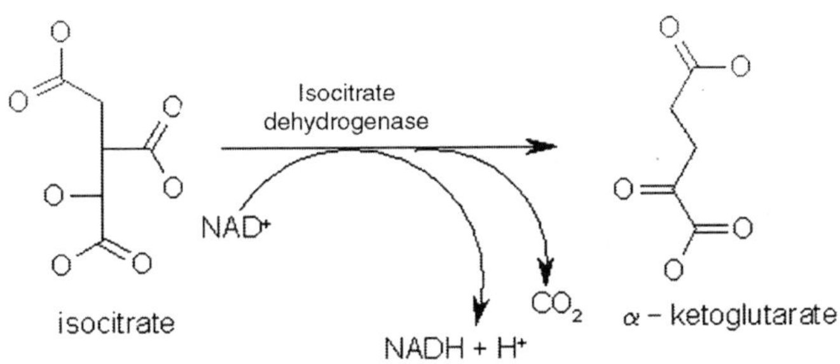

There are two different IDH enzymes. The IDH of the TCA cycle uses NAD^+ as a cofactor, whereas the other IDH uses $NADP^+$ as a cofactor. Unlike the NAD+-requiring enzyme, which is located only in the mitochondrial matrix, the $NADP^+$-requiring enzyme is found in both the mitochondrial matrix and the cytosol. IDH catalyzes the rate-limiting step, as well as the first NADH-yielding reaction of the TCA cycle. The CO_2 produced by the IDH reaction is the original C-1 of the oxaloacetate used in the citrate synthase reaction.

It is generally considered that control of carbon flow through the cycle is regulated at IDH by the powerful negative allosteric effectors NADH and ATP and by the potent positive effectors; isocitrate, ADP and AMP. From the latter it is clear that cell energy charge is a key factor in regulating carbon flow through the TCA cycle.

α-KETOGLUTARATE DEHYDROGENASE COMPLEX

α-ketoglutarate is oxidatively decarboxylated to succinyl-CoA by the *a-ketoglutarate dehydrogenase* (α-KGDH) complex.

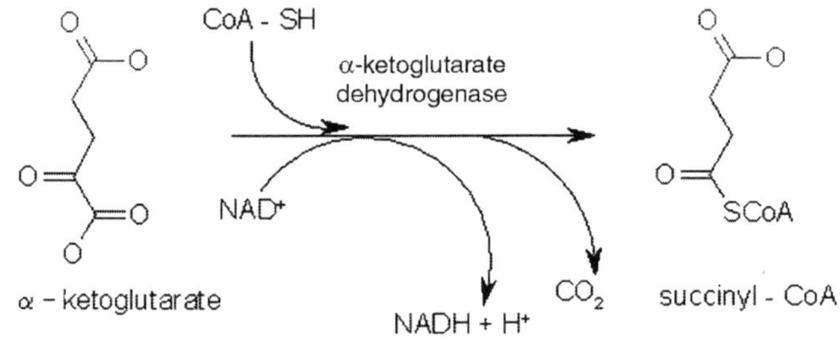

This reaction generates the second TCA cycle equivalent of CO_2 and NADH. This multienzyme complex is very similar to the PDH complex in the intricacy of its protein makeup, cofactors and its mechanism of action. Also, as with the PDH complex, the reactions of the a-KGDH complex proceed with a large negative standard free energy change.

Although the a-KGDH of the complex is not subject to covalent modification, allosteric regulation is quite complex, with activity being regulated by energy charge, the NAD^+/NADH ratio and effector activity of substrates and products. Succinyl-CoA and a-ketoglutarate are also important metabolites outside the TCA cycle.

In particular,a-ketoglutarate represents a key anapleurotic metabolite linking the entry and exit of carbon atoms from the TCA cycle to pathways involved in amino acid metabolism. a-ketoglutarate is also important for driving the malate-aspartate shuttle. Succinyl-CoA, along with glycine, contributes all the carbon and nitrogen atoms required for the synthesis of protoporphyrin heme biosynthesis and for non-hepatic tissue utilization of ketone bodies.

SUCCINYL COA SYNTHETASE

The conversion of succinyl-CoA to succinate by succinyl CoA synthetase involves use of the high-energy thioester of succinyl-CoA to drive synthesis of a high-energy nucleotide phosphate, by a process known as substrate-level phosphorylation.

In this process a high energy succinyl-phosphate intermediate is formed, with the phosphate subsequently being transferred to GDP.

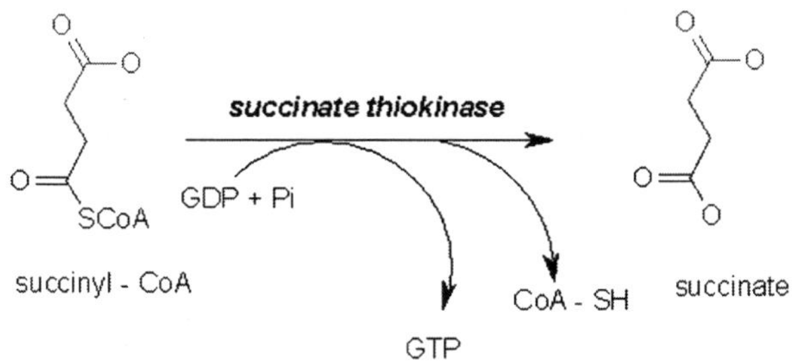

Mitochondrial GTP is used in a trans-phosphorylation reaction catalyzed by the mitochondrial enzyme nucleoside diphospho kinase to phosphorylate ADP, producing ATP and regenerating GDP for the continued operation of succinyl CoA synthetase.

SUCCINATE DEHYDROGENASE (SDH)

Succinate dehydrogenase catalyzes the oxidation of succinate to fumarate with the sequential reduction of enzyme-bound FAD and non-heme-iron.

In mammalian cells the final electron acceptor is coenzyme Q_{10} (CoQ_{10}), a mobile carrier of reducing equivalents that is restricted by its lipophilic nature to the lipid phase of the mitochondrial membrane.

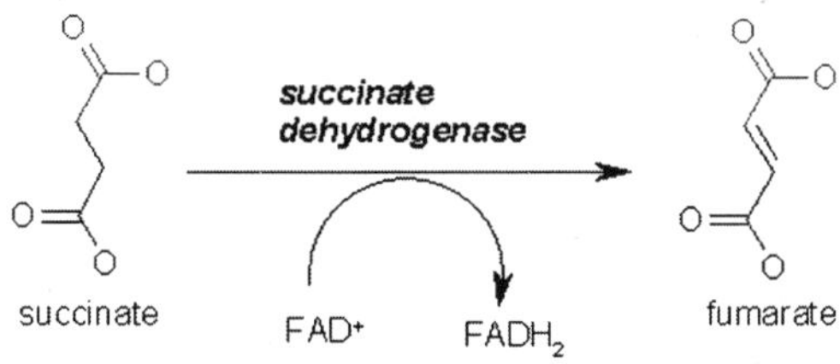

FUMARASE (FUMARATE HYDRATASE)

The fumarase-catalyzed reactions specific for the *trans* form of fumarate.

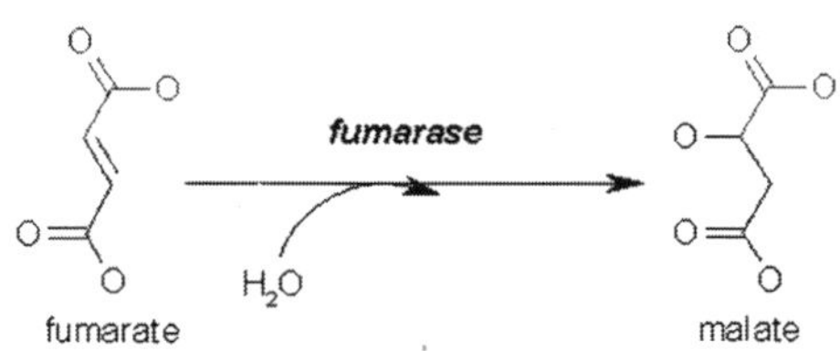

The result is that the hydration of fumarate proceeds stereospecifically with the production of L-malate.

MALATE DEHYDROGENASE (MDH)

L-malate is the specific substrate for MDH, the final enzyme of the TCA cycle. The forward reaction of the cycle, the oxidation of malate yields oxaloacetate(OAA).

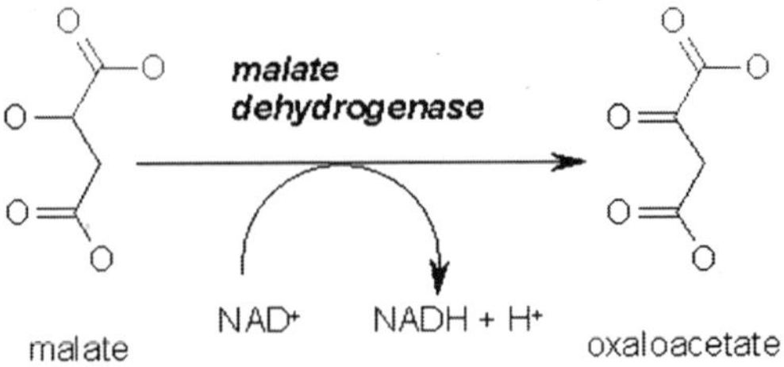

In the forward direction the reaction has a standard free energy of about +7 kcal/mol, indicating the very unfavorable nature of the forward direction. As noted earlier, the citrate synthase reaction that condenses oxaloacetate with acetyl-CoA has a standard free energy of about -8 kcal/mol and is responsible for pulling the MDH reaction in the forward direction. The overall change in standard free energy change is about -1 kcal/mol for the conversion of malate to oxaloacetate and on to succinate.

The overall stoichiometry of the TCA cycle is:

acetyl-CoA + 3NAD+ + FAD + GDP + Pi + 2H2O → 2CO2 + 3NADH + FADH2 + GTP + 2H+ + HSCoA

REGULATION OF THE TCA CYCLE

Regulation of the TCA cycle. like that of glycolysis, occurs at both the level of entry of substrates into the cycle as well as at the key reactions of the cycle. Fuel enters the TCA cycle primarily as acetyl-CoA. The generation of acetyl-CoA from carbohydrates is, therefore, a major control point of the cycle. This is the reaction catalyzed by the PDH complex. By way of review, the PDH complex is inhibited by acetyl-CoA and NADH and activated by non-acetylated CoA (CoASH) and NAD^+.

The pyruvate dehydrogenase activities of the PDH complex are regulated by their state of phosphorylation. This modification is carried out by a specific kinase (PDH kinase) and the phosphates are removed by a specific phosphatase (PDH phosphatase). The phosphorylation of PDH inhibits its activity and, therefore, leads to decreased oxidation of pyruvate. PDH kinase is activated by NADH and acetyl-CoA and inhibited by pyruvate, ADP, CoASH, Ca^{2+} and Mg^{2+}. The PDH phosphatase, in contrast, is activated by Mg^{2+} and Ca^{2+}. Since three reactions of the TCA cycle as well as PDH utilize NAD^+ as co-factor it is not difficult to understand why the cellular ratio of NAD^+/NADH has a major impact on the flux of carbon through the TCA cycle.

Substrate availability can also regulate TCA flux. This occurs at the *citrate synthase* reaction as a result of reduced availability of oxaloacetate. Product inhibition also controls the TCA flux, e.g. citrate inhibits *citrate synthase,* a-KGDH is inhibited by NADH and succinyl-CoA. The key enzymes of the TCA cycle are also regulated allosterically by Ca^{2+}, ATP and ADP.

6

Pyrimidine Biosynthesis

In contrast to the pathways of carbohydrate metabolism discussed above, enzymes of the pyrimidine biosynthesis pathway show a fairly consistent phyletic pattern, although cases of non-orthologous gene displacement can be found here, too.

The whole pathway, with the exception of the last three steps, is missing in the obligate parasitic bacteria with small genomes: rickettsiae, chlamydiae, spirochetes, and mycoplasmas, whereas bacteria and archaea with larger genomes encode all or almost all enzymes of pyrimidine biosynthesis.

CARBAMOYL PHOSPHATE SYNTHASE

In bacteria and archaea, carbamoyl phosphate synthase consists of two subunits, which in eukaryotes are fused into a single multifunctional CAD protein that additionally contains dihydroorotase and aspartate carbamoyltransferase domains. The small subunit, encoded by the carA gene, is a typical glutamine amidotransferase of the Triad family. The large subunit consists of two ATP-grasp domains fused in the same polypeptide chain.

In M. jannaschii and M. thermoautotrophicus, the large subunit is split into two proteins, which are encoded by different, albeit adjacent, genes. In addition to the obligate parasites mentioned above, carbamoyl phosphate synthase is missing in P. horikoshii, P. abyssi,Thermoplasma spp., and H. influenzae. It is present, however, in Pyrococcus furiosus, suggesting a relatively recent loss of this enzyme in the other two pyrococci. In P. abyssi, carbamoyl phosphate biosynthesis is carried out by an unrelated form of the enzyme, which is closely related to carbamate kinase.

This second form is also responsible for the carbamoyl phosphate synthase activity in P. furiosus and might account for this activity in Thermoplasma spp. Although both subunits of carbamoyl phosphate synthase belong to large protein superfamilies and are similar to many proteins with different substrate specificities, the sheer size of the large subunit, which typically contains more than 1,050 amino acid residues, allows an easy identification of this enzyme in genome analyses. However, caution is due with respect to the annotation of any shorter proteins that give statistically significant hits to the large subunit of carbamoyl phosphate synthase: these are likely to be other ATP-grasp superfamily enzymes.

ASPARTATE CARBAMOYLTRANSFERASE

Aspartate carbamoyltransferase, the second enzyme of pyrimidine biosynthesis, has a wide distribution with a phyletic pattern, which is similar to that of carbamoyl phosphate synthase but additionally includes pyrococci and Thermoplasma spp.

This enzyme, however, is lacking in H. influenzae and in its close relativeP. multocida. In eukaryotes, aspartate carbamoyltransferase comprises the C-terminal domain of the multifunctional CAD protein.

Dihydroorotase

The well-characterized form of dihydroorotase (COG0418), encoded by the E. coli pyrC gene and by theURA4 gene in yeast, has a very limited phyletic distribution. In contrast, the second form of this enzyme (COG0044) is almost universal, being present in many bacteria, archaea, and eukaryotes. In eukaryotes, this enzyme forms the middle portion of the multifunctional CAD protein.

In yeast, however, this domain is apparently inactive, most likely because of the presence of the alternative form of dihydroorotase. Simlilar to aspartate carbamoyltransferase, neither form of dihydroorotase is encoded in H. influenzae or P. multocida. Notably, the union of the phyletic patterns for the two forms of dihydroorotase is identical to the phyletic pattern of aspartate carbamoyl-transferase:

------y-------cefg--nu----- COG0418 Dihydroorotase
amopkzyqvdrlbcef--s-uj--- COG0044 Dihydroorotase
amopkzyqvdrlbcefg-snuj---- All froms Dihydroorotase
amopkzyqvdrlbcefg-snuj---- COG0540 Aspartate carbamoltransferase

Dihydroorotate Dehydrogenase

Dihydroorotate dehydrogenase displays the same phyletic pattern as dihydroorotase and aspartate carbamoyltransferase, with the addition of H. influenzae and P. multocida. Both these bacteria encode the enzymes for all downstream steps of pyrimidine biosynthesis.

Orotate Phosphoribosyltransferase

The phyletic pattern of orotate phosphoribosyltransferase differs from that of dihydroorotate dehydrogenase in only one respect, the presence of a pyrE-related gene in C. pneumoniae. The function of the product of this gene in C. pneumoniae is unknown, but given the absence in this organism of the enzymes for the upstream and the downstream steps of the pathway, it is unlikely to function as orotate phospho-ribosyltransferase. Rather, this enzyme might be recruited to catalyze a different phosphoribosyltransferase reaction. In eukaryotes, orotate phosphoribosyltransferase is fused to the next enzyme of the pathway, OMP decarboxylase, forming a two-domain UMP synthase. As a result, orotate phosphoribosyltransferase and OMP decarboxylase are occasionally misannotated as UMP synthases and vice versa.

Orotidine-5'-M onophosphate Decarboxylase

Although the phyletic pattern of OMP decarboxylase is identical to that of dihydroorotate dehydrogenase, a closer look at COG0284 shows that it consists of three distantly related families. Two of these include well-characterized enzymes from E. coli and other bacteria and from yeast and other eukaryotes. The third family includes OMP decarboxylases from archaea and a small number of bacteria, such as M. tuberculosis, M. leprae, and Myxococcus xanthus. Mycobacterial OMP decarboxylases seem to be sufficiently distinct from those of eukaryotes and other bacteria to consider them promising targets for antituberculine drugs.

Uridylate Kinase

There seem to be two distinct forms of uridylate kinase: one specific for UMP and found in bacteria and archaea (COG0528) and another one that phosphorylates both UMP and CMP and is found in eukaryotes.

The eukaryotic form of the enzyme is closely related to bacterial adenylate kinase and could have been recruited from an ancestral prokaryotic adenylate kinase. The prokaryotic form of uridylate kinase is encoded in all bacterial and archaeal genomes sequenced to date, including the 'minimal'genomes of mycoplasmas and Buchnera.

Nucleoside Diphosphate Kinase

Nucleotide diphosphate kinase (COG0105) is highly conserved in most bacteria, archaea, and eukaryotes. Surprisingly, however, this enzyme is not encoded in T. maritima, L. lactis, S. pyogenes, and mycoplasmas. One could imagine that these organisms employ a different nucleotide diphosphate kinase that might have been recruited, just like the eukaryotic uridylate kinase, from the adenylate kinase family (COG0563). This, however, would not solve the problem for T. maritima and mycoplasmas, which encode only a single enzyme of that family. It therefore seems likely that nucleotide diphosphate kinase in these organisms has been recruited from yet another kinase family. Indeed, a phyletic pattern search for a protein that would be encoded in those four genomes, but not in other organisms with relatively small genomes, such as chlamydiae, spirochetes, or H. pylori, easily finds an uncharacterized (predicted) kinase related to dihydro-xyacetone kinase (COG1461), which appears to be a good candidate for the role of nucleoside diphosphate kinase in these organisms:

aompkzyqvdrlbcefghesnujxitw Uridylate kinase (All froms)

aompkzyqv-dr-bcefghesnujxit- COG0105 NDP kinase
-----------v--l-------------------w Missing NDP kinase
-----------vdrlb-----------------w COG1461 Predicted kinase

CTP Synthase

CTP synthase is a two-domain protein, which consists of an N-terminal nucleotide-binding synthetase domain and a C-terminal glutamine amidotransferase domain. This enzyme is extremely highly conserved in bacteria, archaea, and eukaryotes. It is missing only in the genomes of M. genitalium and M. pneumoniae, which apparently make CTP from CDP or CMP in a salvage pathway, rather than from UTP.

CHOLESTROL BIOSYNTHESIS AND DEGRADATION

CHOLESTEROL BIOSYNTHESIS AND CATABOLISM

The formation of cholesterol by the liver, regulated by a type of homeostatic control, is influenced by the hepatic level offree cholesterol or glycogen, or both, and by the availability of the proper enzymes. Here the factors affecting the hepaticbiosynthesis of cholesterol are detailed, including the influence of dietary composition and fasting.

The specific effects oncholesterol synthesis of feeding cholesterol, various steroids, saturated and unsaturated fats and thyroid hormone are reviewed. Also considered are the effects of radiation, glycolysis and various pharmacologic agents.

In the degradation of cholesterol, the chief metabolic end products are bile acids, although cholesterol may also be converted into sex hormones and adrenocortical hormones.

The steps involved in the degradation of cholesterol to bileacids are reviewed, along with the various factors which increase or decrease cholesterol catabolism. Since degradation of cholesterol depends on the levels of bile salts in the liver, any factor which will increase the turnover of bile acids will also accelerate cholesterol catabolism and may in turn lower serum cholesterol.

CHOLESTEROL DESCRIPTION

"Cholesterol is a vital constituent of cell membranes and the precursor for steroid hormones and bile acids. It is clearly essential to life, yet its deposition in arteries is associated with cardiovascular disease and stroke, two leading causes of death in humans. In a healthy organism, an intricate balance is maintained between the biosynthesis, utilization, and transport of cholesterol, keeping its harmful deposition to a minimum."

HOLESTEROL A CRUCIAL COMPONENT OF CELL MEMBRANES

- Cholesterol is the third major lipid in membranes.
- Cholesterol is compact,rigid,hydrophobicmolecule with a polar hydroxyl group.
- The hydroxyl group of cholesterol is orientedtowards the aqueous phase (polar heads of phospholipids) and gives hydrophilic character to this end of cholesterol molecule.
- The cholesterol to polar lipid ratios affect stability, fluidity, permeability, and protein mobility. Cholesterol also controls the microstructure of membranes.
- Cholesterol is especially abundant in myelinated structures of the brain and central nervous system and almost absent from the inner mitochondrial membrane.

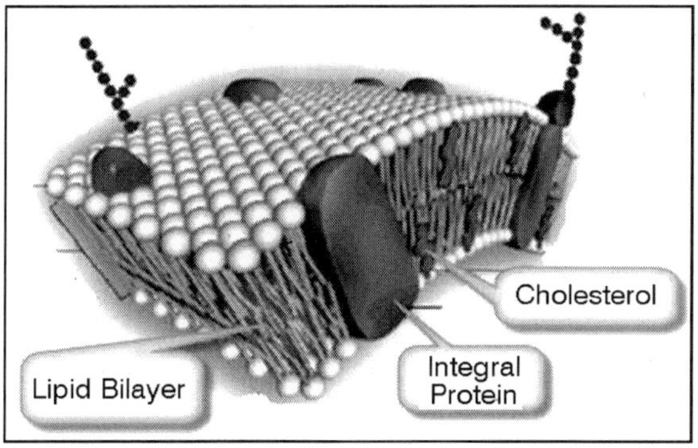

CHOLESTEROL IS PRECURSOR OF BILE SALTS

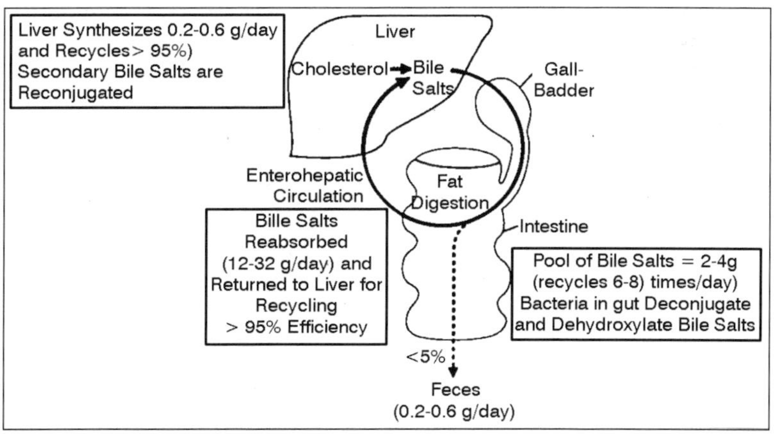

- Cholesterol is abundant in bile.
- The solubilization of hydrophobic molecules of cholesterol is aided

by bile phospholipids and bile salts. They prevent cholesterol from precipitating in gallbladder in form of gallstones.

- Bile salts are metabolites of cholesterol.
- Cholesterol protects gallbladder membranes from irritating and harmful effects of bile acids and bile salts.

Cholesterol is the precursor of steroid hormones

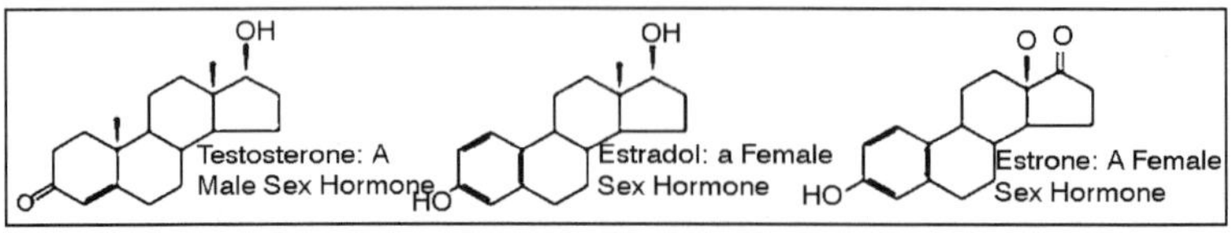

Cholesterol is the precursor of ALL steroid hormones including progesterone, estrogen, testosterone,corticosteroids, and aldosterone.

Thus, cholesterol is crucial for human reproduction, response to the environment, and preservation of human species.

CHOLESTEROL IS PRECURSOR OF VITAMIN D

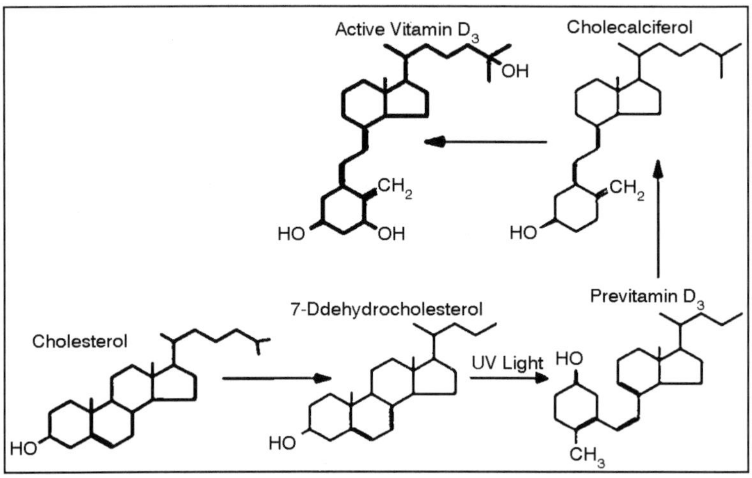

- The various forms of vitamin D, which are really hormones, aresterol derivatives.
- Active vitamin D increases Ca2+concentration in serum by promoting the intestinal absorption of dietary Ca2+. This increases the deposition of Ca2+ in bones and teeth.

BIOSYNTHESIS OF CHOLESTEROL

Slightly less than half of the cholesterol in the body derives from biosynthesis de novo. Biosynthesis in the liver accounts for approximately 10 per cent, and in the intestines approximately 15 per cent, of the amount produced each day. Cholesterol synthesis occurs in the cytoplasm and microsomes from the two-carbon acetate group of acetyl-CoA.

The process has five major steps:

- Acetyl-CoAs are converted to 3-hydroxy-3-methylglutaryl-CoA (HMG-CoA).
- HMG-CoA is converted to mevalonate.
- Mevalonate is converted to the isoprene based molecule, isopentenyl pyrophosphate (IPP), with the concomitant loss of CO_2.
- IPP is converted to squalene.
- Squalene is converted to cholesterol.

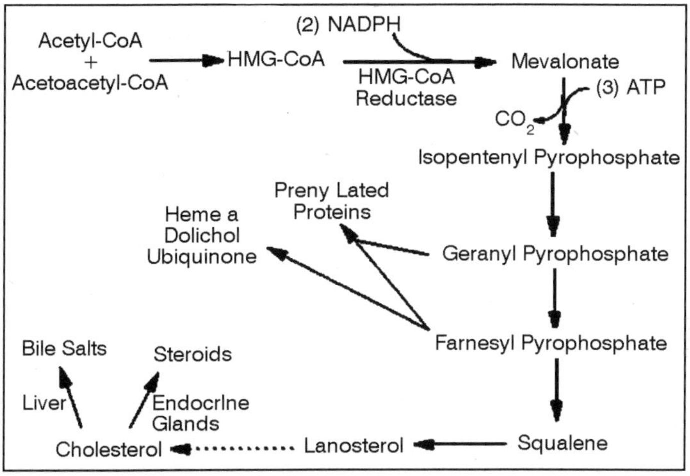

Pathway of cholesterol biosynthesis. Synthesis begins with the transport of acetyl-CoA ffrom the mitochondrion to the cytosol. The rate limiting step occurs at the 3-hydroxy-3 methlgltarl CoA (HMGCoA) redcatase HMGR cataled step The phosphorylation reactions are required to solubilize the isoprenoid intermediates in the pathway. Intermediates in the pathway are used for the synthesis of prenylated proteins, dolichol, coenzyme Q and the side chain of heme a. The acetyl-CoA utilized for cholesterol biosynthesis is derived from an oxidation reaction (eg, fatty acids or pyruvate) in the mitochondria and is transported to the cytoplasm by the same process as that described for fatty acid synthesis. Acetyl-CoA can also be derived from cytoplasmic oxidation of ethanol by acetyl-CoA synthetase. All the reduction reactions of cholesterol biosynthesis use NADPH as a cofactor. The isoprenoid intermediates of cholesterol biosynthesis can be diverted to other synthesis reactions, such as those for dolichol (used in the synthesis of N-linked glycoproteins, coenzyme Q (of the oxidative phosphorylation) pathway or the side chain of heme a. Additionally, these intermediates are used in the lipid modification of some proteins.

Acetyl-CoA units are converted to mevalonate by a series of reactions that begins with the formation of HMG-CoA. Unlike the HMG-CoA formed during ketone body synthesis in the mitochondria, this form is synthesized in the cytoplasm. However, the pathway and the necessary enzymes are the same as those in the mitochondria. Two moles of acetyl-CoA are condensed in a reversal of the thiolase reaction, forming acetoacetyl-CoA.

Acetoacetyl-CoA and a third mole of acetyl-CoA are converted to HMG-CoA by the action of HMG-CoA synthase. HMG-CoA is converted to mevalonate by HMG-CoA reductase, HMGR (this enzyme is bound in the endoplasmic reticulum, ER). HMGR absolutely requires NADPH as a cofactor and two moles of NADPH are consumed during the conversion of HMG-CoA to mevalonate.

The reaction catalyzed by HMGR is the rate limiting step of cholesterol biosynthesis, and this enzyme is subject to complex regulatory controls.

Mevalonate is then activated by three successive phosphorylations, yielding 5-pyrophosphomevalonate. In addition to activating mevalonate, the phosphorylations maintain its solubility, since otherwise it is insoluble in water. After phosphorylation, an ATP-dependent decarboxylation yields isopentenyl pyrophosphate, IPP, an activated isoprenoid molecule. Isopentenyl pyrophosphate is in equilibrium with its isomer, dimethylallyl pyrophosphate, DMPP.

One molecule of IPP condenses with one molecule of DMPP to generate geranyl pyrophosphate, GPP. GPP further condenses with another IPP molecule to yield farnesyl pyrophosphate, FPP. Finally, the NADPH-requiring enzyme, squalene synthase catalyzes the head-to-tail condensation of two molecules of FPP, yielding squalene (squalene synthase also is tightly associated with the endoplasmic reticulum). Squalene undergoes a two step cyclization to yield lanosterol.

The first reaction is catalyzed by squalene monooxygenase. This enzyme uses NADPH as a cofactor to introduce molecular oxygen as an epoxide at the 2,3 position of squalene. Through a series of 19 additional reactions, lanosterol is converted to cholesterol.

REGULATING CHOLESTEROL SYNTHESIS

Normal healthy adults synthesize cholesterol at a rate of approximately 1g/day and consume approximately 0.3g/day. A relatively constant level of cholesterol in the body (150 - 200 mg/dL) is maintained primarily by controlling the level of de novo synthesis. The level of cholesterol synthesis is regulated in part by the dietary intake of cholesterol. Cholesterol from both diet and synthesis is utilized in the formation of membranes and in the synthesis of the steroid hormones and bile acids. The greatest proportion of cholesterol is used in bile acid synthesis.

The cellular supply of cholesterol is maintained at a steady level by three distinct mechanisms:

- Regulation of HMGR activity and levels
- Regulation of excess intracellular free cholesterol through the activity of acyl-CoA:cholesterol acyltransferase, ACAT

- Regulation of plasma cholesterol levels via LDL receptor-mediated uptake and HDL-mediated reverse transport.

Regulation of HMGR activity is the primary means for controlling the level of cholesterol biosynthesis. The enzyme is controlled by four distinct mechanisms: feed-back inhibition, control of gene expression, rate of enzyme degradation and phosphorylation-dephosphorylation.

The first three control mechanisms are exerted by cholesterol itself. Cholesterol acts as a feed-back inhibitor of pre-existing HMGR as well as inducing rapid degradation of the enzyme. The latter is the result of cholesterol-induced polyubiquitination of HMGR and its degradation in the proteosome (see proteolytic degradation below). This ability of cholesterol is a consequence of the sterol sensing domain, SSD of HMGR. In addition, when cholesterol is in excess the amount of mRNA for HMGR is reduced as a result of decreased expression of the gene. The mechanism by which cholesterol (and other sterols) affect the transcription of the HMGR gene is described below under regulation of sterol content.

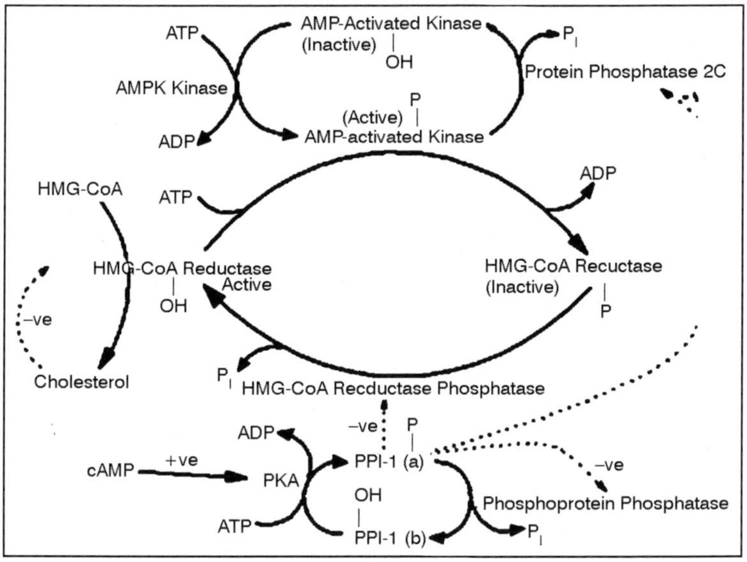

Regulation of HMGR through covalent modification occurs as a result of phosphorylation and dephosphorylation. The enzyme is most active in its unmodified form. Phosphorylation of the enzyme decreases its activity. HMGR is phosphorylated by AMP-activated protein kinase, AMPK (this is not the same as cAMP-dependent protein kinase, PKA). AMPRK itself is activated via phosphorylation. The phosphorylation of AMPK is catalyzed by AMPK kinase.

Regulation of HMGR by covalent modification. HMGR is most active in the dephosphorylated state. Phosphorylation is catalyzed by AMP-activated protein kinase, AMPK, (used to be termed HMGR kinase), an enzyme whose activity is also regulated by phosphorylation. Phosphorylation of AMPK is catalyzed by AMPK kinase. Hormones such as glucagon and epinephrine negatively affect cholesterol biosynthesis by increasing the activity of the inhibitor of phosphoprotein phosphatase inhibitor-1, PPI-1. Conversely, insulin stimulates the removal of phosphates and, thereby, activates HMGR activity. Additional regulation of HMGR occurs through an inhibition of its' activity as well as of its' synthesis by elevation in intracellular cholesterol levels. This latter phenomenon involves the transcription factor SREBP described. The activity of HMGR is additionally controlled by the cAMP signaling pathway. Increases in cAMP lead to activation of cAMP-dependent protein kinase, PKA. In the context of HMGR regulation, PKA phosphorylates phosphoprotein phosphatase inhibitor-1 (PPI-1) leading to an increase in its' activity. PPI-1 can inhibit the activity of numerous phosphatases including protein phosphatase 2C and HMG-CoA reductase phosphatase which remove phosphates from AMPRK and HMGR, respectively. This maintains AMPRK in the phosphorylated and active state, and HMGR in the phosphorylated and inactive state. As the stimulus leading to increased cAMP production is removed, the level of phosphorylations decreases and that of dephosphorylations increases. The net result is a return to a higher level of HMGR activity.

Since the intracellular level of cAMP is regulated by hormonal stimuli, regulation of cholesterol biosynthesis is hormonally controlled. Insulin leads to a decrease in cAMP, which in turn activates cholesterol synthesis. Alternatively, glucagon and epinephrine, which increase the level of cAMP, inhibit cholesterol synthesis.

The ability of insulin to stimulate, and glucagon to inhibit, HMGR activity is consistent with the effects of these hormones on other metabolic pathways. The basic function of these two hormones is to control the availability and delivery of energy to all cells of the body.

Long-term control of HMGR activity is exerted primarily through control over the synthesis and degradation of the enzyme. When levels of cholesterol are high, the level of expression of the HMGR gene is reduced. Conversely, reduced levels of cholesterol activate expression of the gene. Insulin also brings about long-term regulation of cholesterol metabolism by increasing the level of HMGR synthesis.

Proteolytic Regulation of HMG-CoA Reductase

The stability of HMGR is regulated as the rate of flux through the mevalonate synthesis pathway changes. When the flux is high the rate of HMGR degradation is also high. When the flux is low, degradation of HMGR decreases. This phenomenon can easily be observed in the presence of the statin drugs. HMGR is localized to the ER and like SREBP contains a sterol-sensing domain, SSD. When sterol levels increase in cells there is an concommitant increase in the rate of HMGR degradation. The degradation of HMGR occurs within the proteosome, a multiprotein complex dedicated to protein degradation. The primary signal directing proteins to the proteosome is ubiquitination. Ubiquitin is a 7.6kDa protein that is covalently attached to proteins targeted for degradation by ubiqitin ligases. These enzymes attach multiple copies of ubiquitin allowing for recognition by the proteosome. HMGR has been shown to be ubiquitinated prior to its degradation. The primary sterol regulating HMGR degradation is cholesterol itself. As the levels of free cholesterol increase in cells, the rate of HMGR degradation increases.

UTILIZATION OF CHOLESTEROL

Cholesterol is transported in the plasma predominantly as cholesteryl esters associated with lipoproteins. Dietary cholesterol is transported from the small intestine to the liver within chylomicrons. Cholesterol synthesized by the liver, as well as any dietary cholesterol in the liver that exceeds hepatic needs, is transported in the serum within LDLs. The liver synthesizes VLDLs and these are converted to LDLs through the action of endothelial cell-associated lipoprotein lipase.

Cholesterol found in plasma membranes can be extracted by HDLs and esterified by the HDL-associated enzyme LCAT. The cholesterol acquired from peripheral tissues by HDLs can then be transferred to VLDLs and LDLs via the action of cholesteryl ester transfer protein (apo-D) which is associated with HDLs. Reverse cholesterol transport allows peripheral cholesterol to be returned to the liver in LDLs. Ultimately, cholesterol is excreted in the bile as free cholesterol or as bile salts following conversion to bile acids in the liver.

REGULATION OF CELLULAR STEROL CONTENT

The continual alteration of the intracellular sterol content occurs through the regulation of key sterol synthetic enzymes as well as by altering the levels of cell-surface LDL receptors. As cells need more sterol they will induce their synthesis and uptake, conversely when the need declines synthesis and uptake are decreased. Regulation of these events is brought about primarily by sterol-regulated transcription of key rate limiting enzymes and by the regulated degradation of HMGR. Activation of transcriptional control occurs through the regulated cleavage of the membrane-bound transcription factor sterol regulated element binding protein,

Sterol control of transcription affects more than 30 genes involved in the biosynthesis of cholesterol, triacylglycerols, phospholipids and fatty acids. Transcriptional control requires the presence of an octamer sequence in the gene termed the sterol regulatory element, SRE-1. It has been shown that SREBP is the transcription factor that binds to SRE-1 elements. It turns out that there are 2 distinct SREBP gene, SREBP-1 and SREBP-2. In addition, the SREBP-1 gene encodes 2 proteins, SREBP-1a and SREBP-1c as a consequence of alternative exon usage. All 3 proteins are proteolytically regulated by sterols. Full-length SREBPs have several domains and are embedded in the membrane of the endoplasmic reticulum (ER). The N-terminal domain contains a transcription factor motif of the basic helix-loop-helix (bHLH) type that is exposed to the cytoplasmic side of the ER. There are 2 transmembrane spanning domains followed by a large C-terminal domain also exposed to the cytosolic side. When sterols are scarce, cleavage of the full-length SREBP takes place with the result being that the N-terminal bHLH motif is released into the cytosol. The bHLH domain then migrates to the nucleus to direct transcription. Conversely, when sterols are abundant, cleavage of SREBP is inhibited. To control the level of SREBP-mediated transcription, the soluble bHLH domain is itself subject to rapid proteolysis.

The cleavage of SREBP is carried out by 2 distinct enzymes, one of which is regulated by sterols. The regulated cleavage occurs in the lumenal loop between the 2 transmembrane domains. This cleavage is catalyzed by site-1 protease, S1P. High sterol content blocks the activity of S1P. The second cleavage, catalyzed by site-2 protease, S2P, occurs in the first transmembrane span, leading to release of active SREBP.

Additional studies on sterol-regulated gene expression demonstrated that cleavage of SREBP by S1P is controlled by the level and action of an additional protein termed, SREBP cleavage-activating protein, SCAP. SCAP is a large protein also found in the ER membrane and contains at least 8 transmembrane spans. The C-terminal portion, which extends into the cytosol, has been shown to interact with the C-terminal domain of SREBP. This C-terminal region of SCAP contains 4 motifs called WD40 repeats.

The WD40 repeats are required for interaction of SCAP with SREBP. Interestingly, the N-terminus of SCAP, including membrane spans 2-6, resembles HMGR which itself is subject to sterol-stimulated degradation. This shared motif is called the sterol sensing domain, SSD. Several proteins whose functions involve sterols also contain the SSD. These include patched, an important development regulating receptor whose ligand, hedgehog, is modified by attachment of cholesterol and the Neimann Pick C1 (NPC1) protein which is involved in cholesterol transport in the secretory pathway. NPC1 is one of several genes whose activities, when disrupted, lead to severe neurological dysfunction.

The function of SCAP is to positively stimulate S1P-mediated cleavage of SREBP. The function of sterols is to inhibit this positive action of SCAP. The activity of SCAP involves movement from the ER to the Golgi and back. Because the C-terminus of SCAP interacts with SREBP, movement of SCAP takes SREBP along for the ride. When sterols are low, SCAP and SREBP move to the Golgi. This transit is required for SREBP cleavage as S1P is Golgi-localized. When sterols are high, movement of SCAP is halted. Thus, the overall effect of sterols is to regulate the ability of SCAP to present SREBP to S1P.

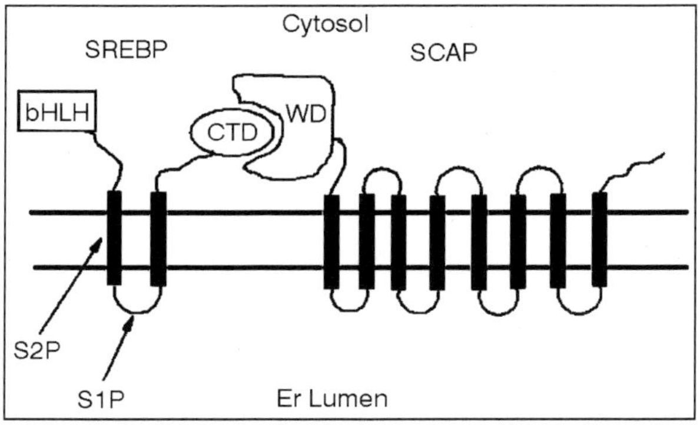

Diagramatic representation of the interactions between SREBP and SCAP in the membrane of the ER CTD = C-terminal domain

BILE ACIDS SYNTHESIS AND UTILIZATION

The end products of cholesterol utilization are the bile acids, synthesized in the liver. Synthesis of bile acids is one of the predominant mechanisms for the excretion of excess cholesterol. However, the excretion of cholesterol in the form of bile acids is insufficient to compensate for an excess dietary intake of cholesterol.

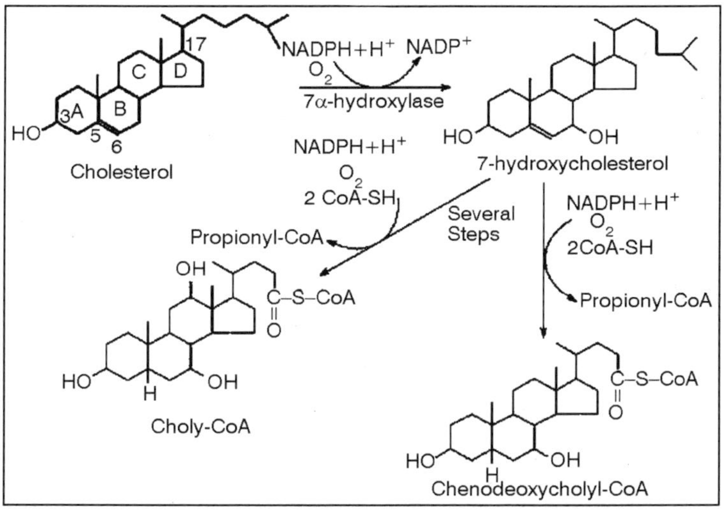

Synthesis of the 2 primary bile acids, cholic acid and chenodeoxycholic acid. The reaction catalyzed by the 7α-hydroxylase is the rate limiting step in bile acid synthesis. Conversion of 7α-hydroxycholesterol to the bile acids requires several steps not shown in detail in this image. Only the relevant co-factors needed for the synthesis steps are shown.

The most abundant bile acids in human bile are chenodeoxycholic acid (45 per cent) and cholic acid (31 per cent). These are referred to as the primary bile acids. Within the intestines the primary bile acids are acted upon by bacteria and converted to the secondary bile acids, identified as deoxycholate (from cholate) and lithocholate (from chenodeoxycholate). Both primary and secondary bile acids are reabsorbed by the intestines and delivered back to the liver via the portal circulation.

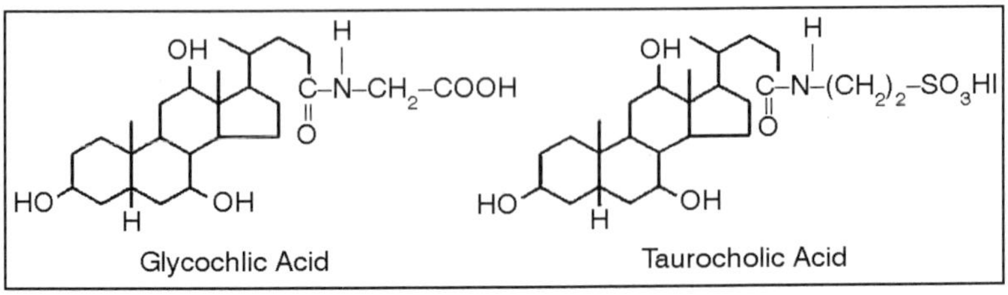

STRUCTURE OF CONJUGATED CHOLIC ACIDS

Within the liver the carboxyl group of primary and secondary bile acids is conjugated via an amide bond to either glycine or taurine before their being resecreted into the bile canaliculi. These conjugation reactions yield glycoconjugates and tauroconjugates, respectively. The bile canaliculi join with the bile ductules, which then form the bile ducts. Bile acids are carried from the liver through these ducts to the gallbladder, where they are stored for future use.

The ultimate fate of bile acids is secretion into the intestine, where they aid in the emulsification of dietary lipids. In the gut the glycine and taurine residues are removed and the bile acids are either excreted (only a small Percentage) or reabsorbed by the gut and returned to the liver. This process of secretion from the liver to the gallbladder, to the intestines and finally reabsorbtion is termed the enterohepatic circulation.

CLINICAL SIGNIFICANCE OF BILE ACID SYNTHESIS

Bile acids perform four physiologically significant functions:

- Their synthesis and subsequent excretion in the feces represent the only significant mechanism for the elimination of excess cholesterol.
- Bile acids and phospholipids solubilize cholesterol in the bile, thereby preventing the precipitation of cholesterol in the gallbladder.
- They facilitate the digestion of dietary triacylglycerols by acting as emulsifying agents that render fats accessible to pancreatic lipases.
- They facilitate the intestinal absorption of fat-soluble vitamins.

GENERAL NOTES ON PYRIMIDINE BIOSYNTHESIS EVOLUTION

Comparison of the phyletic patterns for the enzymes of pyrimidine biosynthesis reveals two important evolutionary trends. First, there appears to be a tendency towards decreasing the genome size by losing genes that have ceased to be essential. Indeed, ample evidence indicates that mycoplasmas evolved from a Gram-positive ancestor by way of massive gene loss associated with their adaptation to parasitism. While bacilli, lactococci, and many other Gram-positive bacteria carry the full set of genes of pyrimidine biosynthesis, most of the pyr genes have been lost in the mycoplasmal lineage. Similarly, many pyr genes apparently have been lost in other parasitic bacteria with small genomes, such as spirochetes, rickettsiae, and chlamydiae.

The trend towards gene loss is much more pronounced for the initial steps of the pyrimidine biosynthesis pathway than it is for the distal steps. Thus, genes for the first three steps of pyrimidine biosynthesis from bicarbonate and ammonia to dihydroorotate (carA, carB, pyrB, and pyrC) are missing in H. influenzae, but the genes for all the subsequent steps of pyrimidine biosynthesis, from dihydroorotate to CTP, are present. This means that, although H. influenzae is incapable of de novo pyrimidine biosynthesis, it still can synthesize UTP and CTP from dihydroorotate, orotate, or OMP. Spirochetes, chlamydiae, rickettsiae, and mycoplasmas show an even deeper loss of pyrimidine biosynthesis genes but nevertheless retain genes for the last three steps of the pathway, the conversion of UMP into CTP. Thus, while depending on the host for the supply of essential nutrients, this strategy allows the parasite to preserve at least some metabolic plasticity. In particular, every organism seems to encode enzymes to synthesize its own nucleoside triphosphates (NTPs). For thermodynamic reasons, bacteria cannot import NTPs directly, although intracellular bacterial parasites do encode ATP/ADP translocases, which are capable of exchanging ADP generated by the parasite for cytoplasmic ATP.

7

Fats and Lipids: Structure and Properties

FAT (LIPIDS)

Despite some recent literature wrongly espousing the benefits of high fat intakes (i.e., intakes of 30 Per cent or more of total calories from fat), fat is a highly concentrated fuel that does nothing to improve athletic performance, body composition, or weight when taken in excess. The adult AMDR for total fat intake is 20 to 35 Per cent of total calories, and there is no scientific information suggesting that more than 25 Per cent of total calories from fat is generally better for athletes. However, for athletes who may have difficulty sustaining weight because of a massive energy expenditure (such as cross-country skiers) or who must sustain high weights (such as linemen on football teams), higher fat intakes (up to the AMDR limit of 35 Per cent) may be necessary.

Few Americans consume less than 35 Per cent of total calories from fat, so consumption of less fat is not easy, and unless steps are taken to provide sufficient energy from other substrates (mainly from more complex carbohydrates) to replace the eliminated fat, athletes may place themselves in an energy-deficit state that is, in itself, a detriment to performance. Therefore, while a reduction in fat intake is generally useful, a conscious effort should be made to provide enough total energy when fat intake is reduced. Since fat is more than twice as concentrated in calories than either protein or carbohydrate (9 calories per gram versus 4 calories per gram), more than twice as much food must be consumed to make up the difference in reduced fat.

Cholesterol, oils, butter, and margarine are all fats, or lipids, but each has slightly different characteristics. The one common attribute shared by lipids is that they are soluble in organic solvents but not soluble in water. (Anyone who has tried to mix Italian dressing knows this to be true. The oil in the dressing eventually rises to the top, no matter how hard the bottle is shaken.) The term fat is usually applied to lipids that are solid at room temperature, and the term oil is applied to lipids that are liquid at room temperature. The most commonly consumed form of lipid is triglyceride, which consists of three fatty acids and one glycerol molecule (thus the name tri glyceride). Despite the numerous forms of lipids, we can obtain them all from the food supply, and we are also capable of making many types of lipids by combining carbon units from other substances. Nearly every cell in the body has the capacity to make cholesterol, which is why a person can have a high blood cholesterol leveleven when on a low-cholesterol diet.

We can also manufacture phospholipids, triglycerides, and oils. In fact, it is this ability to effectively manufacture different types of lipids that limits the necessity to consume large amounts of lipids.

STRUCTURE AND FUNCTION

FAT FUNCTIONS

A certain amount of fat, between 20 and 35 Per cent of total consumed calories, is necessary to ensure a sufficient energy and nutrient intake. The fat-soluble vitamins-vitamins A,D,E, and K-must be delivered in a fat package. The essential fatty acids, which are needed for specific body functions but that we are incapable of synthesizing, must also be consumed. Some dietary fat is also needed to give us a feeling of satiety during the meal, creating the important physiological signal that it is time to stop eating.

Dietary fats have a longer gastric emptying time than do carbohydrates, which contributes to the feeling of satiety. Of course, fat also makes foods taste good.

LIPIDS

All Lipids are hydrophobic: that's the one property they have in common. This group of molecules includes fats and oils, waxes, phospholipids, steroids (like cholesterol), and some other related compounds.

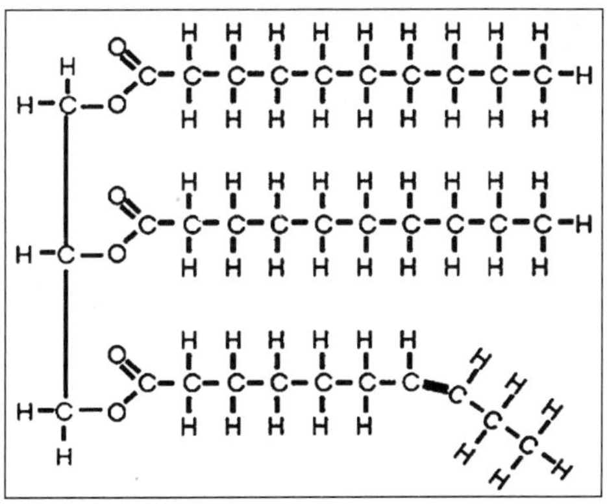

Fats and oils are made from two kinds of molecules: glycerol (a type of alcohol with a hydroxyl group on each of its three carbons) and three fatty acids joined by dehydration synthesis. Since there are three fatty acids attached, these are known as triglycerides."Bread" and pastries from a"bread factory" often contain mono- and diglycerides as"dough conditioners." Can you figure out what these molecules would look like? The main distinction between fats and oils is whether they're solid or liquid at room temperature, and this, as we'll soon see, is based on differences in the structures of the fatty acids they contain.

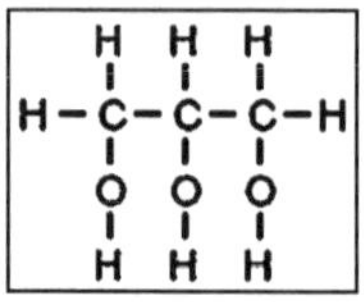

Lipid Structure

Lipids have different levels of saturation, a term that refers to the number of double bonds in the carbon chain. Fatty acids with no double bonds are saturated, those with one double bond are monounsaturated, and those with more than one double bond are polyunsaturated. Single bonds are stronger and less chemically reactive than double bonds, so the greater the number of double bonds, the greater the opportunity for the fatty acid to react with its chemical environment. It is this differential reactive capacity that makes the number of double bonds an important factor in human nutrition.

Saturated fatty acids are most prevalent in fats of animal origin, palm kernel oil, and coconut oil. Monounsaturated fats are highest in olive oil and canola oil but are also present in fats of animal origin. Polyunsaturated fats are highest in vegetable oils (with the exception of olive oil, which is more than 75 Per cent monounsaturated). In the context of a fat intake that does not exceed 35 Per cent of total calories, mono- and polyunsaturated fats should make up the majority of the fats consumed. Saturated fats are associated with higher cholesterol levels, so they should be minimized when possible. This is most easily achieved by reducing the consumption of animal fats, chocolate candies (often high in saturated tropical oils), fried foods, and high-fat dairy products.

STRUCTURE OF FATTY ACIDS

The"tail" of a fatty acid is a long hydrocarbon chain, making it hydrophobic. The"head" of the molecule is a carboxyl group which is hydrophilic. Fatty acids are the main component of soap, where their tails are soluble in oily dirt and their heads are soluble in water to emulsify and wash away the oily dirt. However, when the head end is attached to glycerol to form a fat, that whole molecule is hydrophobic.

The terms saturated, mono-unsaturated, and poly-unsaturated refer to the number of hydrogens attached to the hydrocarbon tails of the fatty acids as compared to the number of double bonds between carbon atoms in the tail. Fats, which are mostly from animal sources, have all single bonds between the carbons in their fatty acid tails, thus all the carbons are also bonded to the maximum number of hydrogens possible. Since the fatty acids in these triglycerides contain the maximum possible amouunt of hydrogens, these would be calledsaturated fats.

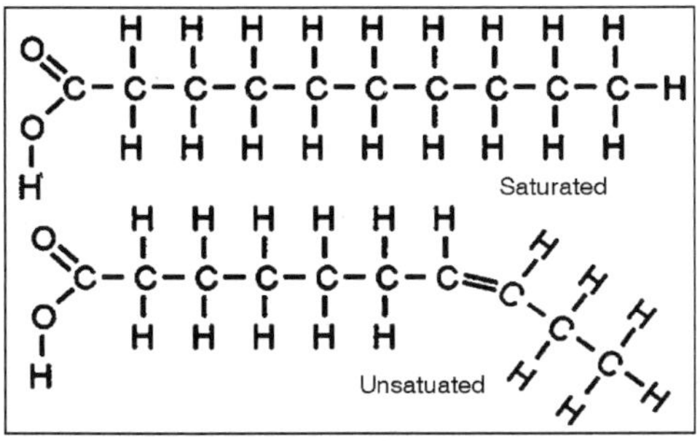

The hydrocarbon chains in these fatty acids are, thus, fairly straight and can pack closely together, making these fats solid at room temperature. Oils, mostly from plant sources, have some double bonds between some of the carbons in the hydrocarbon tail, causing bends or"kinks" in the shape of the molecules. Because some of the carbons share double bonds, they're not bonded to as many hydrogens as they could if they weren't double bonded to each other. Therefore these oils are called unsaturated fats.

Because of the kinks in the hydrocarbon tails, unsaturated fats can't pack as closely together, making them liquid at room temperature. Many people have heard that the unsaturated fats are"healthier" than the saturated ones. Hydrogenated vegetable oil (as in shortening and commercial peanut butters where a solid consistency is sought) started out as"good" unsaturated oil. However, this commercial product has had all the double bonds artificially broken and hydrogens artificially added (in a chemistry lab-type setting) to turn it into saturated fat that bears no resemblance to the original oil from which it came (so it will be solid at room temperature).

In unsaturated fatty acids, there are two ways the pieces of the hydrocarbon tail can be arranged around a C=C double bond. In cis bonds, the two pieces of the carbon chain on either side of the double bond are either both"up" or both"down," such that both are on the same side of the molecule. In trans bonds, the two pieces of the molecule are on opposite sides of the double bond, that is, one"up" and one"down" across from each other.

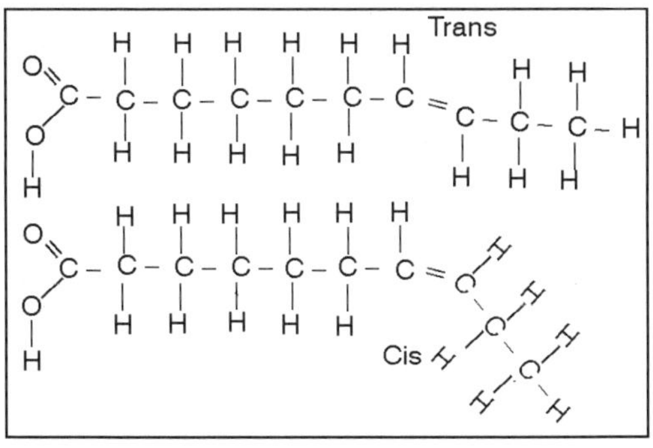

Naturally-occurring unsaturated vegetable oils have almost all cis bonds, but using oil for frying causes some of the cis bonds to convert to trans bonds. If oil is used only once like when you fry an egg, only a few of the bonds do this so it's not too bad. However, if oil is constantly reused, like in fast food French fry machines, more and more of the cis bonds are changed to trans until significant numbers of fatty acids with trans bonds build up. The reason this is of concern is that fatty acids with trans bonds are carcinogenic, or cancer-causing.

The levels of trans fatty acids in highly-processed, lipid-containing products such as margarine are quite high, and I have heard that the government is considering requiring that the amounts of trans fatty acids in such products be listed on the labels.

We need fats in our bodies and in our diet. Animals in general use fat for energy storage because fat stores 9 KCal/g of energy. Plants, which don't move around, can afford to store food for energy in a less compact but more easily accessible form, so they use starch (a carbohydrate, NOT A LIPID) for energy storage. Carbohydrates and proteins store only 4 KCal/g of energy, so fat stores over twice as much energy/gram as fat. By the way, this is also related to the idea behind some of the high-carbohydrate weight loss diets. The human body burns carbohydrates and fats for fuel in a given proportion to each other. The theory behind these diets is that if they supply carbohydrates but not fats, then it is hoped that the fat needed to balance with the sugar will be taken from the dieter's body stores. Fat is also is used in our bodies to a) cushion vital organs like the kidneys and b) serve as insulation, especially just beneath the skin.

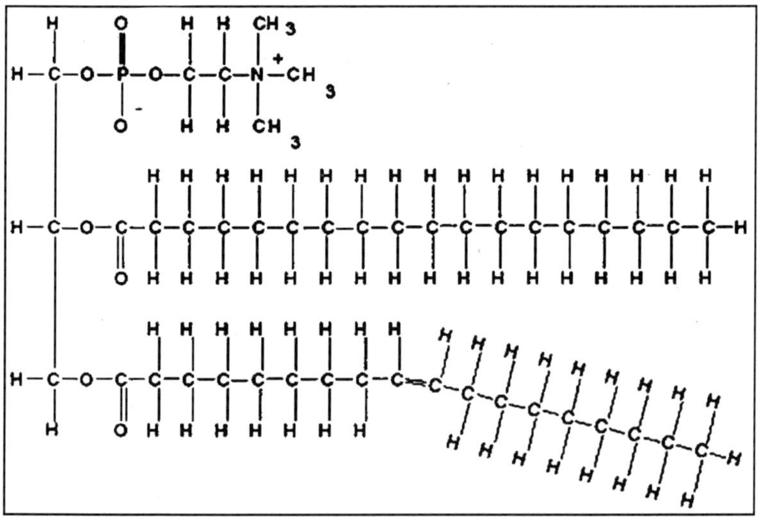

PHOSPHOLIPIDS

Phospholipids are made from glycerol, two fatty acids, and (in place of the third fatty acid) a phosphate group with some other molecule attached to its other end. The hydrocarbon tails of the fatty acids are still hydrophobic, but the phosphate group end of the molecule is hydrophilic because of the oxygens with all of their pairs of unshared electrons. This means that phospholipids are soluble in both water and oil.

An emulsifying agent is a substance which is soluble in both oil and water, thus enabling the two to mix. A"famous" phospholipid is lecithin which is found in egg yolk and soybeans.

Egg yolk is mostly water but has a lot of lipids, especially cholesterol, which are needed by the developing chick. Lecithin is used to emulsify the lipids and hold them in the water as an emulsion. Lecithin is the basis of the classic emulsion known as mayonnaise.

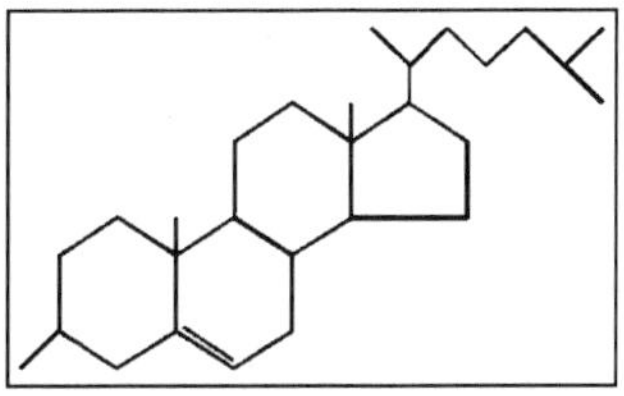

Our cell membranes are made mostly of phospholipids arranged in a double layer with the tails from both layers"inside" (facing toward each other) and the heads facing"out" (toward the watery environment) on both surfaces.

STEROIDS

The general structure of cholesterol consists of two six-membered rings side-by-side and sharing one side in common, a third six-membered ring off the top corner of the right ring, and a five-membered ring attached to the right side of that. The central core of this molecule, consisting of four fused rings, is shared by all steroids, including estrogen (estradiol), progesterone, corticosteroids such as cortisol (cortisone), aldosterone, testosterone, and Vitamin D. In the various types of steroids, various other groups/molecules are attached around the edges. Know how to draw the four rings that make up the central structure.

Cholesterol is not a"bad guy!" Our bodies make about 2 g of cholesterol per day, and that makes up about 85 per cent of blood cholesterol, while only about 15 per cent comes from dietary sources. Cholesterol is the precursor to our sex hormones and Vitamin D. Vitamin D is formed by the action of UV light in sunlight on cholesterol molecules that have"risen" to near the surface of the skin. At least one source I read suggested that people not shower immediately after being in the sun, but wait at least ½ hour for the new Vitamin D to be absorbed deeper into the skin. Our cell membranes contain a lot of cholesterol (in between the phospholipids) to help keep them"fluid" even when our cells are exposed to cooler temperatures. Many people have hear the claims that egg yolk contains too much cholesterol, thus should not be eaten. An interesting study was done at Purdue University a number of years ago to test this. Men in one group each ate an egg a day, while men in another group were not allowed to eat eggs. Each of these groups was further subdivided such that half the men got"lots" of exercise while the other half were"couch potatoes." The results of this experiment showed no significant difference in blood cholesterol levels between egg-eaters and non-egg-eaters while there was a very significant difference between the men who got exercise and those who didn't.

Lipoproteins are clusters of proteins and lipids all tangled up together. These act as a means of carrying lipids, including cholesterol, around in our blood. There are two main categories of lipoproteins distinguished by how compact/dense they are. LDL or low density lipoprotein is the"bad guy," being associated with deposition of"cholesterol" on the walls of someone's arteries. HDL or high density lipoprotein is the"good guy," being associated with carrying"cholesterol" out of the blood system, and is more dense/more compact than LDL.

Triglycerides

The majority of consumed lipids are triglycerides, which contain three fatty acids and a glycerol molecule. Fat is stored in the form of triglycerides, which we manufacture when excess energy is consumed. We store triglycerides in adipose tissues (groups of fat cells) and inside muscle cells (intramuscular tri glyceride), both of which are available as an energy source when needed. When fat is burned as a source of energy, the stored triglycerides are taken out of storage, and each molecule is cleaved into its component fatty acids and glycerol molecule. Each fatty acid can then be broken apart (two carbon units at a time) and thrown into the cellular furnaces for the creation of ATP to form heat and provide the energy for muscular work. This process is referred to as the beta-oxidative metabolic pathway because burning fat, besides requiring some carbohydrate for its complete oxidation, also requires oxygen.

Glycerol is a unique lipid that is burned like a carbohydrate rather than a fat and is also an effective humectant (it holds water). Some long-endurance athletes find that adding glycerol to water helps them retain more water (i.e., to superhydrate) than if they consumed water alone. In an extremely hot and humid environment, water loss is likely to be higher than the athlete's fluid-replacement capacity, so beginning a competition in a superhydrated state may provide some advantages.

In studies of tennis and Olympic distance triathlon, athletes consuming glycerol in water prior to competition experienced some protective hyperhydration benefits when exercising in high heat.17,18 However, being superhydrated is likely to impart a degree of discomfort that requires some adaptation. Athletes who consume fluids containing glycerol often describe the feeling of holding extra body water as making them feel"like a water bag," or"heavy," or"stiff." Still, they maintain that how they feel at the end of a race is more important than how they feel at the beginning, so adding glycerol to precompetition fluids has become a standard protocol for some athletes.

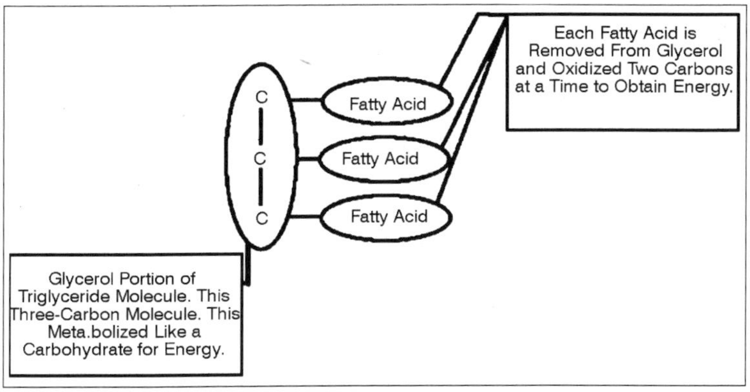

FATTY ACIDS

Linoleic (omega-6) and linolenic (omega-3) fatty acids are the essential fatty acids; although they are needed for metabolic processes, we are incapable of synthesizing them. The"omega-6" classification means that these fatty acids, which are polyunsaturated, have the last double bond six carbons from the end of the carbon chain.

The"omega-3" classification means that these fatty acids have the last double bond three carbons from the end of the carbon chain. Linoleic acid is an essential part of lipid membranes and is required for normal skin health. Linolenic acid is necessary for neural function and growth. The AMDR for omega-6 andomega-3 fatty acids is 5 to 10 and.6 to 1.2 grams per day, respectively.11Both fatty acids are easily obtained from vegetable oils (corn, safflower, canola, and so on) and the oils of cold-water fish.

Fish liver oils that are high in omega-3 acids have recently received much attention. These oils have been shown to reduce the ability of red blood cells to congregate, thereby decreasing the chance for an unwanted blood clot to form. This reduces the risk of a heart attack, which is most commonly caused by a clot formation in one of the major heart arteries. The oils from cold-water fish are the main source of the omega-3 fatty acids eicosapentanoic acid (EPA) and docosohexanoic acid (DHA).

Even a once-weekly consumption of cold-water fish (salmon, albacore tuna, Atlantic herring) is sufficient to significantly reduce the risk of heart attack and stroke.11Despite these findings, excessive intake of these fish oils may cause problems, including an increase in cellular oxidative damage. The best rule of thumb is to make fish consumption a regular part of the weekly diet so that supplemental intake of omega-3 fatty acids is unnecessary.

FAT REQUIREMENTS

From an exercise standpoint, there is little reason to believe that increasing fat consumption results in improving athletic performance, unless the increase in fat intake is the only reasonable means for the athlete to obtain sufficient energy. For the athlete who needs more than 4,000 calories each day to meet the combined demands of growth, exercise, and maintenance, moderate increases in dietary fat (preferably from plant and fish sources) may be needed. Since fat is a more concentrated form of energy than either carbohydrate or protein, more energy can be consumed in a smaller food package if the foods contain more fat.

LIPIDS AND PHYSICAL ACTIVITY

Even the leanest, healthiest athletes have a substantial energy pool of stored lipids. The average storage in adipose tissue ranges between 50,000 and 100,000 calories, or enough energy to walk or run 500 to 1,000 miles19 (800 to 1,600 kilometers) without a refueling stop. In addition, athletes store approximately 2,000 to 3,000 calories of lipids inside the muscle tissue.20 These lipids, which are stored in the form of triglycerides, are available as a fuel under the proper conditions of oxygen availability. Maximal fat oxidation occurs at 60 to 65 Per cent VO_2 max, but at higher levels of VO2max there is insufficient oxygen to derive the majority of energy used from fat catabolism.

Triglycerides stored in adipose tissue are cleaved into their component molecules of glycerol and fatty acids and transported to the blood plasma. The glycerol is available to all tissues for energy metabolism, and the free fatty acids are transported to working muscles where they are oxidized for energy. The triglycerides stored in working muscles are cleaved into glycerol and fatty acids, and the fatty acids can be oxidized for energy where they are resident. The glycerol can also be burned for energy in the working muscle or can be transported to the blood plasma as a source of energy for other tissues.

The lower the exercise intensity, the greater the proportion of fat burned to satisfy energy needs. As exercise intensity increases, the proportion of fat burned decreases and the proportion of carbohydrate burned increases. It is this basic reality that is behind why so many people do low-intensity activity to burn fat and lower body fat levels. However, the proportion of fat burned should not be confused by the total amount of fat burned at different intensities of physical activity. As exercise intensity increases, the total number of calories burned per unit of time also increases. Although there may be a decrease in the proportion of fat burned to satisfy total energy needs in higher-intensity activity, the total volume of fat burned is greater because the total energy requirement is higher.

The take-away message from this metabolic reality is that athletes interested in lowering body fat should exercise at least as high as 65 Per cent of VO2 max for the duration of their workouts to optimize the total mass of fat that is burned. Exercising at lower intensities burns a greater proportion of fat but less total fat than exercising at higher intensities.

BIOSYNTHESIS OF FATTY ACID

Identification of Genes in Biosynthesis of Fatty Acids and Cholesterol Pathways Using Microarrays Sterol regulatory binding proteins (SREBP) are a family of basic helix-loop-helix leucine zipper transcription factors that regulate biosynthesis of fatty acids and cholesterol. To identify candidate genes in the fatty acid synthesis pathway, transgenic mice over expressing SREBP-1a and SREBP-2 isofroms in liver were used. Labeled cRNA was prepared from liver and hybridized to mouse expression microarrays interrogating ~30,000 genes/ESTs. Some of the known genes, ACC, FAS, and ATP citrate lyase regulated by over expression of SREBP were identified by microarray analysis.

A novel gene, long chain fatty acyl elongase (LCE) regulated by the SREBP1a and 2 was also identified. LCE was over expressed with a fold change of 25 and 9 in SREBP-1a and SREBP-2 mice respectively. The confirmation of LCE was done by northern blot analysis, sequence homology searches and biochemical tests. This enzyme specifically catalyzes the elongation of 12, 14 and 16 carbons but does not elongate the C-18 fatty acids or very long chain fatty acids.

DEGRADATION OF FATTY ACIDS

Multi-million pounds of viscera are generated from the catfish filleting process. Our previous study showed that catfish viscera contained 30-35 per cent crude fat, and catfish visceral oil contained health-promoting fatty acids. Catfish visceral oils show potential for use as nutrient supplement and as cooking oil. Information on thermal stability and degradation of catfish visceral oils is not available.

The objective of this study was to determine weight loss (%) of catfish visceral oils and fatty acids after having been subjected to heat treatment. Oil samples included crude, degummed, neutralized, bleached, and deodorized oils.

The oils and individual fatty acids, FAs (C14:0, C16:0, C18:0, C20:0, C16:1, C18:1, C18:2, C18:3, C20:2, C20:4, C22:6) were subjected to the Thermogravimetric Analyser. Samples (0.5-1 mg) was placed in the furnace and heated up to 600°C at the rate of 5°C/min. Weight loss (%) of oils and fatty acids were calculated based on the original weight of non-heated samples. Duplicate experiments were conducted.

Over 90 per cent weight loss of all FAs (except C18:1 and C20:2) was observed at 250°C. Less than 2 per cent remaining weight of FAs (except C20:2) was observed at 350°C. Saturated FAs completely decomposed above 400°C.

Comparing C18:1 with C18:2 and C18:3, weight loss increased with increased double bonds. Between 200-550°C, weight loss of oils slightly increased with increased temperature, regardless of the purification process. There was, however, no distinct weight loss of all oils below 300°C. Oils with less impurity had greater weight loss especially at 450-500°C. At 550°C, all oil samples completely decomposed.

Knowledge of thermal stability or degradation (weight loss through eVapouration or decomposition) of catfish oils during purification or cooking is essential for the design of processing units, analysis of production cost, and cooking quality if to be used as cooking or seasoning oil.

FATTY ACID COMPOSITION OF FISH

ENVIRONMENTAL INFLUENCES

Salinity

The difference between fatty acid compositions of marine and freshwater fish has been noted by several authors. Some examples of fatty acid patterns are given in Table. Although these fish lipids are higher in ω 3 fatty acids, it is clear that freshwater fish have higher levels of ω 6 fatty acids than marine species.

The average ω6/ω3 ratios are 0.37 and 0.16 for freshwater and marine fish, respectively. Fish in general contain more ω 3 than ω 6 polyunsaturated fatty acids (PUFA) and should have a higher dietary requirement for ω 3 PUFA; thus the dietary EFA requirement of marine fish for ω 3 PUFA may be higher than that of freshwater fish. The same type of difference in the ω6/ω3 ratio between freshwater and seawater is seen when some species of fish migrate from oceans to streams or vice versa.

The PUFA ratio of sweet smelt (Plecoglosus altivelis) changes drastically in only one month as they migrate from the sea to a freshwater river. A similar but reverse change occurs in the masu salmon (Oncorhynchus masu) as they migrate from freshwater to seawater. Even within the same species of fish, the salinity of the water seems to cause a dramatic change in the fatty acid pattern. The difference between marine and freshwater fish may be due simply to differences in the fatty acid content in the diet or it may be related to a specific requirement in fish related to physiological adaptations to the environments.

The phospholipids are generally considered to be structural or functional lipid, being incorporated to a large extent in the membrane structure of cell and subcellular particles. The triglycerides are more often storage lipids and reflect the fatty acid composition of the diet to a greater extent than do the phospholipids.

In Table the fatty acid compositions of the triglyceride and phospholipid fractions of fish lipids are presented. It can be seen that the effect of changing environment on the fatty acid composition of the phospholipid is as great in the case of salmon, and considerably greater in the case of the sweet smelt, than it is on the triglyceride composition. Rainbow trout on diets containing either corn oil, which is high in ω6 but low in ω6 PUFA, showed a higher mortality and growth reduction in seawater than in freshwater over the twelve-week feeding period.

Temperature

There are several other factors besides the salinity of the water which affect the fatty acid composition and especially the PUFA of fish. In Tables 1 and 2 it can be seen that the salmonids, even in freshwater, tend to have a higher total PUFA of the 20 and 22 carbon chain length, and a lower ω6/ω3 ratio than the other fish. The salmonids are mostly cold-water fish.

The fatty acids from a number of marine animals from temperate and arctic waters show some significant differences in the general pattern; unfortunately analysis included fatty acids longer than 20:1. There are a number of other experiments demonstrating the effect of environmental temperature on fatty acid composition of aquatic animals. The general trend toward higher content of long chain PUFA at lower temperatures is quite clear.

The ω6/ω3 ratio decreases with a decrease in temperature. If the trends in fatty acid composition can be taken as clues to the EFA requirements of fish, the ω3 requirement would be greater for fish raised at lower temperatures. Fish raised in warmer waters, such as common carp, channel catfish, and tilapia may do better with a mixture of ω6 and ω3 fatty acids.

Effects of Diet

Some of the fatty acid compositions listed in Table may be seriously affected by the dietary lipids. The mosquito fish and guppies were fed trout pellets which had an ω6/ω3 ratio of 2.75. The catfish were fed diets supplemented with either beef tallow or menhaden oil, with ω6/ω3 ratios of 18.13 and 0.15, respectively.

These fish were able to alter the dietary ω6/ω3 ratio in favour of ω3 fatty acid incorporation into the flesh lipids even at the highest temperature. Commercially available trout pellets are often low in ω3 PUFA and high in ω6 fatty acids. It is important not to ignore the effect of dietary lipid composition on fatty acid composition of fish fed artificial diets.

It is clear from the data in Table that the ω6/ω3 ratio of the fish lipids is greatly affected by the ω6/ω3 ratio of the dietary lipids. When the dietary ratio is very high in ω6 fatty acids supplied by animal lard or vegetable oils, there is a tendency for fish to alter the ratio of PUFA incorporated in favour of ω3 fatty acids. When the dietary oil is a fish oil high in to 3 fatty acids, there is little change in the ω6/ω3 ratio of lipids incorporated into the fish. This is further suggestive evidence of an EFA requirement of fish for ω3 PUFA.

Seasonal Variation

Seasonal variations in the fatty acid composition of fish species have often been reported. Seasonal changes have been observed in total lipid and iodine values of herring oils. The iodine value or degree of unsaturation of the oil was minimal in April and maximal in June. The great increase in unsaturation corresponded to the onset of feeding in spring. The absence of a gas liquid chromatograph (GLC) at the time precluded identification of changes in individual fatty acids. Flesh and viscera lipid content of the sardine Sardinops melanosticta vary from 3.9 to 10.77 per cent and from 10.9 to 38.3 per cent, respectively.

The fatty acids of principal interest with respect to EFA metabolism are 20:4ω6, 20:5ω3, and 22:6ω3. There was considerable variation in all of these fatty acids in both neutral and polar lipid from both tissues.

In the flesh, the 20:4ω6 was consistently higher in the neutral lipid than in the polar lipid. The total 20:5 3 plus 22:6 3 was consistently higher in polar lipid than in the neutral lipid. Thus, in spite of the major fluctuations in fatty acids caused by changes in diet and temperature throughout the seasons, there was a consistent preferential incorporation of PUFA of the ω3 series into the polar or phospholipid fraction of the lipids.

One of the best clues to the EFA requirements of a species can be gained from the fatty acid composition of the lipids incorporated into the offspring or egg. The act of reproduction or spawning also has a significant effect on the seasonal fluctuation of lipids in fish.

Fatty acid composition of fish egg lipids is probably distinctive for each species and contains increased levels of 16:0, 20:4ω6, 20:5ω3 and 22:6ω3 compared to the liver lipids of the same female fish (Ackman, 1967). Elevated levels of 16:0, 20:5 3, and 22:6ω3 and reduced 18:1 in the ovary occurred compared to mesenteric fat of Pacific sardine fed a natural copepod diet.

The blood fatty acids of the sardine fed the natural diet were similar to those of the ovary. When the sardines were fed trout food, both the blood and mesenteric fat responded to the diet with elevated 18:2ω6 and reduced 20:5ω3 arid 22:6ω3.

The effect of the diet on ovary fatty acid content was considerably less, as relatively high levels of 20:5ω3 and 22:6ω3 were retained. The ovary lipids of the sweet smelt show an increase in 16:0, and a reduction in the PUFA, especially in the phospholipids, compared to the lipids from the flesh of fish caught at the same time of year. The ω6/ω3 ratio of the ovary was lower than that of the flesh lipids, 0.21 and 0.17 for the ovary compared to 0.31 and 0.20 for the triglycerids and phospholipids of the flesh, respectively.

The hatchability of eggs from common carp fed several different formulated feeds is greatly reduced when the 22:6ω3 of the egg lipids is less than 10 per cent. Further, the muscle, plasma, and erythrocyte fatty acid compositions are more affected by dietary lipid than those of the eggs. The

EFA requirements of a number of species of fish have been investigated in nutritional studies. The fish themselves have given ample evidence for EFA preference by the types of fatty acids they incorporate into their lipids. Fish, in general, tend to utilize ω3 over ω6. This is especially observed when the dietary lipids are high in ω6, as the fish tend to alter the ω6/ 3 ratio toward the ω3 fatty acids in the tissue lipids.

The lipids of the egg must satisfy the EFA requirement of the embryo until it is able to feed. The fatty acid composition data suggest that the ω3 requirement is greater in seawater than in freshwater and higher in cold water than in warm water.

FATTY ACID SYNTHESIS

One might predict that the pathway for the synthesis of fatty acids would be the reversal of the oxidation pathway. However, this would not allow distinct regulation of the two pathways to occur even given the fact that the pathways are separated within different cellular compartments.

The pathway for fatty acid synthesis occurs in the cytoplasm, whereas, oxidation occurs in the mitochondria. The other major difference is the use of nucleotide co-factors. Oxidation of fats involves the reduction of $FADH^+$ and NAD^+. Synthesis of fats involves the oxidation of NADPH. However, the essential chemistry of the two processes are reversals of each other. Both oxidation and synthesis of fats utilize an activated two carbon intermediate, acetyl-CoA. However, the acetyl-CoA in fat synthesis exists temporarily bound to the enzyme complex as malonyl-CoA.

The synthesis of malonyl-CoA is the first committed step of fatty acid synthesis and the enzyme that catalyzes this reaction, acetyl-CoA carboxylase (ACC), is the major site of regulation of fatty acid synthesis. Like other enzymes that transfer CO_2 to substrates, ACC requires a biotin co-factor.

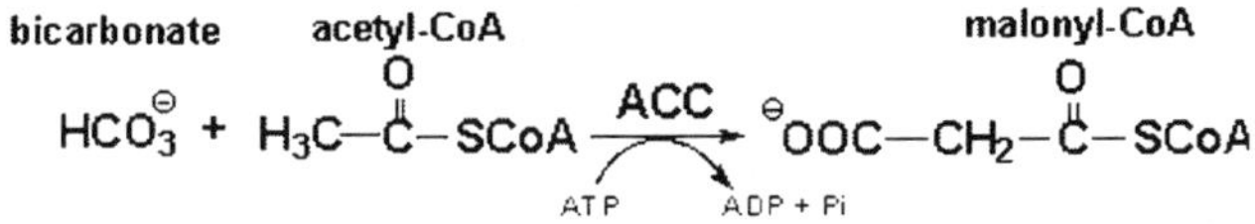

The rate of fatty acid synthesis is controlled by the equilibrium between monomeric ACC and polymeric ACC. The activity of ACC requires polymerization. This conformational change is enhanced by citrate and inhibited by long-chain fatty acids. ACC is also controlled through hormone mediated phosphorylation.

The acetyl groups that are the products of fatty acid oxidation are linked to CoASH. As you should recall, CoA contains a phosphopantetheine group coupled to AMP. The carrier of acetyl groups (and elongating acyl groups) during fatty acid synthesis is also a phosphopantetheine prosthetic group, however, it is attached a serine hydroxyl in the synthetic enzyme complex. The carrier portion of the synthetic complex is called acyl carrier protein, ACP. This is somewhat of a misnomer in eukaryotic fatty acid synthesis since the ACP portion of the synthetic complex is simply one of many domains of a single polypeptide. The acetyl-CoA and malonyl-CoA are transferred to ACP by the action of acetyl-CoA transacylase and malonyl-CoA transacylase, respectively.

The synthesis of fatty acids from acetyl-CoA and malonyl-CoA is carried out by fatty acid synthase, FAS.

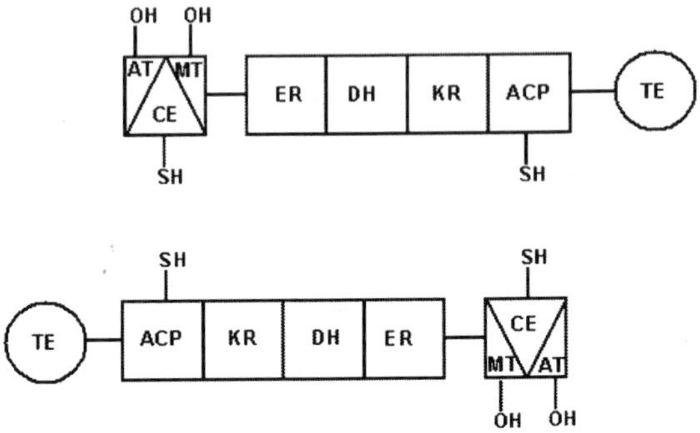

AT: Acetyl transacylase
MT: Malonyl transacylase
CE : Condensing enzyme
ACP: Acyl Carrier Protein
KR : 3-Ketoacyl reductase
DH : 3-Hydroxyacyl dehydratase
ER : Enoyl reductase
TE : Thioesterase

The active enzyme is a dimer of identical subunits.

All of the reactions of fatty acid synthesis are carried out by the multiple enzymatic activities of FAS. Like fat oxidation, fat synthesis involves 4 enzymatic activities. These are, b-keto-ACP synthase (condensing enzyme), b-keto-ACP reductase, 3-OH acyl-ACP dehydratase and enoyl-ACP reductase(give a look to the FAS reaction mechanism). The two reduction reactions require NADPH oxidation to $NADP^+$.

The primary fatty acid synthesized by FAS is palmitate. Palmitate is then released from the enzyme by ***thioesterase*** reaction and can then undergo separate elongation and/or unsaturation to yield other fatty acid molecules.

ORIGIN OF CYTOPLASMIC ACETYL

Acetyl-CoA is generated in the mitochondria primarily from two sources, the pyruvate dehydrogenase (PDH) reaction and fatty acid oxidation. In order for these acetyl units to be utilized for fatty acid synthesis they must be present in the cytoplasm. The shift from fatty acid oxidation and glycolytic oxidation occurs when the need for energy diminishes. This results in reduced oxidation of acetyl-CoA in the TCA cycle and the oxidative phosphorylation pathway. Under these conditions the mitochondrial acetyl units can be stored as fat for future energy demands.

Acetyl-CoA enters the cytoplasm in the form of citrate via the tricarboxylate transport system as diagrammed. In the cytoplasm, citrate is converted to oxaloacetate and acetyl-CoA by the ATP driven *ATP-citrate lyase* reaction. This reaction is essentially the reverse of that catalyzed by the TCA enzyme *citrate synthase* except it requires the energy of ATP hydrolysis to drive it forward. The resultant oxaloacetate is converted to malate by *malate dehydrogenase* (MDH).The malate produced by this pathway can undergo oxidative decarboxylation by *malic enzyme*. The co-enzyme for this reaction is $NADP^+$ generating NADPH. The advantage of this series of reactions for converting mitochondrial acetyl-CoA into cytoplasmic acetyl-CoA is that the NADPH produced by the *malic enzyme* reaction can be a major source of reducing co-factor for the *fatty acid synthase* activities.

REGULATION OF FATTY ACID METABOLISM

One must consider the global organismal energy requirements in order to effectively understand how the synthesis and degradation of fats (and also carbohydrates) needs to be exquisitely regulated. The blood is the carrier of triacylglycerols in the form of VLDLs and chylomicrons, fatty acids bound to albumin, amino acids, lactate, ketone bodies and glucose.

The pancreas is the primary organ involved in sensing the organisms dietary and energetic states via glucose concentrations in the blood. In response to low blood glucose, glucagon is secreted, whereas, in response to elevated blood glucose insulin is secreted.

The regulation of fat metabolism occurs via two distinct mechanisms. One is short term regulation which is regulation effected by events such as substrate availability, allosteric effectors and/or enzyme modification. ACC is the rate limiting (committed) step in fatty acid synthesis. This enzyme is activated by citrate and inhibited by palmitoyl-CoA and other long chain fatty acyl-CoAs. ACC activity also is affected by phosphorylation. Glucagon stimulated increases in PKA activity result in phosphorylation of certain serine residues in ACC leading to decreased activity of the enzyme. On the other hand insulin leads to phosphorylation of ACC at sites distinct from glucagon that result in increased ACC activity. These forms of regulation are all defined as short term regulation.

Control of a given pathways' regulatory enzymes can also occur by alteration of enzyme synthesis and turn-over rates. These changes are long term regulatory effects. Insulin stimulates ACC and FAS synthesis, whereas, starvation leads to decreased synthesis of these enzymes. Adipose tissue lipoprotein lipase levels also are increased by insulin and decreased by starvation. However, in contrast to the effects of insulin and starvation on adipose tissue, their effects on heart lipoprotein lipase are just the inverse. This allows the heart to absorb any available fatty acids in the blood in order to oxidize them for energy production. Starvation also leads to increases in the levels of fatty acid oxidation enzymes in the heart as well as a decrease in FAS and related enzymes of synthesis. Adipose tissue contains hormone-sensitive lipase, that is activated by PKA-dependent phosphorylation leading to increased fatty acid release to the blood. This leads to increased fatty acid oxidation in other tissues such as muscle and liver. In the liver the net result (due to increased acetyl-CoA levels) is the production of ketone bodies. This would occur under conditions where insufficient carbohydrate stores and gluconeogenic precursors were available in liver for increased glucose production. The increased fatty acid availability in response to glucagon or epinephrine is assured of being completely oxidized since PKA also phosphorylates (and as a result inhibits) ACC, thus inhibiting fatty acid synthesis. Insulin, on the other hand, has the opposite effect to glucagon and epinephrine leading to increased glycogen and triacylglyceride synthesis. One of the many effects of insulin is to lower cAMP levels which leads to increased dephosphorylation through the enhanced activity of protein phosphatases such as PP-1. With respect to fatty acid metabolism this yields dephosphorylated and inactive hormone sensitive lipase. Insulin also stimulates certain phosphorylation events. This occurs through activation of several cAMP-independent kinases. Insulin stimulated phosphorylation of ACC activates this enzyme. Regulation of fat metabolism also occurs through malonyl-CoA induced inhibition of carnitine acyltransferase I. This functions to prevent the newly synthesized fatty acids from entering the mitochondria and being oxidized.

ELONGATION AND DESATURATION

The fatty acid product released from FAS is palmitate (via the action of palmitoyl thioesterase) which is a 16:0 fatty acid, i.e. 16 carbons and no sites of unsaturation. Elongation and unsaturation of fatty acids occurs in both the mitochondria and endoplasmic reticulum (microsomal membranes). The predominant site of these processes is in the ER membranes. Elongation involves condensation of acyl-CoA groups with malonyl-CoA. The resultant product is two carbons longer (CO2 is released from malonyl-CoA as in the FAS reaction) which undergoes reduction, dehydration and reduction yielding a saturated fatty acid. The reduction reactions of elongation require NADPH as co-factor just as for the similar reactions catalyzed by FAS. Mitochondrial elongation involves acetyl-CoA units and is a reversal of oxidation except that the final reduction utilizes NADPH instead of FADH2 as co-factor. Desaturation occurs in the ER membranes as well and in mammalian cells involves 4 broad specificity fatty acyl-CoA desaturases (non-heme iron containing enzymes). These enzymes introduce unsaturation at C4, C5, C6 or C9.

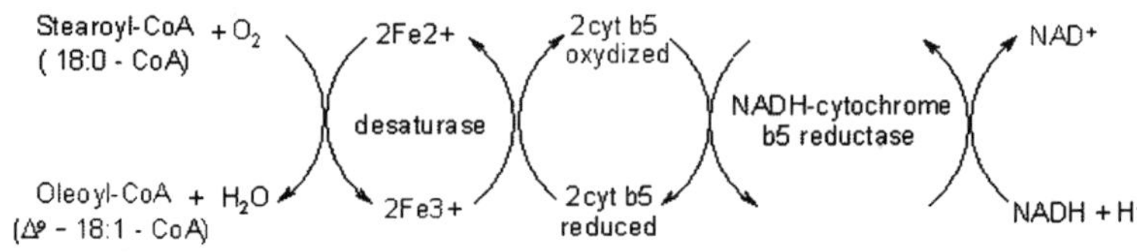

The electrons from the reduced fatty acid (saturated) and from the desaturase are transferred to molecular oxygen to form water and oxydized fatty acid (unsaturated). NADH is the donor of two electrons which keep the desaturase in the reduced form through the action of the enzyme NADH-cytochrome b5 reductase. These electrons are un-coupled from mitochondrial oxidative-phosphorylation and, therefore, do not yield ATP.

Since these enzymes cannot introduce sites of unsaturation beyond C9 they cannot synthesize either linoleate (18:2 9, 12) or linolenate (18:3 9, 12, 15). These fatty acids must be acquired from the diet and are, therefore, referred to as essential fatty acids. Linoleic is especially important in that it required for the synthesis of arachidonic acid. As we shall encounter later, arachindonate is a precursor for the eicosanoids (the prostaglandins and thromboxanes). It is this role of fatty acids in eicosanoid synthesis that leads to poor growth, wound healing and dermatitis in persons on fat free diets. Also, linoleic acid is a constituent of epidermal cell sphingolipids that function as the skins water permeability barrier.

LIPID METABOLISM

Fats (lipids) are an important source of energy for the body. The body's store of fat is constantly broken down and reassembled to balance the body's energy needs with the food available. Groups of specific enzymes help the body break down and process fats. Certain abnormalities in these enzymes can lead to the buildup of specific fatty substances that normally would have been broken down by the enzymes. Over time, accumulations of these substances can be harmful to many organs of the body. Disorders caused by the accumulation of lipids are called lipidoses. Other enzyme abnormalities prevent the body from converting fats into energy normally. These abnormalities are called fatty acid oxidation disorders.

GAUCHER'S DISEASE

Gaucher's disease is caused by a buildup of glucocerebrosides in tissues. Children who have the infantile form usually die within a year, but children and adults who develop the disease later in life may survive for many years.

In Gaucher's disease, glucocerebrosides, which are a product of fat metabolism, accumulate in tissues. Gaucher's disease is the most common lipidosis. The disease is most common among Ashkenazi (Eastern European) Jews. Gaucher's disease leads to an enlarged liver and spleen and a brownish pigmentation of the skin. Accumulations of glucocerebrosides in the eyes cause yellow spots called pingueculae to appear. Accumulations in the bone marrow can cause pain and destroy bone.

RARE HEREDITARY DISORDERS OF LIPID METABOLISM

Wolman's disease results when specific types of cholesterol and glycerides accumulate in tissues. This disease causes enlargement of the spleen and liver. Calcium deposits in the adrenal glands cause them to harden, and fatty diarrhea (steatorrhea) also occurs. Infants with Wolman's disease usually die by 6 months of age.

Cerebrotendinous xanthomatosis occurs when cholestanol, a product of cholesterol metabolism, accumulates in tissues. This disease eventually leads to uncoordinated movements, dementia, cataracts, and fatty growths (xanthomas) on tendons. The disabling symptoms often appear after age 30. If started early, the drug chenodiol helps prevent progression of the disease, but it cannot undo any damage already done.

In sitosterolemia, fats from fruits and vegetables accumulate in blood and tissues. The buildup of fats leads to atherosclerosis, abnormal red blood cells, and xanthomas on tendons. Treatment consists of reducing the intake of foods that are rich in plant fats, such as vegetable oils, and takingcholestyramine resin. In Refsum's disease, phytanic acid, which is a product of fat metabolism, accumulates in tissues. A buildup of phytanic acid leads to nerve and retinal damage, spastic movements, and changes in the bone and skin. Treatment involves avoiding eating green fruits and vegetables that contain chlorophyll. Plasmapheresis, in which phytanic acid is removed from the blood, may be helpful.

Type 1, the chronic form of Gaucher's disease, is the most common. It results in an enlarged liver and spleen and bone abnormalities. Most commonly diagnosed during adulthood, type 1 Gaucher's disease may lead to severe liver disease, including increased risk of bleeding from the stomach and esophagus and liver cancer. Nervous system problems can also occur.

Type 2, the infantile form, usually causes death in the first year of life. Affected infants have an enlarged spleen and severe nervous system problems.

Type 3, the juvenile form, can begin at any time during childhood. Children with type 3 disease have an enlarged liver and spleen, bone abnormalities, and slowly progressive nervous system problems. Children who survive to adolescence may live for many years.

Many people with Gaucher's disease can be treated with enzyme replacement therapy, in which enzymes are given by vein, usually every 2 weeks. Enzyme replacement therapy is most effective for people who do not have nervous system complications.

TAY-SACHS DISEASE

Tay-Sachs disease is caused by a buildup of gangliosides in the tissues. This disease results in early death.

In Tay-Sachs disease, gangliosides, which are products of fat metabolism, accumulate in tissues. The disease is most common among families of Eastern European Jewish origin. At a very early age, children with this disease become progressively intellectually disabled and appear to have floppy muscle tone.

Spasticity develops and is followed by paralysis, dementia, and blindness. These children usually die by age 3 or 4. The disease cannot be treated or cured. Before conception, parents can find out whether they carry the gene that causes the disease. During pregnancy, Tay-Sachs disease can be identified in the fetus by chorionic villus sampling or amniocentesis.

NIEMANN-PICK DISEASE

Niemann-Pick disease is caused by a buildup of sphingomyelin or cholesterol in the tissues. This disease causes many neurologic problems.

In Niemann-Pick disease, the deficiency of a specific enzyme results in the accumulation of sphingomyelin (a product of fat metabolism) or cholesterol. Niemann-Pick disease has several forms, depending on the severity of the enzyme deficiency, which determines how much sphingomyelin or cholesterol accumulates. The most severe forms tend to occur in Jewish people. The milder forms occur in all ethnic groups.

In the most severe form (type A), children fail to grow normally and have several neurologic problems. These children usually die by age 3. Children with type B disease develop fatty growths in the skin, areas of dark pigmentation, and an enlarged liver, spleen, and lymph nodes. They may be intellectually disabled. Children with type C disease develop symptoms during childhood, with seizures and neurologic deterioration. Some forms of Niemann-Pick disease can be diagnosed in the fetus by chorionic villus sampling or amniocentesis. After birth, the diagnosis can be made by a liver biopsy (removal of a tissue specimen for examination under a microscope). None of the types of Niemann-Pick disease can be cured, and children tend to die of infection or progressive dysfunction of the central nervous system. Currently, some therapies that may slow or halt the progression of symptoms in types B and C are being studied.

FABRY'S DISEASE

Fabry's disease is caused by a buildup of glycolipid in tissues. This disease causes skin growths, pain in the extremities, poor vision, recurrent episodes of fever, and kidney or heart failure.

In Fabry's disease, glycolipid, which is a product of fat metabolism, accumulates in tissues. Because the defective gene for this rare disorder is carried on the X chromosome, the full-blown disease occurs only in males (Genetics: X-Linked Inheritance).

The accumulation of glycolipid causes noncancerous skin growths (angiokeratomas) to form on the lower part of the trunk. The corneas become cloudy, resulting in poor vision. A burning pain may develop in the arms and legs, and children may have episodes of fever. Children with Fabry's disease eventually develop kidney failure and heart disease, although most often, they live into adulthood.

Kidney failure may lead to high blood pressure, which may result in stroke. Fabry's disease can be diagnosed in the fetus by chorionic villus sampling or amniocentesis. The disease cannot be cured or even treated directly, but researchers are investigating a treatment in which the deficient enzyme is replaced by transfusion. Treatment consists of taking analgesics to help relieve pain and fever or anticonvulsants. People with kidney failure may need a kidney transplant.

FATTY ACID OXIDATION DISORDERS

Fatty acid oxidation disorders are caused by a lack or deficiency of the enzymes needed to break down fats, resulting in delayed mental and physical development. Several enzymes help break down fats so that they may be turned into energy. An inherited defect or deficiency of one of these enzymes leaves the body short of energy and allows breakdown products, such as acyl-CoA, to accumulate. The enzyme most commonly deficient is medium chain acyl-CoA dehydrogenase (MCAD). Other enzyme deficiencies include short chain acyl-CoA-dehydrogenase deficiency (SCAD), long chain-3-hydroxyacyl-CoA-deficiency (LCHAD), and trifunctional protein deficiency (TFP).

MCAD Deficiency: This disorder is one of the most common inherited disorders of metabolism, particularly among people of Northern European descent.

Symptoms usually develop between birth and age 3. Children are most likely to develop symptoms if they go without food for a period of time (which depletes other sources of energy) or have an increased need for calories because of exercise or illness. The level of sugar in the blood drops significantly, causing confusion or coma. Children become weak and may have vomiting or seizures. Over the long term, children have delayed mental and physical development, an enlarged liver, heart muscle weakness, and an irregular heartbeat. Sudden death may occur.

BODY LIPID COMPOSITION AND DIETARY LIPID REQUIREMENTS

Detailed information is still lacking on the dietary lipid requirements of many species of fish, but there is an abundance of information on the fatty acid composition of fish oils. Information on the lipid composition of fish can be used to make some guesses about dietary lipid requirements.

Linolenic acid (18:3ω3) resulted in some sparing action and growth promotion in rats, and fatty acids of the ω6 EFA prevented all of the EFA-deficiency symptoms. Research with homeothermic land-dwelling animals showed that the ω6 series of fatty acids are the "essential fatty acids", while the ω3 series are considered to be non-essential or only have a partial sparing action on EFA-deficiency. The ω6 series of fatty acids have been shown to be essential to enough species that it began to become accepted that these are the essential fatty acids for all animals. It was assumed by many that fish also required ω6 fatty acids. Many researchers began by supplementing fish diets with vegetable oils, such as corn, peanut, or sunflower oil, which were rich in linoleic acid. The main sympton observed during the development of EFA deficiency in chinook salmon fed fat-free diets was a marked depigmentation that can be prevented by addition of 1 per cent trilinolein, but not by 0.1 per cent linolenic acid.

Although the ω6 fatty acids are considered to be essential, one of the general characteristics of fish oils is the low levels of ω6 series fatty acids and the higher levels of ω3 type fatty acids. There is evidence that polyunsaturated fatty acids (PUPA) of the ω3 series, which are present in relatively large concentrations in fish oil, play the role of essential fatty acid for fish.

POLYUNSATURATED FATTY ACID REQUIREMENTS OF FISH

When a test diet containing 13 per cent corn oil and 2 per cent cod-liver oil was fed to rainbow trout, subsequent deletion of the cod-liver oil from the diet depressed growth and produced some kidney degeneration which might be attributed to a lack of sufficient ω3 PUFA present in significant quantities in cod-liver oil (McLaren et al., 1947).

Dietary fish oil is superior to corn oil in promoting growth of rainbow trout (Salmo gairdneri) and the yellow-tail (Seriola guingueradiata). Dietary linolenic acid or ethyl linolenate (18:3ω3) gives a positive growth response for rainbow trout which may be attributed to a dietary requirement for ω3 fatty acids.

ESSENTIAL FATTY ACID REQUIREMENTS OF FISH

One of the most widely accepted theories explaining the presence of such high levels of 20:5ω3 and 22:6ω3 fatty acids in fish oils is related to the effect of unsaturation on the melting point of a lipid.

The greater degree of unsaturation of fatty acids in the fish phospholipids allows for flexibility of cell membrane at lower temperatures. The ω3 structure allows a greater degree of unsaturation than the ω6 or ω9. This theory is consistent with the fact that cold water fish have a greater nutritional requirement for ω3 fatty acids, while the EFA requirement of some warm water fish can be satisfied by a mixture of ω6 plus ω3.

Rainbow Trout

Rainbow trout, a cold water fish, requires ω3 fatty acids as EFA in the diet. The EFA requirement can be met by 1 per cent 18:3ω3 in the diet. Inclusion of 18:2ω6 in the diet may result in some improvement in growth and feed conversion compared to EFA deficient diets; however, the ω6 fatty acids will not prevent some EFA deficiency symptoms such as the "shock syndrome". Although it is clear that rainbow trout require ω3 fatty acids, it remains to be shown conclusively whether some dietary level of ω6 fatty acid is essential. In all the above studies with rainbow trout, dietary 18:2ω6 or 18:3ω3 were readily converted to C-20 and C-22 PUFA of the same series, and 18:3ω3 or 22:6ω3 had similar EFA value for rainbow trout.

Either 20:5ω3 or 22:6ω3 is superior to 18:3ω3 in an EFA value for rainbow trout, and the former two fatty acids in combination are superior to either alone. This is consistent with data for mammals, where 20:4ω6 has higher EFA value than 18:2ω6. The superior nutritional value of C-20 and C-22 carbon ω3-PUFA is further supported by the excellent growth promoting effects of dietary fish oils such as pollock liver oil and salmon oil for rainbow trout.

Channel Catfish

One of the most important warm water fish in North America is the channel catfish (Ictalurus punctatus). The quantitative EFA requirement of the catfish has not yet been determined. However, the evidence is convincing that the ω3 requirement is not as great as that of rainbow trout. Analysis of fatty acids of lipids from catfish purchased at five processing plants showed very low levels of 20:4ω6, 20:5ω3, and 22:6ω3; 0.8 - 5.5, 0.2 - 1.3, and 0.6 - 6.1 per cent of the total fatty acids, respectively. It was shown that corn oil added to a semipurified casein based diet initially resulted in a positive growth response and protein sparing, but later growth inhibition was observed. The apparent repressive effects of corn oil may be due to its 18:2ω6 content since 20:5ω3 and 22:6ω3 present in menhaden oil had no apparent detrimental effects. The growth suppressing effects of 18:2ω6 were also noted when 3 per cent corn oil was added to 3 per cent beef tallow and 3 per cent menhaden oil. The growth suppression caused by unsaturated fatty acids does not appear to be limited to ω6 fatty acids. Linseed oil (high in 18:3ω3) in the diet of catfish resulted in growth suppression similar to that caused by corn oil compared with dietary beef tallow, olive oil and menhaden oil.

The Common Carp

The picture for another warm water fish, the common carp (Cyprinus carpio) is much clearer than that for the channel catfish. This fish has an EFA requirement for both ω3 and ω6 fatty acids. The best weight gains and feed conversions are obtained in fish receiving a diet containing both 1 per cent 18:2ω6 and 1 per cent 18:3ω3. With the carp, 20:5ω3 and 22:6ω3 at 0.5 per cent of the diet are superior to 1 per cent of 18:3u3. Carp fed a fat-free, or EFA deficient, diet incorporated high levels of 20:3 9 in their lipids, especially in the phospholipids.

The Eel

The eel (Anguilla japonica), another warm water fish, has a requirement for both ω3 and ω6 fatty acids. Corn oil (high in ω6) and cod liver oil (high in ω3)in a 2:1 mixture are most favourable for the growth of eels. The eel requires ω6 and ω3 in the same proportion as the carp, but at a lower level in the diet; namely, 0.5 per cent of each, rather than 1.0 per cent of each PUFA.

The Plaice

The plaice becomes depleted of both ω3 and ω6 PUFA when fed a fat-free diet. The addition of 12:0 and 14:0 to the diet result in the synthesis of saturated and monoenoic fatty acids of chain lengths up to C18; however, increased levels of 20:3ω9 noted in trout and mammals have not been reported in plaice. Plaice fed dietary 18:2ω6 and 18:3ω3 will not produce significant amounts of 20:4 ω6, 20:5ω3, or 22:6ω3.

The Turbot

The growth of turbot (Scophthalmus matimus) is much better with ω3 PUFA than with ω6 or saturated fat (hydrogenated coconut oil) in the diet. The turbot also appears to be unable to convert dietary 18:2ω6 to 20:4ω6 when fed corn oil, or to convert endogenous 18:1ω9 to 20:3ω9 when fed the EFA deficient diet. Although it appears to have an EFA requirement for ω3 fatty acids such as are present in cod liver oil, this requirement is not satisfied by 18:3ω3. The chain elongation and desaturation of 18:lω9, 18:2ω6, or 18:3ω3 has been found to be very limited (3-15 per cent) in turbot compared to the rainbow trout where 70 per cent of the 18:3ω3 was converted to 22:6ω3. The required level of long-chain ω3 fatty acids for turbot is at least 0.8 per cent of the diet.

The Red Sea Bream

The red sea bream (Chrysophyrys major) grows better when the dietary lipid is of marine origin (pollock residual oil) rather than a vegetable oil (such as corn oil). The EFA requirement of the red sea bream is not satisfied by either linoleic acid of corn oil or supplemented linolenate. A mixture of 20:5ω3 and 22:6ω3 supplemented to the corn oil diet has been shown to be effective in improving growth and condition of these fish. Thus, even in warm water, marine fish seem to require not just ω3 fatty acids but 0)3 fatty acids of 20 to 22 carbon-chain length. A direct correlation between feed efficiency and the 18:1 level in the lipids of the red sea bream has been postulated.

Other Species

Among warm water marine fish, mullet and fundulus possess the ability to chain, elongate, and desaturate 18:2ω6 or 18:3ω3 PUFAs. The process is, however, inhibited in fundulus by high levels (about 5 per cent) of these PUFAs of 18:2ω6 or 18:3ω3 in the diet.

FATTY ACID METABOLISM IN FISH

It appears that high levels of 18-carbon ω6 or ω3 fatty acids inhibit the synthesis and metabolism of 18:lω9. It is interesting to note that the channel catfish, which also exhibits negative growth response to dietary 18:2ω6 or 18:3ω3, incorporates very high levels of 18:1 into its body lipids. The inclusion of either 18:2ω6 or 18:3ω3 in the diet reduces the levels of 18:1 fatty acids in body lipids. A similar reduction has also been observed in red sea bream liver phospholipid when either of the PUFAs is added to the diet.

The competitive inhibition of chain elongation and desaturation of members from one series of fatty acids for members of another series is well established, with ω3 > 6 > 9 being the usual order of potency for inhibition.

The pathways of fatty acid metabolism have been reviewed by Mead and Kayama (1967). Fish are able to synthesize, de novo from acetate, the even-chain, saturated fatty acids, as shown in Figure. Radio tracer studies have shown that fish can convert 16:0 to the ω7 monoene and 18:0 to the ω9 monoene. The ω5, 11 and ω3 monoenes are proposed based on the identification of these isomers in the monoenes of herring oil.

Fish are unable to synthesize any fatty acids of the w6 and u3 series unless a precursor with this structure is present in the diet. Fish are able to desaturate and elongate fatty acids of the w9, 6, or w3 series as outlined in Fig. There is competitive inhibition of the elongation desaturation of fatty acids of one series by members of the other series. The w3 fatty acids are the most potent inhibitors, the w9 are the least.

The ability to elongate and desaturate fatty acids is not the same in all species of fish, as was noted earlier. The turbot was able to desaturate and elongate only 3-15 per cent of 18:1w9, 18:2w6, or 18:3w3, when given the C14 labelled fatty acid; in the rainbow trout, 70 per cent of the label from 18:3w3 (C14) was found in 22:6w3. The essential fatty acids are not unique in their ability to supply energy. The -oxidation of fatty acids in fish is basically the same as in mammals. The EFA and saturated and monoenoic fatty acids are all equally utilized by fish for energy production.

Increased swelling rates of liver mitochondria occur in rainbow trout fed diets deficient in ω3 fatty acids. It is possible that EFA plays an important role in the permeability as well as the plasticity of membranes. The role of ω3 fatty acids in membrane permeability may be one of the factors accounting for differences in content of this family of fatty acids between freshwater and marine fish.

Fish mitochondria with high levels of the ω3 PUFA and very low levels of ω6 fatty acids are very similar to mammalian mitochondria with respect to cytochrome content, -oxidation of fatty acids, operation of the tricarboxylic acid cycle, electron transport, and oxidative phosphorylation. The ω3 PUFA may play the same role in fish that the ω6 fatty acids play in rats. The EFA play another role in the mitochondria. In addition to their importance in membrane structure, the EFA are important in some enzyme systems.

Unsaturated fatty acids play an important role in the transportation of other lipids. It has been repeatedly shown that feeding PUFA will lower the cholesterol levels in animals with above-normal blood lipid and cholesterol levels. Fish oils are more effective in lowering cholesterol levels than are most dietary lipids. The major portion of the fatty acids absorbed across the intestinal mucosa are transported as protein-lipid complexes stabilized by phospholipids. The low body temperature in fish probably results in a greater importance for unsaturation in transport of lipids than in homeothermic animals.

NEGATIVE ASPECTS OF LIPIDS IN FISH NUTRITION

The requirement by fish for PUFA of the ω3 series creates problems with respect to feed storage. These types of fatty acids are very labile on oxidation. The products of lipid oxidation may react with other nutrients such as proteins, vitamins, etc., and reducing the available dietary levels or the oxidation products may be toxic. The effect of oxidized lipids on dietary proteins, enzymes and amino acids have been demonstrated.

The use of oxidized menhaden oil in the diets of swine and rats caused decreased appetite, reduced growth, yellowish-brown pigmentation of depot fat, and decreased haemoglobin and haematocrit levels.

The negative effects of the oxidized fish oils were reversed by the addition of alpha-tocopherol acetate or ethoxyoquin to the diet.

Much of the use of vegetable oils in fish diets in the 1950s and 1960s might, in part, have been based on their greater stability in prepared diets.

It has been demonstrated that rancid herring and hake meals in fish feeds caused dark colouration, anaemia, lethargy, brown-yellow pigmented liver, abnormal kidneys, and small gill clubbing in chinook salmon.

The symptons can be alleviated by addition of alpha-tocopherol to the diets containing rancid fish meals. The addition of vitamin E would prevent the toxic or negative effects of adding 5 per cent highly oxidized salmon oil to the diet of rainbow trout. This same sparing effect of alpha-tocopherol can also apply to rancid carp feed. The positive nutritional value of ω3 fatty acids in fish lipids for fish feeds can become a negative factor if adequate care is not taken in the preparation and storage of feeds. Only fresh oils with low peroxide values should be used in feeds. Fish feed ingredients such as fish meals should be protected against oxidation. The level of vitamin E added to the diet should be increased as the PUFA level is increased. The finished feed, if possible, should be stored in air tight containers at reduced temperatures with minimum exposure to UV radiation and other factors accelerating the rate of lipid oxidation. The problems of rancidity or antioxidation of lipids in fish feeds should not be ignored.

LIPOPROTEINS

NTESTINAL UPTAKE OF LIPIDS

In order for the body to make use of dietary lipids, they must first be absorbed from the small intestine. Since these molecules are oils, they are essentially insoluble in the aqueous environment of the intestine. The solubilization (or emulsification) of dietary lipids is therefore accomplished by means of bile salts, which are synthesized from cholesterol in the liver and then stored in the gallbladder; they are secreted following the ingestion of fat.

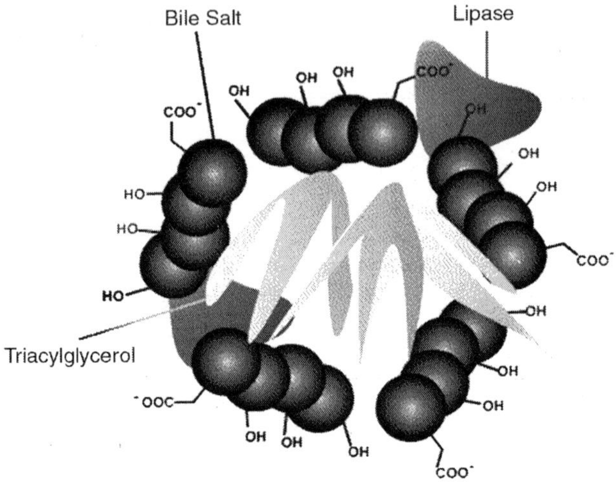

Mixed micelle formed by bile salts, triacylglycerols and pancreatic lipase.

The emulsification of dietary fats renders them accessible to pancreatic lipases (primarily lipase and phospholipase A2). These enzymes, secreted into the intestine from the pancreas, generate free fatty acids and a mixtures of mono- and diacylglycerols from dietary triacylglycerols. Pancreatic lipase degrades triacylglycerols at the 1 and 3 positions sequentially to generate 1,2-diacylglycerols and 2-acylglycerols. Phospholipids are degraded at the 2 position by pancreatic phospholipase A2 releasing a free fatty acid and the lysophospholipid. The products of pancreatic lipases then diffuse into the intestinal epithelial cells, where the re-synthesis of triacyglycerols occurs.

Dietary triacylglycerols and cholesterol, as well as triacylglycerols and cholesterol synthesized by the liver, are solubilized in lipid-protein complexes. These complexes contain triacylglycerol lipid droplets and cholesteryl esters surrounded by the polar phospholipids and proteins identified as apolipoproteins. These lipid-protein complexes vary in their content of lipid and protein.

INTERMEDIATE DENSITY LIPOPROTEINS, IDLS

IDLs are formed as triacylglycerols are removed from VLDLs. The fate of IDLs is either conversion to LDLs or direct uptake by the liver.

Conversion of IDLs to LDLs occurs as more triacylglycerols are removed. The liver takes up IDLs after they have interacted with the LDL receptor to form a complex, which is endocytosed by the cell. For LDL receptors in the liver to recognize IDLs requires the presence of both apo-B-100 and apo-E (the LDL receptor is also called the apo-B-100/apo-E receptor). The importance of apo-E in cholesterol uptake by LDL receptors has been demonstrated in transgenic mice lacking functional apo-E genes. These mice develop severe atherosclerotic lesions at 10 weeks of age.

LOW DENSITY LIPOPROTEINS, LDLS

The cellular requirement for cholesterol as a membrane component is satisfied in one of two way: either it is synthesized de novo within the cell, or it is supplied from extra-cellular sources, namely, chylomicrons and LDLs. As indicated above, the dietary cholesterol that goes into chylomicrons is supplied to the liver by the interaction of chylomicron remnants with the remnant receptor. In addition, cholesterol synthesized by the liver can be transported to extra-hepatic tissues if packaged in VLDLs. In the circulation VLDLs are converted to LDLs through the action of lipoprotein lipase. LDLs are the primary plasma carriers of cholesterol for delivery to all tissues.

The exclusive apolipoprotein of LDLs is apo-B-100. LDLs are taken up by cells via LDL receptor-mediated endocytosis, as described above for IDL uptake. The uptake of LDLs occurs predominantly in liver (75/ per cent), adrenals and adipose tissue. As with IDLs, the interaction of LDLs with LDL receptors requires the presence of apo-B-100. The endocytosed membrane vesicles (endosomes) fuse with lysosomes, in which the apoproteins are degraded and the cholesterol esters are hydrolyzed to yield free cholesterol. The cholesterol is then incorporated into the plasma membranes as necessary. Excess intracellular cholesterol is re-esterified by acyl-CoA-cholesterol acyltransferase (ACAT), for intracellular storage. The activity of ACAT is enhanced by the presence of intracellular cholesterol.

Insulin and tri-iodothyronine (T3) increase the binding of LDLs to liver cells, whereas glucocorticoids (e.g., dexamethasone) have the opposite effect. The precise mechanism for these effects is unclear but may be mediated through the regulation of apo-B degradation. The effects of insulin and T3 on hepatic LDL binding may explain the hypercholesterolemia and increased risk of athersclerosis that have been shown to be associated with uncontrolled diabetes or hypothyroidism.

An abnormal form of LDL, identified as lipoprotein-X (Lp-X), predominates in the circulation of patients suffering from lecithin-cholesterol acyl transferase (LCAT, see HDL discussion for LCAT function) deficiency or cholestatic liver disease. In both cases there is an elevation in the level of circulating free cholesterol and phospholipids.

HIGH DENSITY LIPOPROTEINS, HDLS

HDLs are synthesized de novo in the liver and small intestine, as primarily protein-rich disc-shaped particles. These newly formed HDLs are nearly devoid of any cholesterol and cholesteryl esters. The primary apoproteins of HDLs are apo-A-I, apo-C-I, apo-C-II and apo-E. In fact, a major function of HDLs is to act as circulating stores of apo-C-I, apo-C-II and apo-E.

HDLs are converted into spherical lipoprotein particles through the accumulation of cholesteryl esters. This accumulation converts nascent HDLs to HDL2 and HDL3. Any free cholesterol present in chylomicron remnants and VLDL remnants (IDLs) can be esterified through the action of the HDL-associated enzyme, lecithin-cholesterol acyl transferase, LCAT. LCAT is synthesized in the liver and so named because it transfers a fatty acid from the C-2 position of lecithin to the C-3-OH of cholesterol, generating a cholesteryl ester and lysolecithin. The activity of LCAT requires interaction with apo-A-I, which is found on the surface of HDLs.

Cholesterol-rich HDLs return to the liver, where they are endocytosed. Hepatic uptake of HDLs, or reverse cholesterol transport, may be mediated through an HDL-specific apo-A-I receptor or through lipid-lipid interactions. Macrophages also take up HDLs through apo-A-I receptor interaction. HDLs can then acquire cholesterol and apo-E from the macrophages; cholesterol-enriched HDLs are then secreted from the macrophages. The added apo-E in these HDLs leads to an increase in their uptake and catabolism by the liver.

HDLs also acquire cholesterol by extracting it from cell surface membranes. This process has the effect of lowering the level of intracellular cholesterol, since the cholesterol stored within cells as cholesteryl esters will be mobilized to replace the cholesterol removed from the plasma membrane.

The cholesterol esters of HDLs can also be transferred to VLDLs and LDLs through the action of the HDL-associated enzyme, cholesterol ester transfer protein (CETP, also identified as apo-D). This has the added effect of allowing the excess cellular cholesterol to be returned to the liver through the LDL-receptor pathway as well as the HDL-receptor pathway.

LDL RECEPTORS

LDLs are the principal plasma carriers of cholesterol delivering cholesterol from the liver (via hepatic synthesis of VLDLs) to peripheral tissues, primarily the adrenals and adipose tissue. LDLs also return cholesterol to the liver. The cellular uptake of cholesterol from LDLs occurs following the interaction of LDLs with the LDL receptor (also called the apo-B-100/apo-E receptor). The sole apoprotein present in LDLs is apo-B-100, which is required for interaction with the LDL receptor.

The LDL receptor is a polypeptide of 839 amino acids that spans the plasma membrane. An extracellular domain is responsible for apo-B-100/apo-E binding. The intracellular domain is responsible for the clustering of LDL receptors into regions of the plasma membrane termed coated pits. Once LDL binds the receptor, the complexes are rapidly internalized (endocytosed). ATP-dependent proton pumps lower the pH in the endosomes, which results in dissociation of the LDL from the receptor. The portion of the endosomal membranes harboring the receptor are then recycled to the plasma membrane and the LDL-containing endosomes fuse with lysosomes. Acid hydrolases of the lysosomes degrade the apoproteins and release free fatty acids and cholesterol. As indicated above, the free cholesterol is either incorporated into plasma membranes or esterified (by ACAT) and stored within the cell.

The level of intracellular cholesterol is regulated through cholesterol-induced suppression of LDL receptor synthesis and cholesterol-induced inhibition of cholesterol synthesis. The increased level of intracellular cholesterol that results from LDL uptake has the additional effect of activating ACAT, thereby allowing the storage of excess cholesterol within cells. However, the effect of cholesterol-induced suppression of LDL receptor synthesis is a decrease in the rate at which LDLs and IDLs are removed from the serum. This can lead to excess circulating levels of cholesterol and cholesteryl esters when the dietary intake of fat and cholesterol exceeds the needs of the body. The excess cholesterol tends to be deposited in the skin, tendons and (more gravely) within the arteries, leading to atherosclerosis.

CLINICAL SIGNIFICANCES OF LIPOPROTEIN METABOLISM

Fortunately, few individuals carry the inherited defects in lipoprotein metabolism that lead to hyper- or hypolipoproteinemias. Persons suffering from diabetes mellitus, hypothyroidism and kidney disease often exhibit abnormal lipoprotein metabolism as a result of secondary effects of their disorders. For example, because lipoprotein lipase (LPL) synthesis is regulated by insulin, LPL deficiencies leading to Type I hyperlipoproteinemia may occur as a secondary outcome of diabetes mellitus. Additionally, insulin and thyroid hormones positively affect hepatic LDL-receptor interactions; therefore, the hypercholesterolemia and increased risk of athersclerosis associated with uncontrolled diabetes or hypothyroidism is likely due to decreased hepatic LDL uptake and metabolism.

Of the many disorders of lipoprotein metabolism, familial hypercholesterolemia (FH) may be the most prevalent in the general population. Heterozygosity at the FH locus occurs in 1:500 individuals, whereas, homozygosity is observed in 1:1,000,000 individuals. FH is an inherited disorder comprising four different classes of mutation in the LDL receptor gene.

The class 1 defect (the most common) results in a complete loss of receptor synthesis. The class 2 defect results in the synthesis of a receptor protein that is not properly processed in the Golgi apparatus and therefore is not transported to the plasma membrane. The class 3 defect results in an LDL receptor that is incapable of binding LDLs. The class 4 defect results in receptors that bind LDLs but do not cluster in coated pits and are, therefore, not internalized. FH sufferers may be either heterozygous or homologous for a particular mutation in the receptor gene. Homozygotes exhibit grossly elevated serum cholesterol (primarily in LDLs). The elevated levels of LDLs result in their phagocytosis by macrophages. These lipid-laden phagocytic cells tend to deposit within the skin and tendons, leading to xanthomas. A greater complication results from cholesterol deposition within the arteries, leading to atherosclerosis, the major contributing factor of nearly all cardiovascular diseases.

PHARMACOLOGIC INTERVENTION

Drug treatment to lower plasma lipoproteins and/or cholesterol is primarily aimed at reducing the risk of athersclerosis and subsequent coronary artery disease that exists in patients with elevated circulating lipids. Drug therapy usually is considered as an option only if non-pharmacologic interventions (altered diet and exercise) have failed to lower plasma lipids.

- Mevinolin, Mevastatin, Lovastatin: These drugs are fungal HMG-CoA reductase inhibitors. The net result of treatment is an increased cellular uptake of LDLs, since the intracellular synthesis of cholesterol is inhibited and cells are therefore dependent on extracellular sources of cholesterol. However, since mevalonate (the product of the HMG-CoA reductase reaction) is required for the synthesis of other important isoprenoid compounds besides cholesterol, long-term treatments carry some risk of toxicity.
- Nicotinic acid: Nicotinic acid reduces the plasma levels of both VLDLs and LDLs by inhibiting hepatic VLDL secretion, as well as suppressing the flux of FFA release from adipose tissue by inhibiting lipolysis. Because of its ability to cause large reductions in circulating levels of cholesterol, nicotinic acid is used to treat Type II, III, IV and V hyperlipoproteinemias.
- Clofibrate, Gemfibrozil, Fenofibrate: These compounds are derivatives of fibric acid and promote rapid VLDL turnover by activating lipoprotein lipase. They also induce the diversion of hepatic free fatty acids from esterification reactions to those of oxidation, thereby decreasing the liver's secretion of triacylglycerol- and cholesterol-rich VLDLs.
- Probucol: Probucol increases the rate of LDL metabolism and may block the intestinal transport of cholesterol. The net result is a significant reduction in plasma cholesterol levels.
- Cholestyramine or colestipol (resins): These compounds are nonabsorbable resins that bind bile acids which are then not reabsorbed by the liver but excreted. The drop in hepatic reabsorption of bile acids releases a feedback inhibitory mechanism that had been inhibiting bile acid synthesis. As a result, a greater

amount of cholesterol is converted to bile acids to maintain a steady level in circulation. Additionally, the synthesis of LDL receptors increases to allow increased cholesterol uptake for bile acid synthesis, and the overall effect is a reduction in plasma cholesterol. (This treatment is ineffective in homozygous FH patients, since they are completely deficient in LDL receptor

MEMBRANE LIPIDS AND BILAYER MEMBRANES

The simpler lipids are used primarily for energy storage by living organisms, but more complex lipids are used to form biological membranes. The most important and abundant class of biological membrane lipids is the class of phospholipids. Most phospholipids are derived from glycerol and so they are called phosphoglycerides. The amino alcohol sphingosine gives rise to sphingomyelin, the only significant membrane phospholipid which is not a phosphoglyceride.

In phosphoglycerides, the hydroxy groups at C-1 and C-2 of glycerol are esterified to fatty acids while the hydroxy group at C-3 of glycerol is esterified to phosphoric acid; this forms diacylglycerol-3-phosphates. The two fatty acids found in a diacylglycerol-3-phosphate molecule need not be the same. Diacylglycerol-3-phosphates are only a minor constituent of membranes. The major constituents of membranes are derivatives of diacylglycerol-3-phosphates in which the esterified phosphate also forms an ester linkage with an alcohol. The alcohols commonly found in phosphoglycerides include glycerol, ethanolamine, and choline.

The glycolipids are lipids which contain sugars. Glycolipids are derived from sphingosine, as is the phospholipid sphingomyelin. Glycolipids contain no esterified phosphate.

Both phospholipids and glycolipids are amphipathic molecules, that is, molecules in which one end is hydrophobic and one end is hydrophilic. The fatty acid chains form the hydrophobic end of the molecule while the polar glycerol-phosphate-alcohol or sphingosine-sugar portion forms the hydrophilic end of the molecule.

BILAYER MEMBRANES

When amphipathic molecules are placed in aqueous solution, their hydrophobic tails attempt to orient themselves toward each other. The two ways in which they do so are to form small spherical micelles or to form a planar lipid bilayer. A micelle is a small structure, generally less than two micrometer in diameter, while a lipid bilayer typically has a thickness of about 0.5 micrometer and can have an area of several square millimeters. Sonication or other treatment of lipids can produce spherical liposomes, or lipid vesicles, which contain aqueous solution within an enclosing lipid bilayer. Liposomes are considerably larger than micelles.

Both liposomes and planar bilayer membranes can be prepared from simple solutions of phospholipids such as the diacylglycerol-3-phosphatidyl cholines. Studies of such artificially prepared membranes have given us much insight into the functions and operation of membranes in living cells. Simple bilayer membranes are highly permeable to water molecules, while ions such as sodium ion and potassium ion can traverse them only more slowly (by nine orders of magnitude!). For small molecules, the rates of permeation through simple bilayer membranes increase with their solubility in nonpolar solvents relative to their solubility in water. These observations strongly suggest that transfer of substances through a membrane requires desolvation, transfer through the anhydrous membrane interior, and then resolvation on the other side of the membrane.

The differences in transport behaviour between biological membranes and simple bilayer membranes are now believed to be due to the incorporation of protein molecules or other specifically transporting molecules directly into or onto the surface of a bilayer cell membrane. Considerable research in biology and in chemistry is being directed toward the transport properties of membranes.

LIPIDS AND FATTY ACIDS

Lipids are the generic names assigned to a group of fat soluble compounds found in the tissues of plants and animals,: and are broadly classified as:

- Fats,
- Phospholipids,
- Sphingomyelins,
- Waxes,
- Sterols.

Fats are the fatty acid esters of glycerol and are the primary energy depots of animals. These are used for long-term energy requirements during periods of extensive exercise or during periods of inadequate food and energy intake. Fish have the unique capability of metabolizing these compounds readily and, as a result, can exist for long periods of time under conditions of food deprivation. A typical example is the many weeks of migration by salmon in their return upstream to spawn; stored lipid deposits are burned for fuel to enable body processes to continue during the strenuous journey.

Phospholipids are the esters of fatty acids and phosphatidic acid. These are the main constituent lipids of cellular membranes allowing the membrane surfaces to be hydrophobic or hydrophylic depending on the orientation of the lipid compounds into the intra or extracellular spaces. Sphingomyelins are the fatty acid esters of sphingosine and are present in brain and nerve tissue compounds. Waxes are fatty acid esters of long-chain alcohols. These compounds can be metabolized for energy and to impart physical and chemical characteristics through the stored lipids of some plant and several animal compounds. Sterols are polycyclic, long-chain alcohols and function as components of several hormone systems, especially in sexual maturation and sex-related physiological functions.

Fatty acids can exist as straight chain or branch chain components; many of the fish fats contain numerous unsaturated double bonds in the fatty acid structures. A short bond designation for. fatty acids will be used throughout where the number identifies the position of the first double bond counting from the methyl end. Linolenic acid would be written 18:3ω3.

The first number identifies the number of carbons; the second number, the number of double bonds; and the last number, the position of the double bonds. Many reviews of fish nutrition have been published which contain information on lipid requirements. Most work on lipid requirements of fish has been with salmonids. Rainbow trout have an essential fatty acid (EFA) requirement for the linolenic of ω3[1] series rather than for linolenic or ω6 as required by most mammals. The main emphasis on lipid requirements has been on EFA and on the energy value of lipids.

8

Biomolecules

INTRODUCTION

This BioCoach activity will help you review the characteristics of the different classes of biomolecules: lipids, carbohydrates, proteins, and nucleic acids.

BUILDING BIOMOLECULES

This BioCoach activity will help you review molecular formulas and models, major elements and their valences, and the important functional groups of biological molecules. It will also let you try your hand at building certain molecular models.

GENE TO PROTEIN: TRANSCRIPTION

This BioCoach activity will help you review transcription, the synthesis of RNA from DNA. You will find information about the organization of genes, the structure of RNA, the transcription process, and the differences between mRNA in prokaryotes and eukaryotes. Animations show the transcription process in a dynamic way. You can test your understanding of transcription by using the Self-Quiz at the end of the module. By continuing on to the activity on translation, you can then learn about how mRNA is translated into protein. For the Transcription and Translation activities, it may be helpful to review DNA structure and DNA replication.

GENE TO PROTEINS

This BioCoach activity will help you review translation, the synthesis of protein from mRNA. You will find information about the structure of proteins, the genetic code, the molecular components of translation, and the process of assembling a protein chain as a ribosome moves along an mRNA molecule. Animations show the translation process in a dynamic way. You can test your understanding of translation by using the Self-Quiz at the end of the module. It may be helpful to review other topics, such as DNA structure, DNA replication, transcription, and biomolecules.

LEWIS STRUCTURES OF MOLECULES

An atom of an element has a characteristic number of electrons, equal to its atomic number Z. These electrons are arranged in shells about the nucleus, some being close to the nucleus, others far away. It is the electrons in the outermost shell, furthest from the nucleus, and therefore least tightly held, that are involved in chemical bonding. These are the valence electrons. The electrons close to the nucleus, called inner or core electrons, are much too tightly held to be lost to or shared with another atom. As we have seen, the number of valence electrons that an atom has is equal to the last digit of its group number. Thus Na, in group 1, has one valence electron; Al in group 13 has three valence electrons; Br in group 17 has seven valence electrons; and Fe in group 8 has eight valence electrons. The Lewis symbol for an element shows the valence electrons as dots arranged about the symbol for the element, for Na, Al, and Br. Dots are placed as if the symbol of the element were surrounded by an invisible square.

The first dot can be placed on any of the 4 sides, but the second, third, and fourth dots are then placed singly on the remaining 3 sides of the square; electrons are not placed in pairs (i.e., 2 dots on the same side of the square) until 4 have been placed singly. Dots that singly occupy the side of a square are referred to as unpaired electrons. G.N. Lewis developed this approach to representing the valence electrons of an element before the quantum view of the atom was complete.

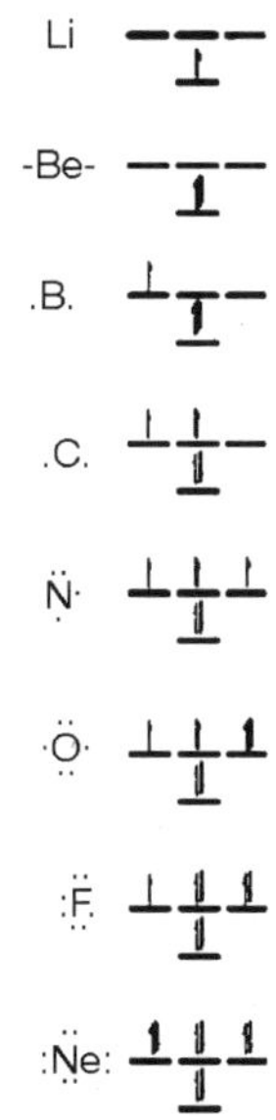

Fig. Valence Electrons for Period2

It is of interest from our perspective, though, to examine the correspondence between the Lewis symbol for an element and its valence electron configuration. In particular, we might expect the number of unpaired electrons in the Lewis symbol to correspond with the number of electrons that singly occupy valence orbitals of the atom. The Lewis symbols and valence orbital occupations of the atoms of period 2.

What we see is interesting and a bit puzzling. First, most of the atoms show the intuitively expected correspondence. Thus Li, N, O, F, and Ne exhibit the same number of unpaired electrons in the Lewis symbol and in the quantum orbital occupation diagram.

Further, the numbers of unpaired electrons for these elements correspond with their usual valences; i.e., with the number of bonds that they form.

However, for Be, B, and C in groups 2, 13, and 14, the numbers of unpaired electrons in the Lewis symbols do not correspond with the orbital occupation diagrams. Thus the Lewis symbol for carbon leads us to expect 4 unpaired electrons; the orbital occupation diagram predicts 2. The observed valence for carbon is (almost) invariably 4, in accord with the Lewis symbol. Similarly, the valences of Be and B are as expected from the Lewis symbols, but in conflict with the orbital occupation diagrams.

Yet the success of the quantum theory of the atom has been so profound that we believe completely that the orbital occupation diagrams are correct. How are we to explain the bonding properties of elements in groups 2, 13, and 14 in terms of the orbital occupation diagrams?

The orbital diagrams can be brought into correspondence with the observed valences by a simple expedient: the downspin electron of the pair in the s orbital is placed in an empty p orbital, with its spin parallel to those of the other electrons. When this is done for Be, B, and C, the numbers of unpaired electrons correspond with the predictions of the Lewis symbol, and with the observed valences. Thus the elements of groups 2, 13, and 14 behave as if their orbital occupation.

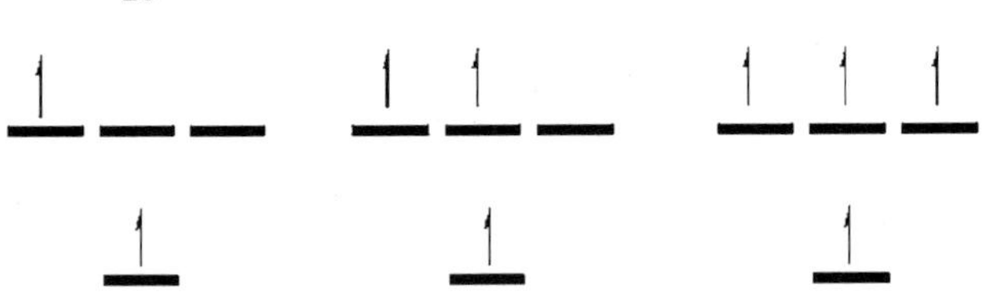

Fig. Apparent Configurations for Be, B, and C

The elements in group 18, called the Noble Gases, have 8 valence electrons, arranged in pairs on the 4 sides of the Lewis symbol square. The one exception is helium, He, which has 2 valence electrons. As their name indicates, these elements occur in nature as very unreactive gases; this implies that 8 valence electrons, called an octet, is in some sense a stable situation. As we saw that, 8 electrons is sufficient to fill the s and p orbitals of a shell. When the atoms of the elements undergo chemical reactions, the frequent result is a filled set of valence s and p orbitals for each atom involved — i.e., each atom acquires the octet of valence electrons possessed by the noble gases. The sodium atom usually loses its single valence electron upon reaction, forming Na^+.

The sodium ion has the same number of electrons as the noble gas, Ne. The Al atom must lose 3 valence electrons to achieve a noble gas configuration, so it forms the Al^{3+} ion. In contrast, Br may much more readily gain a single electron to achieve the same number of electrons as Kr than lose 7 electrons. It therefore is found frequently in compounds as the Br- ion.

Elements with from 3 to 7 valence electrons often achieve an octet of electrons—a full set of valence s and p orbitals—by sharing, rather than gaining or losing, electrons. For example, the Br atom can achieve an octet if it can gain 1 electron by sharing. A second Br atom can supply this electron, and achieve the octet for itself in the process. Shared electrons contribute to the octets of both of the two sharing atoms.

The shared pair of electrons, located between the bromine atoms, is circled. Note that the single unpaired electron of one Br atom is paired with the single unpaired electron of the second Br atom. This pair is placed between them to indicate the sharing.

The shared electron pair constitutes the bond between the two atoms, because it is attracted to both nuclei and therefore holds them together. The bond is called a covalent bond, because it involves cooperation between the 2 atoms so that each may achieve an octet. It is customary and convenient to represent electron pairs with lines rather than dots,. Keep in mind that each line represents 2 electrons an electron pair. Frequently, the number of covalent bonds that an atom forms is the same as the number of unpaired electrons in its Lewis symbol (we will see subsequently that many atoms can be represented by more than one Lewis symbol, giving them some flexibility in the number of covalent bonds formed). Therefore Br, and the other halogens, commonly form one covalent bond. Oxygen has two unpaired electrons and can achieve an octet by sharing both of these with another oxygen atom.

The two oxygen atoms then share a total of four electrons, each thereby achieving the octet. Two pairs of electrons shared between two atoms gives two bonds; the pair of bonds together is called a double bond, and is represented by a double line as in the structure for O_2. You should satisfy yourself that two nitrogen atoms can achieve an octet by sharing three pairs of electrons to form a triple bond. Carbon, with 4 unpaired electrons, forms 4 covalent bonds. Similarly, sulfur forms 2, phosphorus 3, silicon 4, hydrogen 1, and so on, based on the dot. Hydrogen is an exception to the octet rule, because it can accommodate only 2 valence electrons. In acquiring 2, it achieves the electron configuration of the noble gas, helium.

Lewis Structures for Molecules

The realization that when elements react to form compounds the atoms frequently achieve stable electron arrangements octets was first made by G.N. Lewis, a great American chemist. In tribute to the contributions made by Lewis to the theory of chemical bonding, molecular structures in which valence electrons are represented by dots or lines are called Lewis structures. The Lewis approach to structure, called Lewis theory, is our most-used theory of molecular structure. It is simple, yet sophisticated enough to provide an understanding of molecular properties such as shape, bond length, bond angle, and bond strength.

Further, it allows explanation of chemical reactivity and mechanism in terms of structure for many chemical reactions, and suggests a general approach to acids and bases that is used widely. In due course, we will address all of these matters. In this section, we develop a systematic approach for obtaining the Lewis structures of molecules. Some Lewis structures are easily obtained from the dot pictures of the constituent atoms. Dot pictures for carbon and fluorine. Clearly carbon is able to form four bonds, and each fluorine is able to form one. Therefore, one C atom bonds to four F atoms. The Lewis structure for CF_4, tetrafluoromethane. Similarly, the Lewis structures for ammonia and water are simply obtained. In these molecules, each atom forms a number of covalent bonds equal to the number of unpaired electrons in its dot picture, and each atom has an octet (a duet in the case of hydrogen, which gives it the configuration of He, with 2 valence electrons) in the final structure.

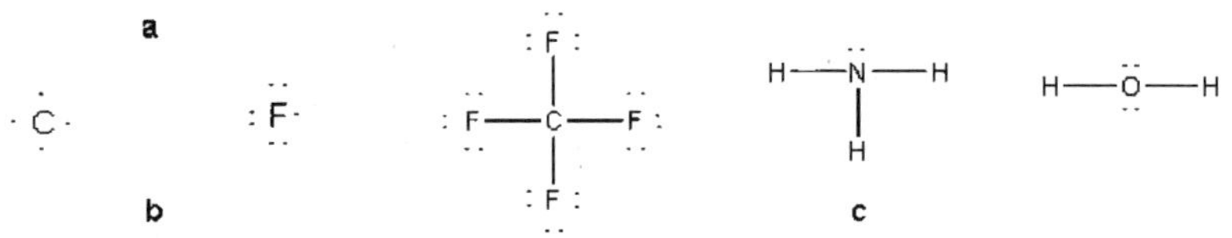

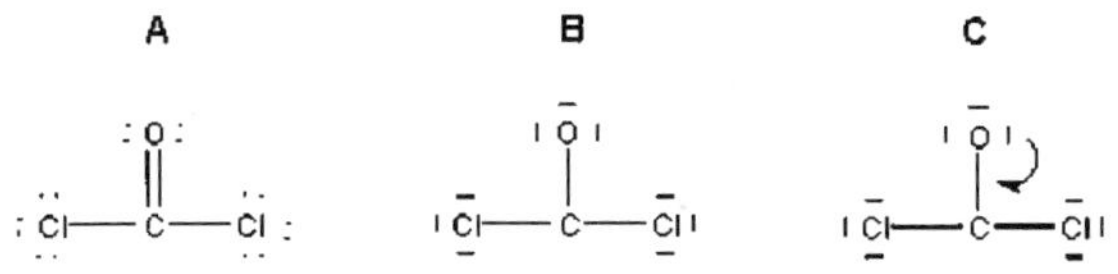

The Lewis structure is not always so readily obtained, however. For example, that for phosgene, $COCl_2$. Although inspection of this structure verifies that each atom has an octet and has formed the expected number of bonds, arriving at this structure from scratch requires some thought.

Note the presence of a double bond between the carbon and oxygen atoms. How are we to know when double bonds and triple bonds, collectively called multiple bonds, are appropriate? To address this question, we now present a systematic and straightforward approach to the writing of Lewis structures.

- *Step:* Arrange the atoms in the correct layout (i.e., bonding arrangement). The layout is experimentally determined, but can often be deduced from the formula for the molecule (or ion). For many molecules (but not all!), the first listed atom is in the centre, and the others are attached to it. This is the case for phosgene.

- *Step:* Count the number of valence electrons in the molecule by summing the electrons contributed by each atom, based on their group numbers. If the molecule is negatively charged, add the corresponding number of electrons. Subtract the appropriate number of electrons if the molecule is positively charged.
  For phosgene,
  Number of valence electrons = 6(O) + 4(C) + 2 × 7(Cl) = 24.
- *Step:* Develop a trial structure for the molecule by putting a single bond (1 electron pair, represented by a single line) between each pair of bonded atoms. Then complete the octets of as many atoms as possible by adding nonbonding electron pairs (often called lone pairs) to the atoms in turn until all of the available valence electrons (step 2) have been used.
  Situations will arise in which electrons remain to be added to the structure after the octets of all atoms are complete. In these situations, the remaining electrons are placed in pairs on the central atom.
- *Step:* Examine the trial structure. If all atoms have octets (or if the terminal atoms have octets while the central atom has more than 8 electrons), the trial structure is acceptable. If the trial structure shows atoms with fewer than eight electrons, it must be altered by using lone pairs to form multiple bonds until all atoms have octets.

The phosgene structure must clearly be amended because the carbon atom has only six electrons in the trial structure. A lone pair from the oxygen atom is moved in to form a second bond, preserving the octet on oxygen and at the same time establishing the octet for carbon.

The result is an acceptable Lewis structure for phosgene. It is important to state at this point that an acceptable structure obtained by direct application of the procedure above is not necessarily the optimal (best) structure.

A B C

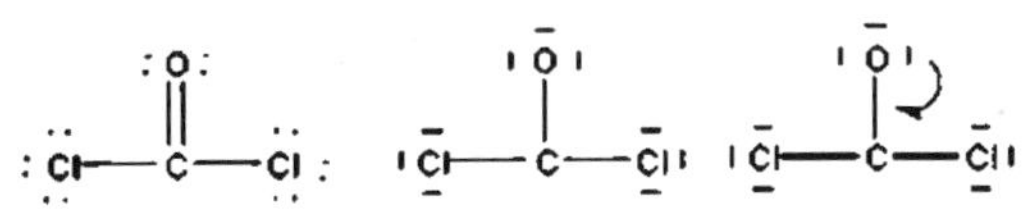

We emphasize that once the correct number of electrons is placed in the structure in step 3, no more may be added, and none may be removed; electrons can, however, be moved. (A lone pair of oxygen is preferred over one from chlorine for reasons dealing with the concept of formal charge.)

Exceeding the Octet

Although we have used the octet rule as the basis for development of a systematic approach to Lewis structures, only the atoms of period 2 of the periodic table obey it (almost) without exception. Any atom having valence shell principal quantum number ^{3} 3 may exceed the octet (violate the octet rule).

When the octet of electrons is exceeded, it is usually at the central atom. An atom can exceed the octet only if it has atomic orbitals available for housing the additional electrons. As we have stated, eight electrons is sufficient to fill the s and p subshells of the valence shell (outermost main shell) of the atom.

To accommodate more than eight electrons in the valence shell, an atom must have available orbitals in addition to the s and p. Atoms for which the valence shell has n ^{3} 3 have a valence d subshell comprised of five d-type atomic orbitals. When an atom exceeds the octet, it makes use of d orbitals to house the additional electrons.

The Se atom in SeF_4 uses a total of 5 valence orbitals (one s, three p, and one d) to accommodate the four bonding pairs and one lone pair of electrons that surround it. The sulfur atom in SF_6 uses 6 valence orbitals (one s, three p, and two d) to accommodate the six bond pairs.

And the Te atom in TeF_8^{2-} uses 8 valence orbitals (one s, three p, and four of the five d) to house eight electron pairs! Note that we would not expect an atom with valence shell n ^{3} 3 to be able to accommodate more than 9 electron pairs in its valence shell because it has only this number of orbitals available in the s, p, and d subshells. Similarly, elements from period 2 can not house more than 4 electron pairs, because the n = 2 shell has no d subshell. The atoms in period 2 are strictly limited by the octet rule.

Many atoms can be represented by more than one Lewis symbol. This is very closely connected with the availability of d orbitals in the valence shell, and we will now examine this matter before moving on. The sulfur atom is a good vehicle for illustration, because it shows a substantial amount of flexibility in the number of covalent bonds that it is able to form. The usual Lewis symbol for S, and the orbital occupation diagram with which it corresponds.

This Lewis symbol allows us to rationalize the many 2-valent compounds that are known for sulfur, among them H_2S, SCl_2, and $(CH_3)_2S$. Formation of two covalent bonds is readily rationalized in terms of the two unpaired electrons in the Lewis symbol (and orbital occupation diagram). Thus in H_2S, each H atom pairs its single unpaired electron with one of the unpaired electrons of S, resulting in completed s and p subshells for S, and completed s subshells for H.

The other two-valent molecules are understood in similar ways. However, sulfur also forms a number of four-valent compounds (for example, SF_4); and six-valent compounds (for example, SF_6), which cannot be readily accounted for in terms of the simple Lewis symbol above. However, useful Lewis symbols for S can be obtained by unpairing electrons, one pair at a time. The resulting Lewis symbols then show four and six unpaired electrons. Using these expanded Lewis symbols, we readily account for the flexivalent nature of the sulfur atom, and in fact for similar flexibility for most atoms with valence shell n ³ 3. The following generalization will prove useful to us in rationalizing many molecular structures:

A p-block atom usually forms compounds with a number of covalent bonds lying in the range between the number of unpaired electrons in its simplest Lewis symbol, and the total number of valence electrons that it has. Further, other common numbers of covalent bonds lie between these extremes, separated by increments of 2.

Applying this to S, we would predict that it should exhibit a minimum of 2 covalent bonds and a maximum of 6, with compounds containing 4 covalent bonds being common. Applying the guideline to chlorine enables us to predict a minimum of 1, a maximum of 7, with 3 and 5 covalent bonds being common. In fact, the compounds HCl (minimum), ClF_7 (maximum), ClF_3, and ClF_5 are well known.

A Formal Charge Refinement—Optimizing the Lewis Structure

We can combine our formal charge criteria for acceptable Lewis structures with the ability of many (central) atoms to exceed the octet to obtain improved structures for molecules in which the central atom is from period n > 3. By way of illustration, consider the structure of sulfuric acid. This structure results from application of the systematic approach to development of Lewis structures presented earlier. Although the sulfur and oxygen atoms obey the octet rule, the formal charges are a bit extreme.

Specifically, S has 2+ formal charge, and the two terminal oxygen atoms not bonded to hydrogen have formal charges of 1-. By using lone pairs on these two oxygens to form double bonds to S, the formal charges of all three atoms can be reduced to zero. This causes the sulfur atom to exceed the octet, but this is acceptable for an atom from period 3. The resonance form called the minimum formal charge structure, results. It is this structure that is usually drawn for sulfuric acid.

The minimum formal charge structure for any molecule can be obtained in a straightforward manner. Starting with the structure generated using the systematic procedure, assign formal charges to all atoms. If the central atom is not from period 2, introduce double bonds between terminal atoms of negative formal charge and the central atom until the central atom formal charge is zero.

One double bond removes one unit of positive formal charge from the central atom, so to achieve zero formal charge requires a number of double bonds equal to the positive formal charge in the initial structure. Initial and minimum formal charge structures for H_4SiO_4, H_3PO_4, and $HClO_4$ are shown.

BIOMOLECULE BIOGRAPHIES

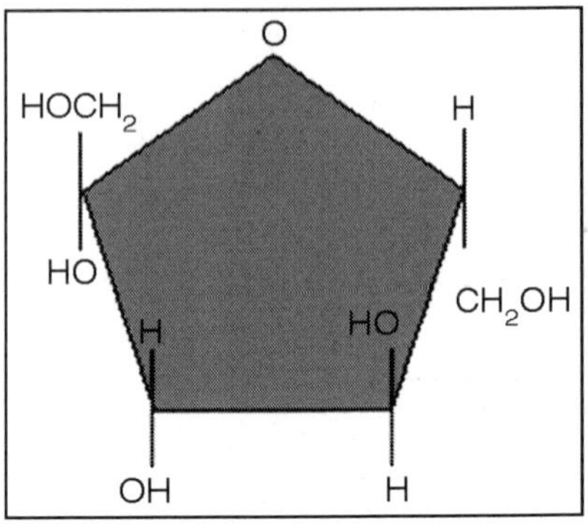

Fig. Fructose, a Monosacharide The Pentagon Structure Represents Five Carbon Atoms.

CARBOHYDRATES

The main function of carbohydrates in animals is to provide an energy source. A few carbohydrates serve as structural units such as deoxyribose. This sugar can be found in DNA. When the energy is not being used by the animal, the carbohydrate is stored as glycogen. Carbohydrates are composed of carbon, oxygen, and hydrogen. In most carbohydrates, there is one carbon atom for each water molecule. Carbohydrates are categorized into three groups based on molecule size.

MONOSACCHARIDES

Monosaccharides are composed of three to seven carbons. Prefixes indicate the number of carbon atoms contained in the molecule. For example, a pentose would consist of five carbon atoms. One of the main energy suppliers to the body happens to be a monosaccharide called glucose, a hexose.

DISACCHARIDES

A disaccharide is the result when two monosaccharides combine by dehydration synthesis. One water molecule also results as a byproduct. For example, the monosaccharides fructose and glucose combine in the chemical equation $C_6H_{12}O_6$(glucose) + $C_6H_{12}O_6$(fructose) $\rightarrow$ $C_{12}H_{22}O_{11}$(sucrose) + H_2O(water) to form sucrose and water. You may notice that glucose and fructose have the same chemical composition; however they differ in the relative positions of their carbon and oxygen atoms. In order to obtain monosaccharides from a disaccharide, all you need to do is add water. This is a process called hydrolysis.

POLYSACCHARIDES

Polysaccharides consist of not only three or four monosaccharides, but tens and hundreds in long chains. These monosaccharides were also joined by dehydration synthesis. The main polysaccharide that is in the human body is glycogen. Glycogen is stored in the liver and skeletal muscles and can be broken down into monosaccharides for use as energy. In order to achieve monosaccharide status, polysaccharides must also go through hydrolysis.

LIPIDS

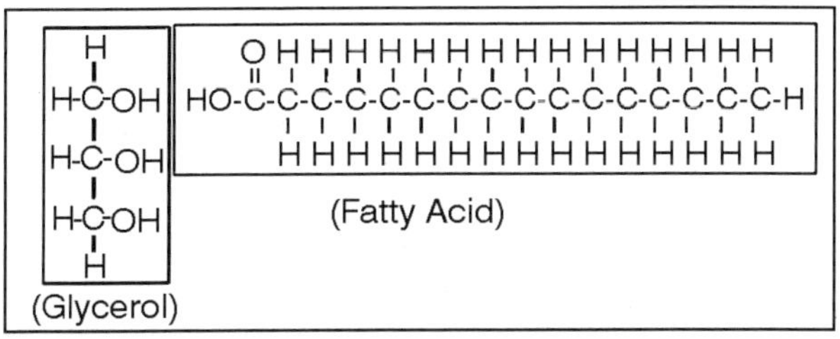

Lipids are very chemically diverse compunds which all share one common trait -- hydrophobia. They possess little or no affinity to water because of their molecular structure - mostly hydrocarbon regions with nonpolar bonds. There are three major categories of lipids:

FATS

large molecules composed of the two smaller molecules glycerol and fatty acids. When these two combine, the product is a triacylglycerol (or tryglyceride). The major function of fat is to supply a reserve of energy which is stored in adipose cells.

PHOSPHOLIPIDS

Similar to fats, but contain two fatty acids and a phosphate group which attaches to the glycerol molecule. The phosphate group is hydrophilic while the hydrocarbons remain hydrophobic. This structure allows phospholipids to act as cell membranes inside the body.

STEROIDS

Consist of four interconnected carbon rings. Cholesterol, an important steroid, is also a part of cell membranes.

BIOMOLECULES: THE CARBOHYDRATES

Carbohydrates are organic molecules with the general formula of CHO in 1:2:1 ratio. Although carbohydrates constitute only 1 to 2 Per cent of cell mass, they provide the raw fuel for cellular energy production. Carbohydrates are classified according to molecular size and solubility. In general, the smaller molecules are more soluble than the larger ones.

Monosaccharides include glucose, fructose, galactose, deoxyribose, and ribose.A single unit of sugar, mono-saccharides are the smallest carbohydrates.Glucose, a 6-carbon sugar (hexose) is the sugar in our blood. Fructose, the sugar that sweetens fruit, and galactose, the sugar found in milk, have the same chemical formula as glucose and are therefore isomers of glucose. Isomers have the same composition but a different arrangement of their atoms and have different properties. Glucose can have a straight chain of carbon atoms or, more commonly, form a ring structure.Two other 5-carbon sugars or monosaccharides (called pentose) used in nucleic acid synthesis are deoxyribose and ribose.

Disaccharides include sucrose, lactose, and maltose.Two monosaccharides are joined together by dehydration synthesis to form a disaccharide molecule.Sucrose (table sugar)Glucose + fructose = sucrose + waterLactose (milk sugar)Glucose + galactose = lactose + waterMaltose (malt sugar)Glucose + glucose = maltose + water.

Polysaccharides include starch, cellulose, and glycogen.These long, chain-like polymers make ideal storage products due to their insolubility.Starch is the storage molecule synthesized from glucose by plants.Cellulose, which is also synthesized by plants for cell wall construction, is indigestible because we lack enzymes for it. Cellulose provides Fibre to promote peristalsis.Glycogen is the carbohydrate storage molecule found in muscle and liver cells. When blood sugar levels drop, liver cells hydrolyze glycogen and release glucose to the blood.

CHARGE DISTRIBUTION IN MOLECULES ELECTRONEGATIVITY

The electronegativity of an atom was defined by Pauling as "the ability of the atom in a molecule to attract electrons to itself." Unlike the molar volume and ionization energy, the electronegativity is not a directly measurable physical property of an atom.

Rather, it is defined in a somewhat vague and qualitative way. Several ways of assigning a numerical value to the electronegativity of an atom have been developed. The most commonly used scale of electronegativity is that of Pauling, based on the experimentally measured values of bond energies. We will not concern ourselves with the details of how the scale was established. We will instead present the scale, make some general statements about the variation of electronegativity with position in the periodic table, and then proceed to use the electronegativity where appropriate. Suffice it to say at this point that the electronegativity of an atom, symbolized c, is roughly proportional to the first ionization energy of the atom. In other words, the more difficult it is to remove the outermost electron from an atom, the more likely it is that the atom will attract additional electrons to itself.

The Pauling electronegativities of the elements are presented. The following trends can be extracted:

- C increases left-to-right across a period. This is just what we expect given the parallel between c and I_1. However, there are no "jogs" in the trend, as there are for I_1.

- C decreases steadily down a family, again paralleling the trend in I_1.
- There is a rough equality of the electronegativities of an element and the element one column to the right and one row below it. Thus the electronegativities of Cl and O are 3.2 and 3.4, respectively. Those of N and S are also quite similar. This results because the increase in c due to the rightward move and the decrease in c due to the downward move roughly cancel.

Electronegativity is the primary concept on which we base our understanding of electron distributions among atoms in a molecule.

Bond Dipoles and Dipole Moment

The distribution of electrons in a molecule determines its shape. One frequent consequence of shape is that the centers of positive and negative charge in the molecule are displaced from one another, so that the molecule has a positive end and a negative end. Such molecules are said to possess total molecular dipole moments. It is important to be able to assess whether or not a molecule is polar; that is, whether it has a total molecular dipole moment. In this section, we will take some time to discuss the origins of total molecular dipole moments so that you will be equipped to do this assessment.

An arrangement in which a positive charge q+ and an equal negative charge q- are separated by a fixed distance, r, is called an electric dipole. The electric dipole is a vector quantity that is by convention taken to point from the positive to the negative end of the dipole. The magnitude (length) of the vector is the product of the charge, q+, and the distance between the positive and negative centers:

$$m_d = qr$$

The electric dipole has units of charge times distance. For a dipole consisting of a charge of 1 electronic charge unit separated from an equal and opposite charge by a distance of 1×10^{-10} m, the dipole moment has the value 1.6×10^{-29} C-m.

The total molecular dipole moment of a molecule is in general the vector sum of various local electric dipoles that exist in the molecule. These local electric dipoles are of two types that we will call bond dipoles and lone pair dipoles. Local bond dipoles are generally larger in magnitude, hence more important, than lone pair dipoles.

They are thus the major contributors to the total molecular dipole moment, and we will discuss them first. A bond between two atoms, X and Y, in which the bond pair is unequally shared is called a polar bond. A polar bond gives rise to a local bond dipole. Unequal sharing occurs when X and Y have different electronegativities; the shared electron pair is on average closer to the more electronegative atom. Consequently, the local bond dipole points along the bond, from the less electronegative atom (positive end) to the more electronegative atom (negative end). Assessment of the polarity of a bond is simple: if the two atoms involved in the bond are different either in identity or environment, the bond is polar.

Only bonds between identical atoms in identical environments are non-polar. A heteronuclear diatomic molecule such as HCl provides an example of a local bond dipole. Heteronuclear means that the two atoms of the molecule are different. Chlorine is more electronegative than hydrogen, so the bond electrons between them are closer to Cl than to H. This makes the Cl atom slightly negative and the hydrogen atom slightly positive. The H-Cl bond is polar, with the electrons biased toward Cl. This polarity gives rise to a local bond dipole that points along the bond from H (positive) to Cl (negative).

The second type of local electric dipole is the lone pair dipole, which exists at any atom having one or more lone electron pairs. A local lone pair dipole points, as expected, from the nucleus of the atom (positive) to the lone pair (negative). Although strictly speaking there is a lone pair dipole generated by each lone pair in a molecule, the major contribution to the total molecular dipole is made by lone pairs on the central atom. Consequently, in examining a molecule for local lone pair dipoles, we will consider only lone pairs on the central atom, ignoring those on terminal atoms.

In addition to its local bond dipole, HCl contains three local lone pair dipoles, one associated with each of the lone pairs on the chlorine atom. The total molecular dipole moment of a molecule is the vector sum of all local electric dipoles (bond dipoles and central atom lone pair dipoles). The total molecular dipole moment of HCl is 0.36×10^{-29} C-m. This is due primarily to the bond dipole; the lone pair dipoles make a minor contribution, as is generally true. Assessment of the total molecular dipole of HCl is quite simple, because there are only two atoms. For larger molecules, we will use the following systematic approach.

- First, examine each bond in the molecule for polarity. For each polar bond, draw a vector along the bond from the positive atom to the negative atom to represent the local bond dipole.
- Combine the local bond dipoles using vector addition methods to obtain the net bond dipole vector. The net bond dipole will usually be the major contributor to the total molecular dipole moment.
- Next, look for local lone pair dipoles on the central atom. Draw a vector from the atom symbol to each lone pair on the central atom. These are the local lone pair dipole vectors.
- Combine the central atom local lone pair dipoles using vector addition methods to obtain the net lone pair dipole vector.
- Combine the net bond dipole vectorially with the net lone pair dipole. The result is the total molecular dipole moment vector.

To illustrate the system, we consider the water molecule, which according to spectroscopy (and VSEPR) is bent, with an angle of 104°. Oxygen is more electronegative than hydrogen, so the O-H bonds are polar with the electrons closer to the oxygen than to the hydrogen atom. Each O-H bond in the water molecule therefore has a local bond dipole that points from H to O. The bond dipoles are along with their vector addition to produce the net bond dipole.

Because the water molecule is bent, the local bond dipoles add together to produce a non-zero net bond dipole that points from the centre of the (imaginary) line connecting the hydrogen atoms (positive) to the oxygen atom (negative). Next, we find two lone pairs on the central oxygen atom. Each lone pair generates a lone pair dipole pointing from the oxygen nucleus to the lone pair.

The result is a non-zero net lone pair dipole that points from the oxygen atom (positive) to the centre of the (imaginary) line connecting the lone pairs (negative). The net lone pair dipole and the net bond dipole point in the same direction. is the vector addition of the net bond and lone pair dipoles to produce the total molecular dipole moment of water.

The magnitude of the total molecular dipole moment is 0.615×10^{-29} C-m. This substantial dipole is a major factor in the ability of water to dissolve a variety of substances, notably including ionic compounds. Although bonds between different atoms are polar to at least some extent, as we saw above, many highly symmetrical molecules have no net dipole moments due to exact cancellation of the bond and lone pair dipoles. An example is the CO_2 molecule, which is linear. Each C-O bond is polar. Each therefore has a local bond dipole pointing directly along the bond from carbon (less electronegative) to oxygen (more electronegative).

However, the two equal local bond dipoles point in opposite directions and cancel exactly. The central carbon atom has no lone pairs, so no lone pair dipoles need be considered. Thus the molecule as a whole has coincident centers of + and–charge. Dipole moments contribute to the attractive forces that molecules exert on one another. To understand the magnitudes of these forces, it is essential that we be able to determine whether or not a molecule possesses a total molecular dipole moment.

Example. Which of the following molecules have (total molecular) dipole moments? SO_3, O_3, CCl_4, NH_3, PF_5.

Solution. Molecular polarity is assessed from molecular shape, which is in turn assessed from the Lewis structure. The method developed earlier leads to the following Lewis structures and shapes for these species. Lone pairs on the terminal atoms are not shown.

SO_3. Each S-O bond is polar, with a local vector bond dipole pointing from S (less electronegative) to O (more electronegative. All S-O bonds are the same when resonance is taken into account. The three local bond dipoles thus point to the corners of an equilateral triangle. When added vectorially, the result is zero. There are no lone pairs on the central S atom, so local lone pair dipoles need not be considered. There is no net molecular dipole; SO_3 is non-polar.

O_3. Although the bonds in this molecule connect pairs of oxygen atoms, the two oxygen atoms of a bond are not in identical environments. Considering the bond between oxygens 1 and 2, we see that oxygen 1 is bonded only to oxygen 2, whereas oxygen 2 is bonded to oxygen 1, but in addition to oxygen 3. Oxygens 1 and 2 are in different environments, so the bond between them is polar. The difference between oxygens 1 and 2 is evident from their formal charges, which are 0 and 1+ in the resonance form shown.

The formal charges also enable us to assign the direction of the local bond dipole, which points from oxygen 2 to oxygen 1.

The molecule is bent, so the two identical local bond dipoles do not add to zero. They cancel horizontally, but not vertically. The local lone pair dipole pointing from oxygen atom 2 to its lone pair only partially cancels the vertical net bond dipole (recall our statement that in general bond dipoles are more important than lone pair dipoles). Ozone has a net molecular dipole running from the centre oxygen through the centre of the molecular angle.

CCl_4. From the Lewis structure for CCl_4, a perfect tetrahedral shape is expected (like methane, CH_4). The local bond dipoles are non-zero, with the negative end toward Cl, but cancel when added vectorially. The central carbon atom has no lone pairs.

There is no molecular dipole moment. The cancellation of local bond dipoles may not be obvious here because of the three-dimensionality of the figure. However, the following argument may be used. Molecules having total molecular dipole moments are inherently less symmetric than molecules with zero molecular dipole moments.

For example, if you were to orient the ammonia molecule with the nitrogen lone pair pointing down rather than up, it is easy to tell that the molecule has been moved. In contrast, if you were to orient carbon tetrachloride with any of the three other chlorine atoms at the top, the molecule would look exactly the same as it does above; you would not be able to tell that it had been moved! Thus it has no dipole moment.

NH_3. NH_3 has a trigonal pyramidal shape. The bonds are polar, with the negative end of the local bond dipole toward nitrogen. These will combine to give a non-zero net bond dipole colinear with an axis passing through the nitrogen atom and the centre of the equilateral triangle defined by the H atoms.

This is reinforced by the nitrogen local lone pair dipole. The molecule therefore has a molecular dipole moment with the nitrogen end of the dipole negative. PF_5. PF_5 is trigonal bipyramidal. The bonds are very polar because F is the most electronegative element. However, the two axial local bond dipoles oppose and cancel, and cancellation occurs in the equatorial plane as well, exactly as in SO_3 above. There are no lone pairs on the central atom. There is no molecular dipole moment.

CARBOHYDRATES MONOSACCHARIDES

- (Monosaccharoses) Dioses ($C_2H_4O_2$) - Glycolose.
- Moses ($C_3H_6O_3$). Aldoses - Glycerose. Ketose - Dioxyacetone.
- Tetroses ($C_4H_8O_4$).
- Aldoses - Erythrose,3 Threose.3 Ketose - Erythrulose.2
- Pentoses ($C_6H_{10}O_5$).
- Aldoses - Arabinose,2 Xylose,2 Ribose,2 Lyxose.8 Ketoses
- - Araboketose3, Xyloketose (ketoxylose).2 [Methyl pentoses ($C_6H_{10}O_5$) - Rhamnose,2 Fucose2].

Some writers use the term polysaccharides to include all carbohydrates other than monosaccharides. Mathews applies it to all carbohydrates more complex than the disaccharides.

Names of a few of the most important carbohydrates are printed in small capitals. Separate mention of the d, l, and dl forms of the various sugars is omitted, since in the study of food and nutrition we are practically concerned only with that one of the three forms which is found in or derived from natural products.

- Hexoses ($C_6H_{12}O_6$).
- Aldoses - Glucose,1 Mannose,2 Galactose,2 Gulose,3 Idose,3Talose,3.
- Allose,2 Altrose.3 Ketoses - Fructose,1 Sorbose,2 Tagatose.3.
- Heptoses ($C_7H_{14}O_7$). Aldose - Mannoheptose.1 Ketose - Sedoheptose.1.

Note: No attempt is here made to summarize the occurrence of any but the tetroses, pentoses, hexoses and heptoses. Glycolose and the trioses if formed in nature are probably too reactive to accumulate sufficiently for identification.

Disaccharides (Disaccharoses)

Dihexoses (Hexobioses) - ($C_{12}H_{22}O_{11}$). Anhydride of glucose + fructose-Sucrose. Anhydrides of glucose + galactose - Lactose, Melibiose. Anhydrides of glucose + glucose- Maltose, Isomaltose, Trehalose, Turanose.

Trisaccharides (Trisaccharoses)

Trihezoses ($C_{18}H_{22}O_{16}$). Anhydride of glucose + galactose +fructose - Raffinose. Anhydride of glucose + glucose + glucose - Melezitose. Anhydride of fructose + fructose + fructose - Secalose.

Tetrasaccharides (Tetrasaccharoses)

Tetrahexoses ($C_{24}H_{42}O_{21}$). Anhydrides of 2 galactose + glucose + fructose - Stachyose, Lupeose.

Polysaccharides (Polysaccharoses)

Pentosans (chief constituents of gums and mucilages). Anhydrides of xylose - Xylans. Anhydrides of arabinose - Arabans.

Hexosans

Anhydrides of glucose-Starch, Cellulose, Glycogen, Dextrin (and other"dextrans"). Anhydrides of mannose - Mannans. Anhydrides of galactose-Galactans (pectins). Anhydrides of fructose - Inulin (and other"levulans").

- Occurs free in nature.
- Not yet found free in nature (or only in small amounts) but obtained by hydrolysis or fermentation of natural product.
- Known only (with certainty) as a Labouratory product.

SIMPLE BIOCHEMICAL TESTS

Materials

- Benedict's solution (not"quantitative")
- Iodine solution (I/KI)
- Waterbaths set at about 100°C
- Boiling tubes Racks
- 1 per cent glucose soln
- 1 per cent fructose soln
- % sucrose soln
- other sugars?
- 1 per cent starch sol solution X

HCl, NaOH, litmus/universal indicator paper filter funnels, papers, flasks? measuring cylinders (10, 25, 50, 100 ml) plastic pipettes glass pipettes (1, 5, 10 ml?), pipette pumps

Objectives

Familiarisation with Benedict's test - for reducing sugars modification of Benedict's test for non-reducing sugars iodine test dilution techniques sensitivity of above tests.

Procedure

See also Roberts Students' manual p. 42 You will be expected to write up and hand in a concise account of your techniques and the results. There is no need to use large amounts of test substances or reagents.

SAFETY

Irritant Solutions Wear Safety Glasses And Lab Coats Hot Water And Steam Take Care: Keep Working Area Clear:

- Carry out Benedict's test on the named sugar solutions and starch as provided.

- Use fixed (noted!) quantities of test substance and reagents.
- Do not contaminate stock solutions shared by the group.
- Note results - colour and intensity - separately from conclusions.
  - *Refinement*: Filter the contents of the tube in which you have been testing glucose (using paper previously labelled with pencil). Consider the colour of the filtrate, and accordingly modify the quantities of reagents used. Repeat filtration process. Explain your conclusions in your write-up.
- Dilute the sugar solutions (say ×10, ×100 etc), and carry out Benedict's test on suitable samples (i.e. amounts same as before).
- Explain the dilution technique you used in your write-up.
- What does this tell you about the sensitivity of the test?
- Using sucrose, carry out the test for non reducing sugars. Having obtained a negative result from ordinary benedict's test, a second test is performed after hydrolysing it to reducing sugars (acid hydrolysis, followed by neutralisation), then performing Benedict's test:-

  Add... drops/ml HCl - leave for... minutes at room temperature (... °C) or... mins at 100 °C - add... drops/ml NaOH (same strength) - check neutrality with litmus (colour......) or universal indicator paper (colour.....).
- Test solution X using a combination of the above tests.
- Test starch solution using iodine solution (I/KI). You could use drops in a recessed tile, but for sensitivity purposes it is probably best to use boiling tubes and drops of iodine.
  - Dilute starch × 10, × 100, etc. and repeat (standardised) test.
  - Warm the last tubes - i.e. the least blue coloured (100 °C waterbath)), and observe colour.
  - Cool (the lower end of) the tube under the tap.

9

Photosynthesis

The ultimate source of energy for life on Earth is the sun. Plants are able to transform the light energy from the sun into chemical energy through a process called photosynthesis. Through absorption of light energy, plants, algae, and a few types of bacteria transform carbon dioxide and water into carbohydrates. Carbohydrates serve as fuel for their growth and metabolism. Other organisms benefit indirectly from photosynthesis. Oxygen is a waste product of photosynthesis. By releasing oxygen into the environment, plants make the air breathable for many other life forms.

The presence of oxygen in relation to plants was first demonstrated by the English chemist, Joseph Priestley in 1772. He noted that air that had been depleted through burning candles could be made breathable again by plants. Seven years later, a Dutch doctor, Jan Ingen- Housz, showed that plants required sunlight in order to make air breathable. Ingen-Housz also demonstrated that it was only the green parts of plants that had this ability. However, neither scientist knew anything of oxygen's existence.

In 1782, Swiss scientist Jean Senebier proved that the plants also needed "fixed air"–what is now called carbon dioxide–in addition to sunlight. His discovery was followed in 1804 by another Swiss scientist, Nicholas Théodore de Saussure, proving that water was also a necessary factor. By the early 19th century, scientists had at least a general understanding of photosynthesis. They knew that green plants, if given carbon dioxide, water, and sunlight, would produce organic material and oxygen. Building on this foundation, tists began to tease apart the details of how the plants accomplished this pocess.

A big step forward was the discovery that chloroplasts were the site of oxygen generation in plant cells. Chloroplasts are cell organelles comparable to the mitochondria of animal cells. These organelles contain a cell's energy-generating mechanisms. Discovering the function of chloroplasts prompted scientists to scrutinize the organelle's structure and chemical makeup. Not all cells capable of photosynthesis have chloroplasts, however. Cells are generally divided into two categories prokaryotes and eukaryotes. Eukaryotes have chloroplasts, but prokaryotes do not. The difference between eukaryotes and prokaryotes is that eukaryotes have a nucleus, while prokaryotes do not. The nucleus is a cellular structure that stores genetic material separately from the rest of the cell interior. Plants and certain types of algae are eukaryotes. Blue- green algae and photosynthetic bacteria are prokaryotes. The following discussion of photosynthesis will concentrate on photosynthetic eukaryotes–i.e., plants and algae.

Chloroplasts contain chlorophyll, a green-coloured pigment. Chlorophyll is the key molecule in photosynthesis. There are several types of chlorophyll, but the predominant form is chlorophyll *a*. Chlorophyll molecules are composed of a central porphyrin ring and side chains. These side chains differ slightly between types of chlorophyll, but the general appearance of the molecule is very similar. The porphyrin ring is actually a complex multi-ring structure composed of carbon, hydrogen, nitrogen, and oxygen atoms.

Light is broken down into units called photons which move in waves. The distance between waves is called the wavelength. Visible light contains wavelengths ranging from 350 nm to 800 nm. Longer wavelengths contain less energy; short wavelengths contain more. Wavelengths correspond with the colours of the rainbow–called the visible light spectrum–and each colour represents a different energy level.

The colours that an object appears to be are those that are reflected; colours that are absorbed are not seen. Chlorophyll absorbs energy from the red portion of the spectrum, approximately 680-700 nm, and reflects the green portion. For that reason, the leaves and stems of plants appear green. When a photon is absorbed by a chlorophyll molecule, its energy causes the chlorophyll to enter an energy-rich excited state.

Within a chloroplast, chlorophyll molecules are densely packed together. Because the chlorophyll molecules are so close to each other, the energy they absorb circulates from molecule to molecule. This energy is eventually transmitted to a reaction centre, thought to be a special arrangement of chlorophyll and protein molecules. There are two types of reaction centres in chloroplasts, each of which corresponds with a photosystem. The P700 reaction centre is associated with photosystem I; P680 is linked with photosystem II. The names of the photosystems arise from the wavelength of light which they most efficiently absorb.

Both photosystems serve as trigger points for electron transport chains. Electron transport chains operate through the reduction and oxidation of chemicals or chemical complexes that make up its links. Although reduction usually means that something is lost, it has the opposite meaning in chemistry. As a chemical term, reduction means that an electron is gained. Oxidation means that an electron is lost.

When P680 absorbs the energy of a photon, it goes from a ground state to an excited state. This change is seen at the atomic level. In the ground state, the electrons of an atom are at a low- energy configuration. In an excited state, an electron jumps from a normal low-energy position, or orbital, to a high-energy orbital. The high-energy position is unstable, and the electron is quickly transferred to another chemical, pheophytin. Pheophytin is the first link in the photosystem II electron transport chain.

As P680 returns to its ground state, it picks up an electron to replace the one transferred to pheophytin. This electron is taken from a water molecule. In chemical terms, P680 is reduced and water is oxidized. The water molecule is split into hydrogen ions (positively charged atoms) and oxygen. For every two water molecules broken down in this manner, four hydrogen ions and one oxygen molecule are produced. The hydrogen ions can be used for other reactions, but the molecular oxygen is released as a useless waste product.

The electron transport chain that picks up an electron from P680, leads to P700. The chain itself comprises several chemical links. The extra electron is passed from one link to the next in a series of reductions and oxidations. As the electron proceeds to P700, it encounters a chemical complex called cytochrome *bf*. The final link in the photosystem II electron transport chain is plastocyanin.

The P700 reacts to a photon in the same manner as P680. When the photon's energy is absorbed, P700 is transformed from its ground state to an excited state. In its excited state, P700 transfers an electron to the photosystem I electron transport chain. In returning to its ground state, P700 is reduced by plastocyanin, courtesy of photosystem II.

The photosystem I electron transport chain leads to an enzyme called ferredoxin-NADP reductase. (Enzymes are protein molecules that trigger speedy chemical reactions.) Once this enzyme has the extra electron available, it is able to transform NADP+ to NADPH and a hydrogen ion. NADPH is the abbreviation for nicotinamide adenine dinucleotide phosphate. This molecule is a high-energy fuel used for certain chemical reactions in the cell.

Under certain circumstances, P700 is raised to its excited state and the electron passes through only the first few links of the photosystem I electron transport chain. Instead of proceeding to the last link, the electron is shunted over to the cytochrome *bf* complex of photosystem II. The end result of this process–called cyclic photophosphorylation–is the production of a different high-energy fuel. This fuel is adenosine triphosphate, or ATP. Both ATP and NADPH are use to drive the chemical reactions that produce carbohydrates.

Both photosystem I and photosystem II require sunlight. As a result, this part of photosynthesis falls under the heading of light reactions. The actual generation of carbohydrates is a dark reaction, because it can occur in the absence of light. Carbohydrate construction, or synthesis, is accomplished through the Calvin cycle. The Calvin cycle is a series of chemical reactions through which carbon dioxide is used to build glucose. Glucose, in turn, is used as an eventual building block for sucrose, starch, and other carbohydrates. The first step of the Calvin cycle is the addition of carbon dioxide to an acceptor molecule. This step was first described by Melvin Calvin in the 1940s, and the entire cycle is named in his honor. The cycle begins with the addition of carbon dioxide to another molecule, ribulose-1,5-bisphosphate. This reaction is triggered, or catalyzed, the enzyme ribulose 1,5- bisphosphate carboxylase/oxygenase–often called by its short name, rubisco. Rubisco is able to add either carbon dioxide or molecular oxygen to a ribulose-1,5-bisphosphate molecule. The addition of carbon dioxide leads into the Calvin cycle; the addition of oxygen leads to a different process called photorespiration.

When rubisco binds carbon dioxide to ribulose-1,5-bisphosphate, a six-carbon molecule is temporarily formed. This molecule quickly splits into two molecules of 3-phosphoglycerate. At this stage, ATP enters the cycle by transferring a phosphate group to the molecule. The resulting molecule is reduced by NADPH to form glyceraldehyde-3-phosphate. At this point, glyceraldehyde-3-phosphate can go in one of two directions. A certain quantity of the molecule is dedicated to completing the cycle by regenerating the ribulose-1,5-bisphosphate stores. The remaining glyceraldehyde-3-phosphate continues on toward glucose, sucrose, starch, and other carbohydrate production.

Some plants have the ability to concentrate carbon dioxide levels within their cells before it enters the Calvin cycle. This ability helps circumvent photorespiration. Photorespiration occurs when rubisco adds oxygen, rather than carbon dioxide, to ribulose-1,5- bisphosphate. This is a wasteful reaction for the plant because it uses up ATP and reduces carbohydrate synthesis. Generally rubisco will use carbon dioxide in preference to oxygen, but in hot, sunny climates, the carbon dioxide near plants is rapidly used up. With lower levels of carbon dioxide in the area, oxygen uptake would increase and photorespiration would occur.

Some plants have evolved a mechanism to reduce photorespiration. These plants are called C4 plants, because they create a four-carbon molecule as an intermediary between carbon dioxide and the start of the Calvin cycle. The C4 plants concentrate carbon dioxide levels prior to the Calvin cycle through the Hatch-Slack pathway. The Hatch-Slack pathway is named for its discovers, Australian biochemists, Marshall Hatch and Roger Slack. Examples of C4 plants are corn, sorghum, and sugar cane, both of which do well in hot, sunny conditions.

Plants which add carbon dioxide directly to the Calvin cycle are called C3 plants, because they create the three- carbon molecule, 3-phosphoglycerate. The C3 plants do best in temperate climates. Wheat is an example of a C3 plant. Regardless of how carbon dioxide enters the Calvin cycle, the products are the same.

Photosynthesis evolved over three billion years ago, shortly after the appearance of the first living organisms. The food we eat and the oxygen we breathe are both formed by plants (including algae) through photosynthesis.

The power to drive this reaction comes from sunlight absorbed by chlorophyll in the chloroplasts of plants. At the present time, no known chemical system can be made to serve as a substitute for this process.

It has been calculated that each CO_2 molecule in the atmosphere is incorporated into a plant structure every 200 years and that all the O_2 in air is renewed by plants every 2000 years. All life depends directly or indirectly on the sun's energy, and only plants are capable of capturing and converting this energy into chemical energy in the form of sugar and other organic compounds. Thus, if plants should suddenly disappear from the earth, so would we.

Our geological heritage of coal, oil, and gas also originated directly or indirectly from photosynthesis, since these fossil fuels were all derived from the remains of living organisms. Our stake in photosynthesis is, therefore, great, since we are not only dependent upon it for the food we eat, but also for many of the goods and most of the energy we use.

AN OVERVIEW OF PHOTOSYNTHESIS

Photosynthesis converts light energy into the chemical energy of sugars and other organic compounds. This process consists of a series of chemical reactions that require carbon dioxide (CO_2) and water (H_2O) and store chemical energy in the form of sugar. Light energy from light drives the reactions. Oxygen (O_2) is a byproduct of photosynthesis and is released into the atmosphere. The following equation summarizes photosynthesis:

$$6\ CO_2 + 6\ H_2O \rightarrow \underset{\text{sugar}}{6(CH_2O)} + 6\ O_2$$

Photosynthesis transfers electrons from water to energy-poor CO_2 molecules, forming energy-rich sugar molecules. This electron transfer is an example of an oxidation-reduction process: the water is oxidized (loses electrons) and the CO_2 is reduced (gains electrons). Photosynthesis uses light energy to drive the electrons from water to their more energetic states in the sugar products, thus converting solar energy into chemical energy.

PHOTOPHOSPHORYLATION

Pq, the acceptor molecule, releases the excited electron into the care of an electron transport system that is sort of like a downhill bucket brigade. The transport system moves electrons extracted from water temporarily to a high-energy storage molecule called nicotinamide adenine dinucleotide phosphate ($NADP^+$).

$NADP^+$ is an electron acceptor for photosystem. The transport chain is essentially iron-containing pigments, cytochromes, a copper containing protein called plastocyanin and other electron transferring molecules. As electrons are passed through the chain and protons are being shuffled through a coupling factor, ATP molecules are assembled from ADP and phosphate in a process called photophosphorylation. A similar series of events occurs in photosystem I. After a photon of light strikes a P_{700} molecule, the resulting excited electron is passed along to an iron-sulphur molecule Fe-S which in turn passes it to another acceptor molecule ferrodoxin, (Fd). The ferrodoxin molecule releases the electron to a carrier molecule called flavin adenine dinucleotide (FAD) and then eventually on to $NADP^+$. A reduction occurs and NADP+ becomes NADPH.

Electrons from photosystem II and the activities of the electron transport system replace any electrons removed from the P_{700}molecule. Because the electrons move in one direction, the movement of electrons from water to photosystem II to photosystem I to $NADP^+$ are said to be part of noncyclic electron flow. Any ATP that is produced are designated noncyclic phosphorylation.

It should be noted that photosystem I can operate independently of photosystem II. When this occurs, the electrons boosted from P_{700} reaction-center molecules (photosystem I) are passed through an intermediary acceptor molecule called P_{430} and then on to the electron transport chain. This is rather then to the ferrodoxin and $NADP^+$.

After being passed through the electron transport chain, the electron is dumped back into the reaction-center of photosystem I. This process demonstrates cyclic electron flow and any ATP generated by cyclic electron flow is termed cyclic phosphorylation. Note, that no water molecules are split and no NADPH or oxygen is produced.

CHEMIOSMOSIS

Earlier we mentioned in passing a coupling factor. The enzyme necessary for mediation of the splitting of water molecules is on the inside of the thylakoid membrane. As a result of this, a proton gradient forms across the membrane and the movement of these protons is thought to be a source of energy for generating ATP. The motion is thought to be similar to molecular movement during osmosis and has hence been termed chemiosmosis. As the protons move across the membrane, they are assisted in crossing by protein channels called ATPase or coupling factor. Because of the proton movement, ADP and phosphate combine which produces ATP.

NITTY-GRITTY OF CARBON-FIXING REACTIONS

Both ATP and NADPH are important products of the light reactions and both of them play roles in the synthesis of carbohydrates from atmospheric carbon dioxide. Although the carbon-fixing reactions do not require daylight, they generally are conducted during daylight hours as there is some indication that some of the enzymes required for the processes in carbon-fixing may require some level of light. These reactions take place in the stroma of the chloroplast. Three known mechanisms of converting carbon dioxide to sugar.

The Calvin Cycle or the 3-carbon pathway

The Calvin cycle is the most common of the three mechanisms and has four main results:

1. With the assistance of the enzyme rubisco (RuBP carboxylase), six molecules of atmospheric carbon dioxide combine with six molecules of ribulose 1, 5-bisphosphate (RuBP)
2. The result of the first step is six unstable 6-carbon complexes, which immediately split into two 3-carbon molecules of 3-phosphoglyceric acid or 3PGA. This is the first stable compound in photosynthesis.
3. NADPH and ATP from the light-reactions, supply the energy required to convert the 3PGA to 12 molecules of glyceraldehydes 3-phosphate (GA3P), which is a 3-carbon sugar phosphate complex.
4. Finally, of the 12 molecules formed; 10 are restructured into six 5-carbon molecules of RuBP—the sugar that the process started with.

The sugars produced can either add to an increase in the sugar content (carbohydrate content) of the plant or they can be used in pathways that lead to the production of lipids and amino acids.

4-Carbon Pathway—C_4 plants

These plants use a 4-carbon molecule called oxaloacetic acid in place of the 3-carbon 3-phosphoglyceric acid used in step two of the Calvin cycle. Oxaloacetic acid is produced from a 3-carbon compound PEP—phosphoenolpyruvate and carbon dioxide. This process is enzyme mediated and occurs in the mesophyll cells of the leaf. Some species will convert the resulting oxaloacetic acid to aspartic, malic or other acids.

Note that the acids do not substitute for 3PGA. The 4-carbon acids migrate to the bundle sheaths surrounding the vascular bundles, where they are further converted to pyruvic acid and carbon dioxide. In returning to the mesophyll cells and interacting with ATP molecules, the pyruvic acids molecules are able to produce additional PEP. In the bundle sheath cells, the carbon dioxide formed converts into 3PGA and other molecules, by combining with RuBP. The other molecules formed are similar to the other ones formed in the Calvin cycle.

The C_4 cycle furnishes carbon dioxide to the Calvin cycle in a more roundabout way than the C_3 pathway, but there is an advantage to this extra pathway. The extra pathway greatly reduces photorespiration in C_4 plants, and this is a good thing because photorespiration is in direct competition with the Calvin cycle and even takes place in the light while the Calvin cycle is functioning. During photorespiration, RuBP reacts with oxygen to create carbon dioxide; in contrast, during photosynthesis RuBP and carbon dioxide are used to form carbohydrates. C_4 plants are able to pick up carbon dioxide in very low concentrations via the mesophyll cells. The Calvin cycle occurs in the bundle sheath where carbon dioxide is readily available. Because of the location, the enzyme rubisco will be in a prime spot to catalyze the reaction between RuBP and carbon dioxide, rather than oxygen. As a result C_4 plants have photosynthetic rates that are two to three times higher than C_3 plants.

There are a few other characteristic features of C_4 plants worth noting:

1. C_4 plants have two types of chloroplasts and an alternate pathway for using carbon dioxide. C_3 plants only have one type of chloroplast and one pathway. Chloroplasts with starch grains and are large with very little grana, and sometimes none, in the bundle sheath cells. In the mesophyll, the small, but numerous chloroplasts have no starch grains and contain highly developed grana.
2. PEP carboxylase is found in high concentration in the mesophyll cells which permits ready conversion of carbon dioxide to carbohydrate at lower concentrations than does rubisco (in bundle sheath cells) of the Calvin cycle.
3. The temperature ranges for C_4 plants are much higher than C_3 plants which enables C_4 plants to live well in conditions that would likely kill a C_3 plant.

CAM Photosynthesis

Crassulacean acid metabolism is a modified photosynthetic system that is somewhat similar to C_4 photosynthesis in that 4-carbon compounds are produced during the carbon-fixing reactions. CAM plants accumulate malic acid in their chlorenchyma tissues at night, which is converted back to carbon dioxide during the day. In the daytime, malic acid diffuses out of the vacuoles and is converted to carbon dioxide for use in the Calvin cycle. PEP carboxylase is responsible for converting the carbon dioxide plus PEP to malic acid at night. This modification allows for a greater amount of carbon dioxide to be converted to carbohydrate during the day than would be otherwise converted given the conditions CAM plants generally grow in. CAM plants generally close their stomata during the day in order to reduce water loss. There are more than 20 families that contain CAM plants, including cacti, stonecrops, orchids, bromeliads and many succulents growing in regions of high light intensity. There are some succulents that do not have CAM photosynthetic capabilities, as well as non-succulents that do have the ability.

There are great resources available that go into even greater detail on these reactions. If you are interested in these titles, please don't be afraid to ask on the forum for direction.

RESPIRATION

Respiration is the group of processes that utilizes the energy that is stored through the photosynthetic processes. The steps in respiration are small enzyme-mediated steps tha release tiny amounts of immediately available energy, the energy released is usually stored in ATP molecules which allow for even more efficient use of an organism's energy. Respiration occurs in the mitochondria and cytoplasm of cells.

There are several forms of respiration: aerobic—which requires oxygen, anaerobic—which occurs in the absence of oxygen, and fermentation—which also occurs in the absence of oxygen.

Aerobic respiration is the most common form of respiration and cannot be completed without oxygen gas. The controlled release of energy is the main event in aerobic respiration.

Certain types of bacteria and other organisms without oxygen gas carry on anaerobic respiration and fermentation. Compared to aerobic respiration the amount of energy released is quite small. The main difference between aerobic respiration and fermentation is in the way hydrogen is released and combined with other substances. Two very common forms of fermentation are summed up by the following equations:

$$C_6H_{12}O_6 \rightarrow \text{(with enzymes)} \rightarrow 2C_2H_5OH + 2CO_2 + \text{energy (ATP)}$$

glucose ethyl alcohol carbon dioxide

$$C_6H_{12}O_6 \rightarrow \text{(with enzymes)} \rightarrow 2C_3H_6O_3 + \text{energy (ATP)}$$

glucose lactic acid

Note the first equation is particularly valuable to the brewing industry.

MAJOR STEPS IN RESPIRATION

Glycolysis

The first step does not require oxygen gas (O_2) and takes place in the cytoplasm. The glycolytic phase is subdivided into three main steps and several smaller ones. Each step is mediated by an enzyme. A small amount of energy is released and hydrogen atoms are removed from compounds derived from glucose.

The main gist of the steps are:

- The glucose molecules goes through several steps and becomes a double phosphorylated fructose molecule.
- The 6-carbon fructose molecule is split into two 3-carbon fragments, each with a phosphate, GA3P
- Hydorgen, energy and water are removed from the GA3P molecules leaving pyruvic acid.

Glycolysis requires two molecules of ATP to get the process started. In the processes, four ATP molecules are created, with a net gain of 2 ATP molecules at the end of glycolysis. The hydrogen ions and electrons that are released are held by an acceptor molecule called NAD—nicotinamide adenine dinucleotide. The overall end products of gylcolysis is: 2-ATP molecules, 2-NADH molecules, and pyruvic acid. The next step depends on the kind of respiration involved—aerobic, true anaerobic or fermentation.

AEROBIC RESPIRATION (WITH OXYGEN PRESENT)

The Krebs Cycle (or Citric Acid Cycle)

The Krebs cycle takes place in the fluid matrix of the cristae compartments of the mitochondria. It is called the citric acid cycle because of all the intermediate acids in the cycle. The pyruvic acid product of glycolysis is restructured, some of the CO_2 is lost and becomes acetyl CoA which then dumps into the Krebs cycle. During the restructuring of pyruvic acid, a molecule of NADH is produced. The Krebs cycle removes energy, CO_2 and hydrogen from acetyl CoA via enzyme mediated reactions of organic acids.

The hydrogen removed during the Krebs cycle is picked up by FAD and NAD acceptor molecules. The end result of the metabolizing of two acetyl CoA molecules in the Krebs cycle is: 2-ATP molecule, oxaloacetic acid (to further drive the cycle), 6-$NADH_2$ molecules, 2-$FADH_2$ molecules and $2CO_2$ molecules.

The NAD and FAD molecules and the hydrogens that they carry will be dumped into the next step in respiration in order to extract the energy from the molecules.

FACTORS REGULATING RATE OF RESPIRATION

Temperature

To a point, the higher the temperature the faster respiration occurs. At some temperature, enzymes will become inactivated, although there are thermophilic (heat-loving) organisms that do quite well in high-temperature environments. Energy from sugar is released faster as the rate of respiration increases which results in a net weight loss. Plants offset the weight loss by increasing photosynthetic production of sugar. Note that during respiration, some of the energy is lost as heat, which results in an overall increase in organism temperature—not necessarily detectible by human hands.

Water

Enzymes generally operate in the presence of water, and reduced water in a plant will reduce the rate of respiration. Seeds usually have a water content of less than 10 per cent, while mature living cells usually are in excess of 90 per cent water. Seeds keep better if they are kept dry as the respiration rate remains quite low.

However, if a seed comes into contact with water and via imbibition swells, the respiration rate will skyrocket. The temperature could increase to the point of killing the seeds. Spontaneous combustion can occur from the respiration generated heat when a fungi or bacterium is permitted to grow on wet seeds. Kind of a neat little trivia fact to tuck away. Oxygen—Oxygen is an important regulator of respiration. If oxygen is drastically reduced, respiration may drop off to the point of retarding growth or death. Low oxygen concentrations can lead to the onset of fermentation processes.

ASSIMILATION AND DIGESTION

Assimilation is the conversion of the sugar produced by photosynthesis to fats, proteins, complex carbohydrates and other substances. While digestion is the breakdown of large insoluble molecules by hydrolysis to smaller soluble forms that can be transported to various parts of the plant.

Summary of key differences between photosynthesis and respiration:

Photosynthesis

- Energy stored in sugar molecules
- Carbon dioxide and water used
- Increases weight
- Requires light
- Occurs in chlorophyll
- In green organisms, produces oxygen
- With light energy, produces ATP

Respiration

- Energy released from sugar molecules
- Carbon dioxide and water released
- Decreases weight
- Can occur in light or darkness
- Occurs in all living cells
- Uses oxygen (aerobic respiration)
- With energy released from sugar, produces ATP.

THE CHEMISTRY OF PHOTOSYNTHESIS

Photosynthesis takes place in the chloroplasts of plant cells and consists of Light-Dependent and Light-Independent reactions. The Light-Dependent reaction occurs in the thylakoids and converts light energy into ATP and $NADPH_2$.

During this process water is split (oxidized) and oxygen is given off. Light-Independent reactions (the Calvin Cycle) incorporate CO into sugar, the basic food source for all organisms. Thylakoid membranes are the sites of the Light-Dependent reactions, whereas the Calvin cycle occurs in the stroma. These reactions can be summarized by the following equations

A DESIGN FLAW IN PHOTOSYNTHESIS PHOTORESPIRATION

Since plants first moved onto land about 425 million years ago, they have been adapting to the problems of terrestrial life, particularly the problem of dehydration. The solutions often involve tradeoffs. An important example is the compromise between photosynthesis and the prevention of excessive water loss from the plant. The CO_2 needed for photosynthesis enters a leaf via microscopic pores called stomata.

However, the stomata are also the main avenues of transpiration, the evapouration of water from leaves. On hot, dry days, most plants close their stomata in order to conserve water. This response limits access to CO_2, thereby reducing photosynthetic yield. Under these conditions, CO_2 concentrations in the air spaces within the leaf begin to decrease and the concentration of oxygen released from photosynthesis begins to increase. This favours what appears to be a wasteful process within the leaf called photorespiration.

In most plants (about 800,000 species), CO_2 is initially fixed via rubisco of the Calvin cycle into ribulose bisphosphate. Because the product of this reaction is a three-carbon compound, 3-phosphoglycerate, such plants are called C_3 plants.

Rice, wheat, and soybeans are among the C_3 plants that are important in agriculture. On hot, dry days when their stomata close, these plants produce less food as CO_2. levels within the leaf decline. Making matters worse, rubisco can use oxygen in place of CO_2. As O_2 increases, rubisco adds oxygen to RuBP instead of CO_2. The product splits, and one piece, a two-carbon compound (glycolate), is exported from the chloroplast. Other organelles (mitochondria and peroxisomes) then break down glycolate back to CO_2.

This process is called photorespiration because it occurs in the light (photo) and consumes oxygen (respiration). However, unlike normal cellular respiration, photorespiration generates no ATP. And unlike photosynthesis, photorespiration produces no food. Photorespiration is probably a metabolic relic from a much earlier time when the atmosphere had less O_2 and more CO_2 than it does today.

When rubisco first evolved, the inability of the enzyme's active site to exclude O_2 would have made little difference. Now, it is considered to be wasteful, since photorespiration drains away as much as 50 per cent of the carbon fixed by the Calvin cycle. If photorespiration could be reduced in certain plant species, without affecting photosynthetic productivity, crop yields and food supplies would increase.

ALTERNATIVE STRATEGIES OF CARBON FIXATION - C_4 AND CAM C_4 PLANTS

The environmental conditions that promote photorespiration are hot, bright, dry days. In these climates, alternate modes of carbon fixation have evolved to minimize photorespiration. The two most important of these photosynthetic adaptations are exhibited by C_4 and CAM C_4 plants. C_4 plants are so named because they form a four-carbon compound as the first product of the nonlight requiring reactions of photosynthesis. Several thousand species in at least 19 families use the C_4 pathway. Agriculturally important C_4 plants are sugarcane and corn, members of the grass family. Leaves of C_4 plants contain two distinct types of photosynthetic cells a cylinder of bundle-sheath cells surrounding the vein, and mesophyll cells located outside the bundle sheath. CO_2 is initially fixed in mesophyll cells by the enzyme PEP carboxylase. A four-carbon compound is formed (malate in this case) which conveys the fixed CO_2 via plasmodesmata (protoplasmic connections) into a bundle sheath cell where the enzymes of the Calvin cycle are located.

In the bundle sheath cell, the malate is converted into pyruvate and CO_2; the latter is now used by rubisco and the Calvin cycle to make sugar. Compared to rubisco, the enzyme PEP carboxylase has a much higher affinity for CO_2. Thus, PEP carboxylase can fix CO_2 efficiently when it is hot and dry and stomata are partially closed. Also, by pumping CO_2 from the mesophyll cells into the bundle sheath, this keeps the CO_2 concentration high enough for rubisco to accept CO_2 rather than O_2. In this way, C_4 photosynthesis minimizes photorespiration and enhances sugar production. This adaptation is especially advantageous in hot climates with intense sunlight and it is where C_4 plants evolved and thrive today.

A second photosynthetic adaptation to arid conditions (as found in deserts) has evolved in succulent (water-storing) plants (including ice plants), many cacti, and representatives of other plant families. These plants close their stomata in the day and open them during the night, just the reverse of other plants. Closing the stomata during the day helps desert plants conserve water, but it also prevents CO_2 from entering the leaves. At night, when the stomata are open, these plants take up CO_2 and initially fix it into four-carbon compounds like malate.

This mode of carbon fixation is called crassulacean acid metabolism, or CAM, after the plant family Crassulaceae, the succulents in which the process was first discovered. The photosynthetic cells of CAM plants store the malate formed in the night in their vacuoles until morning, when the stomata close. In the daytime, when the light reactions can make ATP and $NADPH_2$ for the Calvin cycle, CO_2 is released from the malate made the night before to become fixed into sugar in the chloroplasts.

The diagram compares C_4 and CAM C_4 photosynthesis. Both adaptations are characterized by initial fixation of CO_2 into an organic acid such as malate followed by transfer of the CO_2 to the Calvin cycle. In C_4 plants, such as sugarcane, these two steps are separated spatially; the two steps take place in two cell types. In CAM C_4 plants, such as pineapple, the two steps are separated temporally (time); carbon fixation into malate occurs at night, and the Calvin cycle functions during the day. Both C_4 and CAM C_4 are two evolutionary solutions to the problem of maintaining photosynthesis with stomata partially or completely closed on hot, dry days.

However, it should be noted, that in all plants, the Calvin cycle is used to make sugar from carbon dioxide. On a global scale, this represents a prodigious amount of sugar, about 160 billion metric tons of carbohydrate per year. An example of naturally-occuring biological oxidation-reduction reactions is the process of photosynthesis. It is a very complex process carried out by green plants, blue-green algae, and certain bacteria. These organisms are able to harness the energy contained in sunlight, and via a series of oxidation-reduction reactions, produce oxygen and sugar, as well as other compounds which may be utilized for energy as well as the synthesis of other compounds. The overall equation for the photosynthetic process may be expressed as

$$6\ CO_2 + 6H_2O \rightarrow C_6H_{12}O_6 + 6\ O_2$$

(glucose)

The equation is the net result of two processes. One process involves the splitting of water. This process is really an oxidative process that requires light, and is often referred to as the "light reaction". This reaction may be written as

$$12\ H_2O \rightarrow 6\ O_2 + 24\ H^+ + 24e^-$$

light or radiant energy

The oxidation of water is accompanied by a reduction reaction resulting in the formation of a compound, called nicotinamide adenine dinucleotide phosphate (NADPH). This reaction is illustrated below

$$NADP^+ + H_2O \rightarrow NADPH + H^+ + O$$

(oxidized form) (reduced form) (oxygen)

This reaction is linked or coupled to yet another reaction resulting in the formation of a highly energetic compound, called adenosine triphosphate, (ATP). As this reaction involves the addition of a phosphate group (labeled, as P_i) to a compound called, adenosine diphosphate (ADP) during the light reaction, it is called photophosphorylation.

$$ADP + P_i \rightarrow ATP$$

Think of the light reaction, as a process by which organisms "capture and store" radiant energy as they produce oxygen gas. This energy is stored in the form of chemical bonds of compounds such as NADPH and ATP. The energy contained in both NADPH and ATP is then used to reduce carbon dioxide to glucose, a type of sugar ($C_6H_{12}O_6$). This reaction, shown below, does not require light, and it is often referred to as the "dark reaction".

$$6\ CO_2 + 24\ H^+ + 24\ e^- \rightarrow C_6H_{12}O_6 + 6\ H_2O$$

The chemical bonds present in glucose also contain a considerable amount of potential energy. This stored energy is released whenever glucose is catabolized (broken down) to drive cellular processes.

The carbon skeleton in glucose also serves as a source of carbon for the synthesis of other important biochemical compounds such as, lipids, amino acids, and nucleic acids. In simplest terms, the process of photosynthessis can be viewed as one-half of the carbon cycle. In this half, energy from the sun is captured and transformed into nutrients which can be utilized by higher organisms in the food chain. The release of this energy during the metabolic re-conversion of glucose to water and carbon dioxide represents the second half of the carbon cycle and it may be referred to as catabolism or "oxidative processes".

HUMAN RESPIRATION

There are a number of books listed in the bibliography that have excellent illustrations. Teachers might also consider use of a purchased or homemade chart of the process of respiration. All human beings need air to survive. Actually, we need oxygen contained in the air. While food is also a necessity, we can survive for days without it. Without air, however, we would quickly die. During the process we call respiration we breathe in air into our lungs. When we release or exhale, carbon dioxide and water vapour are released. The oxygen we take in helps to oxidize or "burn food." Plants are the corner stone of this process. All living things depend upon food synthesized from raw materials in the leaves of green plants. Plants depend on their parts to help produce this food. Humans and animals breathe through their lungs. As we breathe in air passes through the nasal passages above the mouth where it is warmed and filtered. The air then moves through the voice box or larynx where two tubes are found. One channels food into the stomach and the other called the windpipe goes into the lungs. When we eat or drink a flap of tissue covers the windpipe so we do not swallow food into our lungs.

From there the windpipe divides into two tubes called bronchi. At the end of each bronchus tube is one of our two lungs. Within the lungs are a series of progressively smaller branches of tubes. At the end of the tiniest tube is a tiny air sac. These sacs swell like tiny balloons when we breathe air into our lungs. Surrounding these sacs are many capillaries. Oxygen breathed into the sacs pass through the sacs' thin walls into the capillaries. Carbon dioxide and water vapour in the capillaries pass the opposite way into the air sacs. Children should remember oxygen in, carbon dioxide out.

OVERALL EQUATION OF PHOTOSYNTHESIS

Photosynthesis is the intracellular anabolic process, characteristic of the green cells of plants in which carbohydrates are synthesized from carbon dioxide and water in the presence of light and chlorophyll. In this process, light energy is converted into chemical energy and stored in carbohydrate molecules, while oxygen is liberated. The overall process of photosynthesis is represented by the following general chemical equation.

$$6CO_2 + 12H_2O \xrightarrow[\text{Chlorophyll-a}]{\text{Light}} C_6H_{12}O_6 + 6H_2O + 6O_2$$

PRIMARY PROCESSES OF PHOTOSYNTHESIS

Nature of Light

As stated earlier, solar radiation is the only natural source of light for all organisms. Green plants can utilize light belonging to the visible spectrum only for photosynthesis. This visible part of the electromagnetic spectrum from the sun is confined between wavelengths of 390 nm (violet) and 760 nm (red).

Light is a form of energy. It appears to travel as a stream of discrete particles called photons. Each photon contains one quantum (unit) of light energy.

The energy quantity of each type of light depends on its wavelength. Light with a shorter wavelength has greater energy; light with a longer wavelength has lesser energy.

Rate of Photosynthesis

Rate of photosynthesis is measured in terms of the amount of carbon dioxide used (reduced) or oxygen released. It has been estimated that 8 quanta (photons) of light are required to reduce each molecule of carbon dioxide (or to release each molecule of oxygen) during photosynthesis (Emerson and Lewis, 1943). This is called the quantum requirement in photosynthesis.

Light Trapping Systems

During the photochemical phase, light is trapped by the photosynthetic pigments present in the quantasomes of the grana thylakoids. These pigments are organized into two pigment systems called pigment system I (PS I) and pigment system II (PS II).

(I) PS I is composed of the following pigment molecules:

Chl-a 700 (P700)	– one molecule	– Reaction center
Chl-a 683	– 200 molecules	– Antenna chlorophyll
Carotenoids	– 50 molecules	– Accessory pigments

(II) PS II is composed of the following pigment molecules:

Chl- a 680	– one molecule	– Reaction center
Chl- a 670	– 200 molecules	– Antenna chlorophyll
Chl- b	– up to 200 molecules	} Accessory pigments
Carotenoids	– up to 50 molecules	

In both the systems, all pigment molecules help in trapping light energy. However, all other molecules transfer their energy to the reaction center to be used for the photochemical reactions in photosynthesis.

Photo-excitation of Chlorophyll-a

When a molecule of chlorophyll-a (acting as the reaction center) receives light energy in the form of photons (quanta), it becomes energy-rich and activated.

This is called the excited state of chlorophyll. In this excited state, the chlorophyll-a expels one electron. With the loss of an electron, the chlorophyll-a develops a positive charge. This is called ionized chlorophyll-a.

The expelled electron contains an extra amount of energy received from light. In other words, the energy-rich expelled electron represents light energy. This electron energy is used for the formation of ATP during photosynthesis.

Photolysis of Water

In the photosynthesis of green plants, water is used as a source of hydrogen required for the reduction of carbon dioxide to form carbohydrates.

In this process, water is oxidized in presence of light and chlorophyll. Hydrogen is removed from water, and oxygen is released as follows:

This is described as photolysis of water, or photochemical oxidation of water. This was first indicated by Van Niel (1931). It was demonstrated with the help of an experiment by R. Hill (1937). Hill suggested that during the photolysis of water, hydrogen combines with some hydrogen acceptor in the plant, while oxygen is released.

(where A = unknown H-acceptor in plants).

Later on, Arnon (1951) discovered this hydrogen acceptor to be a coenzyme called NADP (nicotinamide adenine dinucleotide phosphate).

Substance A = NADP

Similarly, Ruben, Kamen, *et al.* (1941) confirmed that the oxygen evolved during photosynthesis comes from the splitting of water only. They used heavy isotope of oxygen (O_{18}) to demonstrate the fact as follows:

HILL REACTION

R. Hill (1937) suspended isolated chloroplasts in water in a beaker. He added some known hydrogen accepting compound (*e.g.* benzoquinone or ferric salt) to the water and exposed the beaker to the light. Immediately the oxidation of water took place and bubbles of oxygen were evolved. When the hydrogen acceptor in the beaker was examined, it was found to be reduced.

(A = hydrogen acceptor)

This is photolysis or the photochemical oxidation of water and is commonly called the Hill reaction.

Photophosphorylation

The formation of ATP molecules (chemical energy) from ADP and H_3PO_4 in presence of light and chlorophyll-a during the photochemical phase of photosynthesis is called photophosphorylation or photosynthetic phosphorylation.

More commonly, it is described as the conversion of light energy into chemical energy (ATP).

(ADP = adenosine diphosphate)

During the light reactions (the primary process or the photochemical phase), ATP formation takes place through two types of phosphorylation reactions. These are:

The non-cyclic process involves both PS I and PS II. The cyclic process involves only PS I.

- Non-cyclic photophosphorylation

In the light reaction of photosynthesis, the formation of ATP from ADP and H_3PO_4 in presence of light and chlorophyll-a during the non-cyclic transfer of electrons is called non-cyclic photophosphorylation.

Important Features:

- It is the major pathway of the light reaction of photosynthesis.
- It is a photochemical reaction.
- It takes place in the grana of chloroplasts.
- It requires participation of both PS I and PS II.
- The transfer of electrons through the ETS is unidirectional or non-cyclic.
- The non-cyclic process involves:
  - Photophosphorylation (ATP formation)
  - Photolysis of water
  - Formation of assimilatory power (ATP and $NADPH_2$)
  - Liberation of oxygen

The non-cyclic photophosphorylation takes place as follows:

Photoexcitation of PS I

The chlorophyll-a (P 700) of PS I is activated on receiving photons of light and so it expels electrons. As a result of this loss, it becomes ionized chl-a^+.

The electrons from PS I are first accepted by FRS (ferredoxin reducing substance) and are transferred to co-enzyme NADP via Fd (ferredoxin). NADP retains the electrons and is thus reduced.

Photoexcitation of PS II

Similarly, on receiving photons of light, the chl-a (680) of PS II also gets activated and expels electrons. As a result, it becomes ionized chl-a^+. The high energy containing electrons from PS II are first accepted by an electron acceptor PQ (plastoquinone) and are then transferred along stet Electron Transport System consisting of plastoquinone, cytochrome b6, Cytochrome f and plastocyanin. Finally, the electrons are accepted by the ionized chl-a^+ of PS I. In this way, ionized PS I comes back to the ground state and can again participate in the process.

Photolysis of Water

The ionized chlorophyll-a^+ of PS II is brought back to the ground state with the help of the electrons made available through the photolysis of water.

Formation of $NADPH_2$ and Liberation of O_2

During photolysis of water, oxygen is released as a by-product. The 2

NADP receive $4e^-$ from PS I and $4H^+$ from the photolysis of water to form reduced coenzyme $NADPH_2$

Non-Cyclic Electron Transfer

In this process, the movement of electrons is considered to be non-cyclic, or unidirectional. This is because, the final electron acceptor is different from the initial electron donor.

In this way, the light-induced electron flow is maintained continuously in the non-cyclic manner from water to NADP as follows

Because of the zig-zag path of the electrons, the non cyclic process is also called 'Z' scheme electron transport.

ATP Formation

During the non-cyclic transfer of energy-rich electrons through the Electron Transport System from PS II to PS I, energy from the electrons is released. This is used for the formation of ATP from ADP and H_3PO_4. This is called non-cyclic photophosphorylation.

End Products of Non-cyclic Process and the Assimilatory Power

The end products are ATP, $NADPH_2$ and O_2. Of these, O_2 is liberated from the green plants. ATP and $NADPH_2$ are used in the following dark reactions (phase II) for carbon assimilation(*i.e.*, reduction of CO_2 to form carbohydrates). ATP and $NADPH_2$ of light reaction are called 'assimilatory power.

ROLE OF PHOTOSYNTHESIS

Photosynthesis is the physico-chemical process by which plants, algae and photosynthetic bacteria use light energy to drive the synthesis of organic compounds. In plants, algae and certain types of bacteria, the photosynthetic process results in the release of molecular oxygen and the removal of carbon dioxide from the atmosphere that is used to synthesize carbohydrates (oxygenic photosynthesis).

Other types of bacteria use light energy to create organic compounds but do not produce oxygen (anoxygenic photosynthesis). Photosynthesis provides the energy and reduced carbon required for the survival of virtually all life on our planet, as well as the molecular oxygen necessary for the survival of oxygen consuming organisms1.In addition, the fossil fuels currently being burned to provide energy for human activity were produced by ancient photosynthetic organisms. Although photosynthesis occurs in cells or organelles that are typically only a few microns across, the process has a profound impact on the earth's atmosphere and climate. Each year more than 10 per cent of the total atmospheric carbon dioxide is reduced to carbohydrate by photosynthetic organisms. Most, if not all, of the reduced carbon is returned to the atmosphere as carbon dioxide by microbial, plant and animal metabolism, and by biomass combustion. In turn, the performance of photosynthetic organisms depends on the earth's atmosphere and climate.

Over the next century, the large increase in the amount of atmospheric carbon dioxide created by human activity is certain to have a profound impact on the performance and competition of photosynthetic organisms. Knowledge of the physico-chemical process of photosynthesis is essential for understanding the relationship between living organisms and the atmosphere and the balance of life on earth. Several books on photosynthesis are available for the uninitiated or advanced student. Taiz and Zeiger place the photosynthetic process in the context of over all plant physiology, and Cramer and Knaff describe the bioenergetic foundation of photosynthesis.

The overall equation for photosynthesis is deceptively simple. In fact, a complex set of physical and chemical reactions must occur in a coordinated manner for the synthesis of carbohydrates. To produce a sugar molecule such as sucrose, plants require nearly 30 distinct proteins that work within a complicated membrane structure.

Research into the mechanism of photosynthesis centers on understanding the structure of the photosynthetic components and the molecular processes that use radiant energy to drive carbohydrate synthesis. The research involves several disciplines, including physics, biophysics, chemistry, structural biology, biochemistry, molecular biology and physiology, and serves as an outstanding example of the success of multidisciplinary research. As such, photosynthesis presents a special challenge in understanding several interrelated molecular processes.

BRIEF HISTORY

In the 1770s Joseph Priestley, an English chemist and clergyman, performed experiments showing that plants release a type of air that allows combustion. He demonstrated this by burning a candle in a closed vessel until the flame went out. He placed a sprig of mint in the chamber and after several days showed that the candle could burn again. Although Priestley did not know about molecular oxygen, his work showed that plants release oxygen into the atmosphere. It is noteworthy that over 200 years later, investigating the mechanism by which plants produce oxygen is one of the most active areas of photosynthetic research.

Building on the work of Priestley, Jan Ingenhousz, a Dutch physician, demonstrated that sunlight was necessary for photosynthesis and that only the green parts of plants could release oxygen. During this period Jean Senebier, a Swiss botanist and naturalist, discovered that CO_2 is required for photosynthetic growth and Nicolas- The odore de Saussure, a Swiss chemist and plant physiologist, showed that water is required. It was not until 1845 that Julius Robert von Mayer, a German physician and physicist, proposed that photosynthetic organisms convert light energy into chemical free energy. An interesting time line of the history of photosynthesis has been presented by Huzisige and Ke. By the middle of the nineteenth century the key features of plant photosynthesis were known, namely, that plants could use light energy to make carbohydrates from CO_2 and water. The empirical equation representing the net reaction of photosynthesis for oxygen evolving organisms is :

$$CO_2 + 2H_2O + \text{Light Energy} \rightarrow [CH_2O] + O_2 + H_2O, \quad (1)$$

where [CH2O] represents a carbohydrate (e.g., glucose, a six-carbon sugar). The synthesis of carbohydrate from carbon and water requires a large input of light energy. The standard free energy for the reduction of one mole of CO_2 to the level of glucose is +478 kJ/mol. Because glucose, a six carbon sugar, is often an intermediate product of photosynthesis, the net equation of photosynthesis is frequently written as:

$$6CO_2 + 12H_2O + \text{Light Energy} \rightarrow C_6H_{12}O_6 + 6O_2 + 6H_2O. \quad (2)$$

The standard free energy for the synthesis of glucose is +2,870 kJ/mol. Not surprisingly, early scientists studying photosynthesis concluded that the O_2 released by plants came from CO_2, which was thought to be split by light energy. In the 1930s comparison of bacterial and plant photosynthesis lead Cornelis van Niel to propose the general equation of photosynthesis that applies to plants, algae and photosynthetic bacteria. Van Niel was aware that some photosynthetic bacteria could use hydrogen sulfide (H_2S) instead of water for photosynthesis and that these organisms released sulfur instead of oxygen. Van Niel, among others, concluded that photosynthesis depends on electron donation and acceptor reactions and that the O_2 released during photosynthesis comes from the oxidation of water. Van Niel's generalized equation is :

$$CO_2 + 2H_2A + \text{Light Energy} \rightarrow [CH_2O] + 2A + H_2O. \quad (3)$$

In oxygenic photosynthesis, 2A is O_2, whereas in anoxygenic photosynthesis, which occurs in some photosynthetic bacteria, the electron donor can be an inorganic hydrogen donor, such as H2S (in which case A is elemental sulfur) or an organic hydrogen donor such as succinate (in which case, A is fumarate). Experimental evidence that molecular oxygen came from water was provided by Hill and Scarisbrick (1940) who demonstrated oxygen evolution in the absence of CO_2 in illuminated chloroplasts and by Ruben et al. (1941) who used 18O enriched water.

The biochemical conversion of CO_2 to carbohydrate is a reduction reaction that involves the rearrangement of covalent bonds between carbon, hydrogen and oxygen. The energy for the reduction of carbon is provided by energy rich molecules that are produced by the light driven electron transfer reactions. Carbon reduction can occur in the dark and involves a series of biochemical reactions that were elucidated by Melvin Calvin, Andrew Benson and James

Bassham in the late 1940s and 1950s. Using the radioisotope 14C, most of the intermediate steps that result in the production of carbohydrate were identified. Calvin was awarded the Nobel Prize for Chemistry in 1961 for this work.

In 1954 Daniel Arnon and coworkers discovered that plants, and A. Frenkel discovered that photosynthetic bacteria, use light energy to produce ATP, an organic molecule that serves as an energy source for many biochemical reactions. During the same period L.N.M. Duysens showed that the primary photochemical reaction of photosynthesis is an oxidation/ reduction reaction that occurs in a protein complex (the reaction centre). Over the next few years the work of several groups, including those of Robert Emerson, Bessel Kok, L.N.M. Duysens, Robert Hill and Horst Witt, combined to prove that plants, algae and cyanobacteria require two reaction centers, photosystem II and photosystem I, operating in series.

In 1961 Peter Mitchell suggested that cells can store energy by creating an electric field or a proton gradient across a membrane. Mitchell's proposal that energy is stored as an electrochemical gradient across a vesicular membrane opened the door for understanding energy transformation by membrane systems. He was awarded the Nobel Prize in Chemistry in 1978 for his theory of chemiosmotic energy transduction.

Most of the proteins required for the conversion of light energy and electron transfer reactions of photosynthesis are located in membranes. Despite decades of work, efforts to determine the structure of membrane bound proteins had little success. This changed in the 1980s when Johann Deisenhofer, Hartmut Michel, Robert Huber and co-workers determined the structure of the reaction centre of the purple bacterium Rhodospeudomonas viridis.. They were awarded the Nobel Prize for Chemistry in 1988 for their work, which has provided insight into the relationship between structure and function in membrane-bound proteins.

A key element in photosynthetic energy conversion is electron transfer within and between protein complexes and simple organic molecules. The electron transfer reactions are rapid (as fast as a few picoseconds) and highly specific. Much of our current understanding of the physical principles that guide electron transfer is based on the pioneering work of Rudolph A. Marcus, who received the Nobel Prize in Chemistry in 1992 for his contributions to the theory of electron transfer reaction in chemical systems.

PHOTOSYNTHESIS ORIGINATES AND ENERGY

The energy that drives photosynthesis originates in the centre of the sun, where mass is converted to heat by the fusion of hydrogen. Over time, the heat energy reaches the sun's surface, where some of it is converted to light by black body radiation that reaches the earth. A small fraction of the visible light incident on the earth is absorbed by plants. Through a series of energy transducing reactions, photosynthetic organisms are able to transform light energy into chemical free energy in a stable form that can last for hundreds of millions of years (e.g., fossil fuels).

A simplified scheme describing how energy is transformed in the photosynthetic process is presented in this section. The focus is on the structural and functional features essential for the energy transforming reactions. For clarity, mechanistic and structural details are omitted. A more highly resolved description of oxygenic and anoxygenic photosynthesis is given in the remaining sections.

The photosynthetic process in plants and algae occurs in small organelles known as chloroplasts that are located inside cells. The more primitive photosynthetic organisms, for example oxygenic cyanobacteria, prochlorophytes and anoxygenic photosynthetic bacteria, lack organelles. The photosynthetic reactions are traditionally divided into two stages - the "light reactions," which consist of electron and proton transfer reactions and the "dark reactions," which consist of the biosynthesis of carbohydrates from CO_2.The light reactions occur in a complex membrane system (the photosynthetic membrane) that is made up of protein complexes, electron carriers, and lipid molecules. The photosynthetic membrane is surrounded by water and can be thought of as a two-dimensional surface that defines a closed space, with an inner and outer water phase.

A molecule or ion must pass through the photosynthetic membrane to go from the inner space to the outer space. The protein complexes embedded in the photosynthetic membrane have a unique orientation with respect to the inner and outer phase. The asymmetrical arrangement of the protein complexes allows some of the energy released during electron transport to create an electrochemical gradient of protons across the photosynthetic membrane.

Photosynthetic electron transport consists of a series of individual electron transfer steps from one electron carrier to another. The electron carriers are metal ion complexes and aromatic groups. The metal ion complexes and most of the aromatic groups are bound within proteins. Most of the proteins involved in photosynthetic electron transport are composed of numerous polypeptide chains that lace through the membrane, providing a scaffolding for metal ions and aromatic groups. An electron enters a protein complex at a specific site, is transferred within the protein from one carrier to another, and exits the protein at a different site. The protein controls the pathway of electrons between the carriers by determining the location and environment of the metal ion complexes and aromatic groups.

By setting the distance between electron carriers and controlling the electronic environment surrounding a metal ion complex or aromatic group, the protein controls pairwise electron transfer reactions. Between proteins, electron transfer is controlled by distance and free energy, as for intraprotein transfer, and by the probability that the two proteins are in close contact. Protein association is controlled by a number of factors, including the structure of the two proteins, their surface electrical and chemical properties and the probability that they collide with one another. Not all electron carriers are bound to proteins. The reduced forms of plastoquinone or ubiquinone and nicotinamide adenine dinucleotide phosphate (NADPH) or NADH act as mobile electron carriers operating between protein complexes. For electron transfer to occur, these small molecules must bind to special pockets in the proteins known as binding sites. The binding sites are highly specific and are a critical factor in controlling the rate and pathway of electron transfer.

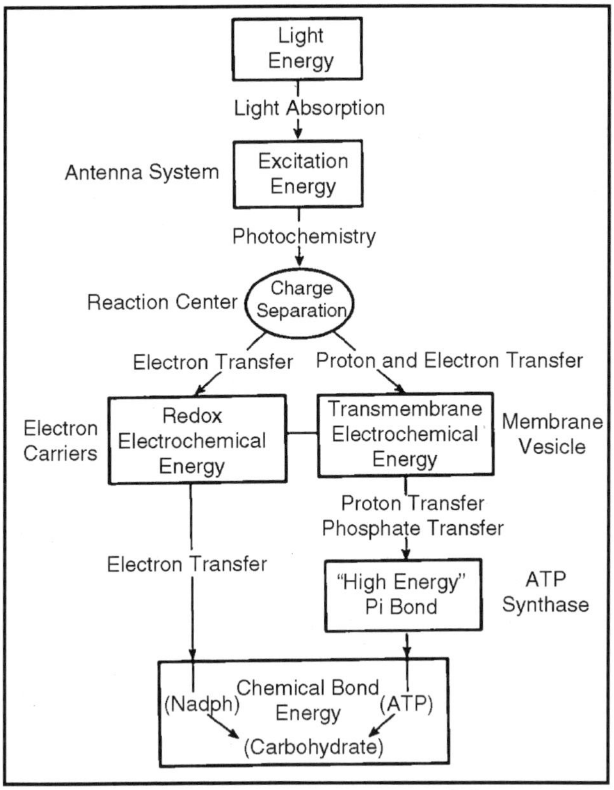

The light reactions convert energy into several forms. The first step is the conversion of a photon to an excited electronic state of an antenna pigment molecule located in the antenna system. The antenna system consists of hundreds of pigment molecules (mainly chlorophyll or bacteriochlorophyll and carotenoids) that are anchored to proteins within the photosynthetic membrane and serve a specialized protein complex known as a reaction centre. The electronic excited state is transferred over the antenna molecules as an exciton.

Some excitons are converted back into photons and emitted as fluorescence, some are converted to heat, and some are trapped by a reaction centre protein. Excitons trapped by a reaction centre provide the energy for the primary photochemical reaction of photosynthesis - the transfer of an electron from a donor molecule to an acceptor molecule. Both the donor and acceptor molecules are attached to the reaction centre protein complex. Once primary charge separation occurs, the subsequent electron transfer reactions are energetically downhill.

In oxygenic photosynthetic organisms, two different reaction centers, known as photosystem II and photosystem I, work concurrently but in series. In the light photosystem II feeds electrons to photosystem I. The electrons are transferred from photosystem II to the photosystem I by intermediate carriers.

The net reaction is the transfer of electrons from a water molecule to $NADP^+$, producing the reduced form, NADPH. In the photosynthetic process, much of the energy initially provided by light energy is stored as redox free energy (a form of chemical free energy) in NADPH, to be used later in the reduction of carbon. In addition, the electron transfer reactions concentrate protons inside the membrane vesicle and create an electric field across the photosynthetic membrane. In this process the electron transfer reactions convert redox free energy into an electrochemical potential of protons.

The energy stored in the proton electrochemical potential is used by a membrane bound protein complex (ATP-Synthase) to covalently attach a phosphate group to adenosine diphosphate (ADP), forming adenosine triphosphate (ATP). Protons pass through the ATP-Synthase protein complex that transforms electrochemical free energy into a type of chemical free energy known as phosphate group-transfer potential (or a high-energy phosphate bond). The energy stored in ATP can be transferred to another molecule by transferring the phosphate group. The net effect of the light reactions is to convert radiant energy into redox free energy in the form of NADPH and phosphate group-transfer energy in the form of ATP. In the light reactions, the transfer of a single electron from water to $NADP^+$ involves about 30 metal ions and 7 aromatic groups.

The metal ions include 19 Fe, 5 Mg, 4 Mn, and 1 Cu. The aromatics include quinones, pheophytin, NADPH, tyrosine and a flavoprotein. The NADPH and ATP formed by the light reactions provide the energy for the dark reactions of photosynthesis, known as the Calvin cycle or the photosynthetic carbon reduction cycle. The reduction of atmospheric CO_2 to carbohydrate occurs in the aqueous phase of the chloroplast and involves a series of enzymatic reactions.

The first step is catalyzed by the protein Rubisco (D-ribulose 1,5-bisphosphate carboxylase/oxygenase), which attaches CO_2 to a five-carbon compound. The reaction produces two molecules of a three-carbon compound. Subsequent biochemical reactions involve several enzymes that reduce carbon by hydrogen transfer and rearrange the carbon compounds to synthesize carbohydrates. The carbon reduction cycle involves the transfer and rearrangement of chemical bond energy. In anoxygenic photosynthetic organisms water is not used as the electron donor. Electron flow is cyclic and is driven by a single photosystem, producing a proton electrochemical gradient that is used to provide energy for the reduction of NAD^+ by an external H-atom or e-donor (e.g., H_2S or an organic acid) in a process known as "reverse electron flow". Fixation of CO_2 occurs via different pathways in different organisms.

10

Structure and Constituents of Cell

THE THEORY OF CELL

The Cell Theory is one of the basic principles of biology. Credit for the formulation of this theory is given to german scientists Theodor Schwann, Matthias Schleiden, and Rudolph Virchow. More directly related to the origin of cell biology was the establishment of the *cell theory,* one of the broadest and most fundamental of all biological generalizations. It shows that all living beings-animals, plants, or protozoa-are composed of cells and cell products. This theory resulted from numerous investigations that started at the beginning of the 19th century, and finally led to the studies of the botanist Schleiden (1838) and the zoologist Schwann, who established the theory in a proper and definite manner.

The cell theory has had wide-ranging effects in all the fields of biological research. It was established immediately that every cell is formed by the division of another cell. Much later, with the progress of biochemistry, it was shown that there are fundamental similarities in the chemical composition and metabolic activities of all cells. The function of the organism as a whole was also recognized to be a result of the sum of the activities and interactions of the cell-units. After that cell theory was applied to pathology by Virchow, and Kölliker extended it to embryology after it was demonstrated that the organism develops from the fusion of two cells — the spermatozoon and the ovum.

A more general conclusion was reached at the same time by investigators such as Brown, who established that the nucleus is a basic and constant component of the cell. Others such as Dujardin, Schultze, Purkinje, von Mohl) concentrated on the description of the cell contents, known as *protoplasm.*

Thus the primitive idea of the *cell* was transformed into the more sophisticated concept of a mass of protoplasm limited in space by a cell membrane *and* possessing a nucleus. The protoplasm surrounding the nucleus became known as the *cytoplasm* to distinguish it from the *karyoplasm,* the protoplasm of the nucleus. The remarkable changes produced in the nucleus at each cell division attracted the attention of a great number of investigators. For example, the phenomena of *amitosis,* or direct division, and of indirect division were discovered by Flemming (in animals) and by Strasburger (in plants). Indirect division was also called *karyokinesis* or *mitosis*. It was proved that fundamental to mitosis is the formation of nuclear filaments, or *chromosomes*, and their equal division between the nuclei. Other discoveries of importance were the fertilization of the ovum and the fusion of the two pronuclei. In the cytoplasm, the aster, the mitochondrion, and the reticular apparatus were discovered. In 1892 O. Hertwig published his monograph *Die Zelle und das Gewebe (The Cell and the Tissue)* in which he attempted to achieve a general synthesis of biological phenomena based on the characteristics of the cell, its structure and function. He thus created cytology as a separate branch of biology.

If one follows the development of cell biology in the present century it is fact that there were two main reasons for the advance of cytologic knowledge: former is the increased resolving power of instrumental analysis, due essentially to the introduction of electron microscopy and x-ray diffraction techniques, and second is the convergence of *cytology* with other fields of biological research, especially genetics, physiology, and biochemistry.

Cytology and Genetics

By the middle of the 19th century the universality of cell division as the central phenomenon in the reproduction of organisms was established, and Virchow expressed it *in* the famous aphorism *"Omnis cellula e cellula."* From this time on, the study of cells and of heredity and evolution converged. Observations on the germ cells made by van Beneden, Flemming, Strasburger, Boveri, and others gave support to the theory of the continuity of germ plasm proposed by Weissmann in 1883. This theory stated that the transference of hereditary factors from one generation to the next takes place through the continuity of what he called *germ plasm,* located in the sex elements (spermatozoon and ovum), and not through somatic cells.

The discovery of fertilization led to the theory that the cell nucleus is the bearer of the physical basis of heredity. Furthermore, Roux postulated that chromatin, the substance of the nucleus that constitutes the chromosomes, must have a linear organization.

In 1965 Gregor or Mendel gave the fundamental law of heredity, but at that time the cytologic changes produced in the sex cells were not sufficiently known to permit an interpretation of the independent segregation of hereditary characters. For this and other reasons, little attention was paid to Mendel's work until the botanists Correns, Tschermack, and De Vries in 1901 independently rediscovered Mendel's laws. At that time cytology had advanced enough that the rnechanism of distribution of the hereditary units postulated by Mendel could be understood and *and* explained. It was known that the somatic cells have a double, or *diploid,* hereditary constitution, whereas in the reproductive cells or gametes this constitution is single, or *haploid*. In addition, cytologists had observed that the cycle the chromosomes undergo in *meiosis* of germ cells was related to hereditary phenomena. McClung suggested that sex determination was related to some special chromosomes; this theory was later corroborated by Stevens and Wilson. The chromosome theory of heredity was finally established by Morgan and his collaborators, Sturtevant and Bridges, who assigned to the *genes,* or hereditary units, definite loci within the chromosomes. After that experimental research on heredity and evolution became a separate branch of biology, which Bateson in 1906 called genetics. Almost from the beginning, however, the science of genetics maintained a close relationship with cytology. From the convergence of cytology and genetics *cytogenetics* originated. In the past decade the study of genetics has become linked to biochemistry and has reached the molecular level. Thus the new fields of *biochemical* and *molecular genetics* were established.

Cell Physiology

Most early cytologic knowledge was based on observations of fixed and stained cells and tissues. By 1899 interest shifted toward the study of living cells, principally as a result of the work of Fischer and Hardy, who showed that several of the structures observed in fixed cells could be reproduced by the action of fixatives on colloidal models. Various types of movement, such as cyclosis, ameboid motion, and ciliary, flagellar, and muscular contraction, were studied at a cellular level. At the end of the 19th century, Overton advanced the theory that the cell membrane was a lipoidal film. Michaelis made membrane models to study the passage of substances and did the first vital staining of mitochondria. The basic concepts of cell irritability and nerve function were established by the middle of the 19th century by Du Bois-Reymond; physiological techniques were then developed to aid in the study of these cells and to measure action potentials and nerve currents.

In 1909 Harrison have opened the important avenue to study the living cell, who demonstrated that nerve cell from an embryo could grow and differentiate in vitro. This gave rise to the technique of tissue culture, which had a great impact on cytology.

Another method for the study of the physicochemical properties of the living cell — *microsurgery* — came from the field of bacteriology. Schouten and Barber used fine micropipets, moved by precision instruments, to isolate and culture single bacteria. In 1911 Kite adapted this method to cytology. Levi, Peterfi, Chambers, and others have perfected techniques of intracellular operations and have obtained data on the viscosity, hydrogen concentration, and other physicochemical properties of the cell.

Among the important phenomena studied in cell physiology are: the nature of the cell membrane and of active transport across membranes, the reaction of cells to changes in environment, the mechanisms of cell excitability and contraction, and cell nutrition, growth, secretion, and other manifestations of cellular activity.

Chemical and Physicochemical Cellular Analysis

Cytochemistry is the another branch of cell biology the result of the convergence of methods and sciences devoted to the chemical and physicochemical analysis of living matter. Other outstanding biochemical studies were those of Fischer and Hofmeister in 1902, they have independently recognized that the protein molecule consists of a small number of amino acids united by peptide bonds. Of similar importance to cell biology were the earlier investigations of Miescher and Kossel they, by the analysis of pus cells, spermatozoa, hemolyzed erythrocytes of birds, and other cell types, isolated the nucleic acids, whose basic role in heredity and protein synthesis has been recognized only recently.

Another great advance was the introduction into biological thinking by Ostwald of the concept of catalytic activity and the discovery that enzymes are the molecular entities used by the cell to produce the various types of energy transformations necessary for maintenance of living activities. The main types of cellular oxidations were discovered by Wieland and by Warburg but keilin has discovered the final mechanism in 1934.

Because of cytology's emphasis on morphology, cytologists were at first very slow in grasping the importance of the biochemical approach; at the same time, biochemists, because of their field's emphasis on organic chemistry, had no interest in cell structure and were concerned principally with isolating chemical components and studying elementary enzyme reactions. The point of convergence of cytology and biochemistry can be traced back to 1934, when Bensley and Hoerr isolated mitochondria from cells in large enough quantities to permit analysis by chemical and physicochemical methods.

Research in this direction was pursued with great success by Claude, Hogeboom, and others they concluded that mitochondria are centres of cellular oxidations.

Advances in cell fractionation have been of the greatest importance to biology and biochemistry, especially with the development of radioactive tracer techniques that permit a dynamic approach to the study of cell metabolism. Another great advance was the use of the electron microscope for the observation and characterization of cell fractions.

Modern cytochemistry has also developed along the lines of microchemical and ultramicrochemical analysis by means of techniques for assay of minute quantities of material and isolation of single cells and even parts of cells. Chemical analysis can be carried out by cytophotometry, which facilitates the study of the localization of nucleic acids and proteins within parts of a single cell.

Another important branch of cytochemistry arose from the application of numerous enzyme reactions that could be observed under the light and electron microscopes. This last approach is of particular interest since it combines cytochemistry and ultrastructure and thus permits study of the localization of enzymes at the level of resolution of the electron microscope. Of similar importance have been the autoradiographic studies on the localization of radioactive tracers in different cellular structures.

Ultrastructure and Molecular Biology

In studying the limits and dimensions in biology, the impact of instrumental analysis was mentioned and the modern fields of ultrastructure and molecular biology were delineated. These are the most progressive branches of biology, in which the merging of cytology with biochemistry, physicochemistry, and especially macromolecular and colloidal chemistry becomes increasingly intimate.

Knowledge of the submicroscopic organization or ultrastructure of the cell is of fundamental importance because practically all the functional and physicochemical transformations take place within the molecular architecture of the cell and at a supramolecular level.

The new advances are having an remarkable impact on cell biology: the discovery that in the protein molecule the exact sequence of amino acids and the three-dimensional arrangement of the polypeptide chain have a direct relationship to definite biological properties; the studies of active groups in different enzymes; the molecular model of DNA suggested by Watson and Crick in 1953; as well as all the knowledge on the stereochemistry of macromolecules. Molecular biology is thus providing valuable information to the fields of genetics (through molecular genetics), biochemistry, and even pathology —the latter through the establishment of molecular diseases.

Both in ultrastructure and molecular biology the integration of morphology and physiology becomes so intimate that it is impossible to separate them; the concepts of *form* and *function* fuse into inseparable unity.

GROWTH OF CELL WALL

Formerly two theories were held regarding the method of cell wall growth in thickness: that of growth by intussusception, where new microfibrils were held to be laid down between existing microfibrils and that of growth by apposition where new microfibrils were laid down on top of the existing ones, forming a new layer.

It is now considered that the formation of both primary and secondary cell walls occurs principally by the mechanism of apposition. It is probable, however, that some growth by intussusception does occur. Deposition of cellulose uniformly over the whole surface of the cell has been demonstrated by the use of the radio-active isotope C^{14}. This was incorporated into the whole length of the primary cell wall.

With respect to longitudinal growth the theory now most widely held is the multi-net theory of cell growth. This accounts also for the observed orientation of the microfibrils in successive layers of the wall. In this theory microfibrils are first deposited transversely to the long axis of the cell, and this layer is later pushed outwards as a result of the formation by apposition of a layer internal to it. During cell elongation the first formed layers of microfibrils are stretched and thus become orientated in a progressively more longitudinal plane.

Recent studies of cell wall formation in fibres and tracheids using the electron microscope and also the technique of autoradiography whereby the path of a radio-active isotope is followed are consistent with the multi-net theory of cell wall growth.

As in early work with the primary cell walls, labelled carbon was found to be deposited more or less uniformly over the secondary wall fibres and tracheids. In fibres, however, the formation of the secondary wall may begin near the centre of the cell and progress towards the tips. The wall is thus thicker near the centre. In some cells, *e.g.* root hairs, pollen tubes, tracheids and fibres, growth occurs only at the tip. This tip growth is regarded as a localised type of multi-net growth. The role of determining whether the wall of the whole cell will grow or only a localised part of it, as for example in root hairs or stellate parenchyma cells, is attributed to the cytoplasm.

SECONDARY WALLS

These are formed after cell expansion has stopped. The secondary walls, particularly of fibres and tracheids, show microscopic layering. The microscopic layers are commonly known as the S_1 (outer), S_2 (middle) and S_3 (inner). The S_3 layer is usually thinner than either the S_1 or S_2 and may be absent altogether. The S_1 layer normally consists of four submicroscopic lamellae, alternate ones having microfibrils in opposed helices. The S_2 layer, the middle of the secondary wall, consists of numerous lamellae in which the orientation of the microfibrils is at only a small angle to the long axis of the cell. There seems to be a tendency for the microfibrils of this very thick and conspicuous wall layer to be aggregated into macrofibrils. The S_3 layer is always poorly developed in contrast to the S_2 layers and there is evidence too that the S_3 layer may differ chemically in some way from S_1 and S_2.

GENERAL CHARACTERISTICS OF THE CELL

The cell is the basic unit common to all living organisms. Two general types of cells exist: prokaryotic and eukaryotic. The main differences between the two groups are the presence or absence of a nuclear membrane, structure enclosing the cell's genetic information, and the use of functional organelles.

Each of us starts life as a single cell, a microscopic package which contains directions for everything that we can become. This cell is defined and separated from the rest of the world by a membrane, a transparent film so thin that it could not be seen under any light microscope. For years scientists debated whether such a membrane truly existed -- perhaps it was just an illusion? Despite its ethereal appearance, however, the surface membrane is not only real but exceedingly powerful, controlling everything that goes in and out of the cell and relaying vital messages. Similar membranes enclose or make up a large number of the cell's organelles. "It takes a membrane to make sense out of disorder in biology," writes Lewis Thomas. "You have to be able to catch energy and hold it, storing precisely the needed amount and releasing it in measured shares. A cell does this and so do the organelles inside it.... you can only transact this business with membranes in our kind of world."

Sheltered by the cell's surface membrane (which is also called the plasma membrane), different configurations of organelles move about in a watery environment. The number and kinds of organelles in each cell depend on its function. Now that researchers have become acquainted with these organelles, they can often guess at a cell's specialty from its structure: A cell that contains many mitochondria, for instance, must be involved in producing large quantities of energy, (It was once calculated that cells deepens as the cells move apart.

A human cell Splits: These SEM (scanning electron microscope) photographs show one cell dividing into two identical daughter cells. The cleavage furrow between the daughter. The hair-like projections on the cell surface are microvilli, cytoplasmic extensions which seem to be involved in cell motion. Bumble-bees cannot fly, since there seemed no way for their short wings to get enough lift -- yet obviously they do fly, despite their weight and one reason is that the cells in their wing muscles contain exceptionally large numbers of mitochondria which supply them with the energy to beat at enormous speed. A cell in which ribosomes are prominent must be synthesizing proteins. A cell with a great many lysosomes must be destroying wastes. And so on.

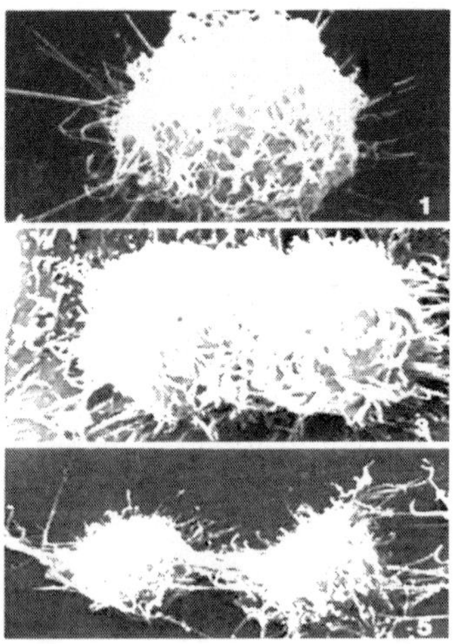

The single cell with which we start life divides again and again. As Daniel Mazia, Professor of Zoology at the University of California, puts it, the story of the cell cycle is "Double or nothing. With few exceptions, a living cell either reproduces or dies; the principle is so simple that no one has bothered to call it a principle. A cell is born in the division of a parent cell. It then doubles in every respect: in every part, in every kind of molecule, even in the amount of water it contains."

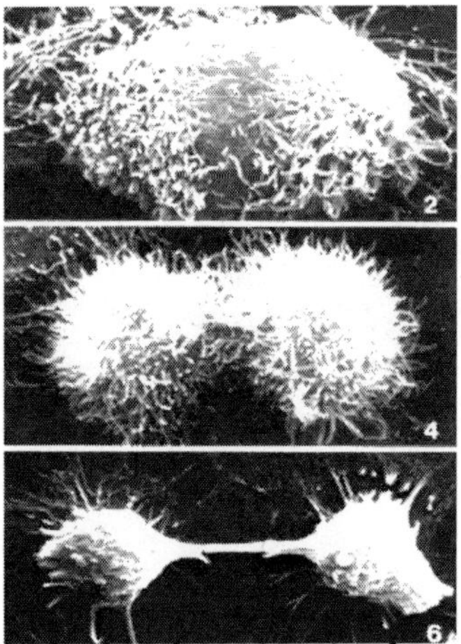

The process is particularly clear in single-celled organisms such as amoebae, which may live forever through doublings. Each cell annihilates its individual existence in the production of two daughter cells, which repeat the process roughly 40 hours later, until limited by the food supply, eaten by another, organism, destroyed in other ways. Some bacteria reproduce every 20 minutes. In more complex organisms, however, the life span of individual cells varies. "The governance of an organism such as a higher animal, which is a society of cells, dictates that some of the cells in the society will reproduce and that others will not," writes Mazia. "In general, the cells of tissues that perform special services for the entire organism, such as cells of the nervous system and the muscular system, do not reproduce at all." We are born with our full complement of brain neurons. Most of these neurons live as long as we do, but when one dies it is not replaced.

By contrast, the white phagocytic (scavenger) cells in our blood are replaced every two or three weeks. The life span of our red blood cells is three to four months. Our skin cells, too, keep being replaced. In such tissues, "The rate of production of new cells is nicely modulated to compensate for the continual loss of old cells," Mazia points out. In case of wounds or infections, when many cells are lost, the remaining cells multiply rapidly until the number of new cells is sufficient to replace what was lost and then stop. It is only in cancer that the cells repeat the cycle without restraint.

Surprisingly, all cells, whether of plant, bacteria, mouse or man, are made of the same fundamental materials: nucleic acids, proteins, lipids, carbohydrates, water and salts. "The uniformity of the earth's life, more astonishing than its diversity, is accountable by the high probability that we derived, originally, from a single cell," notes Lewis Thomas. "It is from the progeny of this parent cell that we take our looks; we still share genes around and the resemblance of the enzymes of grasses to those of whales is a family resemblance."

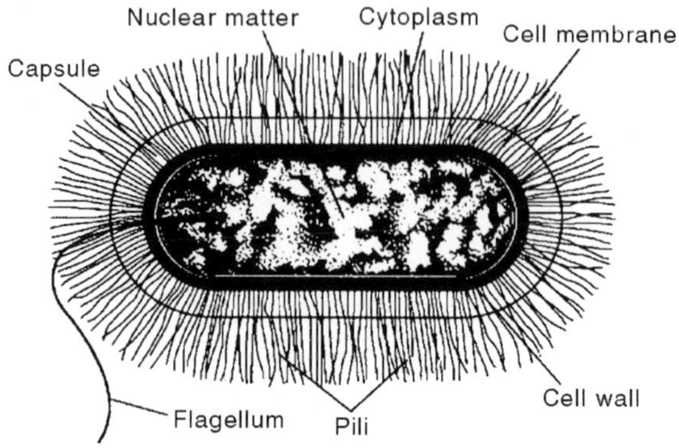

Fig. Bacterial Cells

The genetic material in all these cells is DNA (deoxyribonucleic acid), a large molecule which directs the making of duplicate cells according to a complex code. DNA also directs the building of proteins within the cell. Even the smallest and simplest living cells -- the mycoplasma -contain a relatively large amount of DNA, enough to code for up to a thousand different proteins. Every human cell has about 6 feet of DNA strands and every adult carries about 100 billion miles of ultrathin DNA strands in his body -- a distance greater than the diameter of the solar system. This DNA is concentrated in the cell's nuclear region. But there is a fundamental distinction between two major categories of cells: The procaryotic cells, which include the bacteria, mycoplasma and blue-green algae, do not have any membrane around their nuclear region.

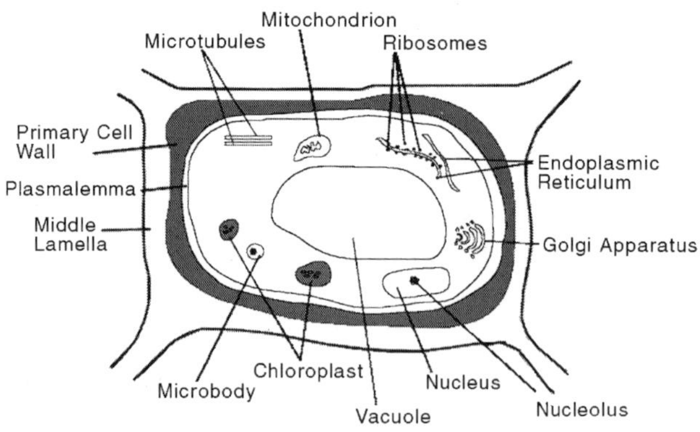

Fig. Plant Cell

In fact, they do not have any membrane-bound organelles at all. The eucaryotic ("proper nucleus") cells do have a double membrane to separate their nucleus from the cytoplasm, as well as many other internal membranes to segregate their organelles. The cells of all animals and plants (except blue-green algae) and those of one-celled protozoa fall in this category. Only the eucaryotic cells are able to combine with one another to form multicellular systems -- an important step Bacterial cells have no nucleus; their chromosomes, smaller and simpler than those of animal cells, are not enclosed in a membrane. As this diagram of an Escherichia coil bacterium shows, bacteria do not have any membrane -- bound organelles at all. They cannot combine with one another to form multicellular systems, as plant or animal cells do.

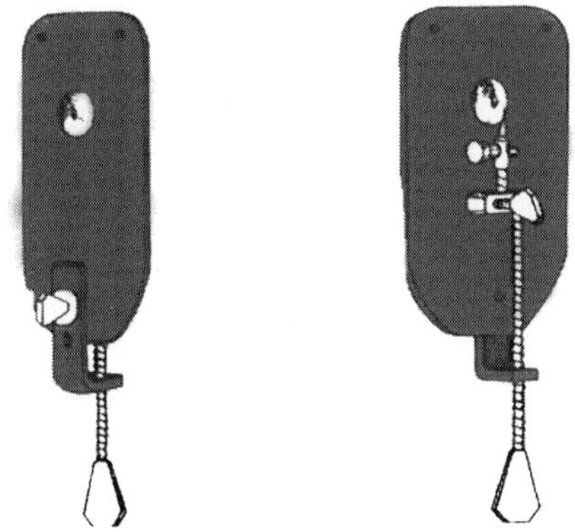

This is the actual size of a typical microscope built by Leeuwenhoek. He peered through the tiny lens opening on one side of a metal plate (left) to see the specimen mounted on the point of a pin on the other side (right). The specimen could be moved into focus by a system of screws.

And while, in general, the procaryotic cells produce only exact duplicates of themselves, the eucaryotic cells are capable of differentiation into many different kinds of cells, at least in higher organisms. This gives the eucaryotic cells certain obvious advantages.

However, the procaryotes have advantages of one thousand times larger than bacterial cells, plant cells have a well-defined nucleus, as well as many membrane-bound organelles. They also have chloroplasts which allow them to utilize solar energy in photosynthesis. A rigid cell wall made of cellulose lies outside their plasma membrane. The large vacuole, containing water and salt, preserves osmotic pressure.

They are not necessarily inferior. As Mazia puts it, the differences between the two types of cells are simply "different ways of making a living." Some of the most important discoveries in modern cell biology have involved the eucaryotic cell and its organelles, beginning with the most prominent organelle of all, the nucleus.

BIOSYNTHESIS OF CELL WALL MATERIAL

The endoplasmic reticulum, Golgi bodies and microtubules have all been implicated in the synthesis and organised deposition of the material of the plant cell wall on the basis of their association in electron micrographs with developing cell walls. Direct chemical evidence of a role for these organelles in the synthesis of wall materials is almost entirely lacking except for a demonstration by Northcote and Picket-Heaps that pectic substances are formed in the Golgi apparatus and are transported to the cell wall in the Golgi vesicles.

The possible role that the endoplasmic reticulum plays in the mechanism for laying down cell walls has been investigated but as yet ribosomes have not been identified in electron micrographs of the cell walls. However, Kivilaon *et al.* have detected trace amounts of ribosomal RNA (rRNA) in a preparation of cell walls of *Zea mays* coleoptiles. Phetheon *et al.* considered that the rRNA found in *Zea mays* was bacterial contamination, but they have isolated highly purified preparations of cell walls from a number of plants and have found that RNA was invariably present.

The possible role of rRNA in the cell wall may be concerned with the synthesis of protein in situ. However, Phetheon *et al.* consider that protein synthesis requires such an elaborate system of enzymes and nucleic acids that it is unlikely that such a synthesis would occur in the wall.

The process of biosynthesis and orientated deposition of cellulose fibrils is considered by Ben-Hayyim and Ohad to consist of four steps:

- Polymerisation (of the activated monomeric precursor) to form a cellulose molecule of high molecular weight;
- Transport of the molecule from the site of synthesis to that of crystallisation;
- Crystallisation or fibril formation;
- Orientation of fibrils during deposition.

These steps are not necessarily isolated from one another. Frei and Preston believe the polymerisation, fibril formation and orientation occur simultaneously in plants. They propose that a catalytically active protein moves around the cell membrane and deposits the formed cellulose fibrils in pre-determined directions, the whole process being under genetic control. Preston and Goodman considered the hypothesis of end synthesis of microfibrils through ordered enzyme-complex granules as a satisfactory explanation for the organisation and biosynthesis of the microfibrils. They suggested that the cytoplasmic surface is covered by layers of granules in cubic packing each containing cellulose synthesising enzymes. When these granules come into contact with a microfibril end they are stimulated and start synthesising the microfibril. The microfibrils are maintained straight both by intra-chain hydrogen bonding and mutual orientation of the granules. The packing of the granules are such that they would restrict the direction of growth of the microfibrils in such a way that only two directions at right angles to another and an occasional departure along the diagonals to these directions would be possible.

It is considered that this hypothesis may explain the observations of Setterfield and Bailey that cellulose synthesis can occur in outer wall layers away from the cytoplasm, as the granules which come to be buried in the wall could continue to synthesise cellulose. Where or how the microfibril ends are produced is not known; however it is proposed that they could be produced by enzyme degradation of the cellulose or by stresses in the wall during cell elongation.

Frey-Wyssling has suggested that extrusions of the plasmolemma inside the cell wall are responsible for the synthesis and orientation of cellulose fibrils in plants.

Opposed to this is evidence that the synthesis, transport and crystallisation may be separate processes in bacterial systems and probably in some plants. Many workers still regard the process of cellulose fibril orientation in the secondary wall of plant cells as a passive one, rearrangement being the result of mechanical strain or stress. Northcoate noted that the synthesis of material and the extent of enlargement of the cell wall during development can be influenced by nutrition of the growing cell. Factors affecting growth are thus important in affecting the structure and growth of the cell wall.

MICROFIBRIL ORIENTATION

The microfibrils are orientated in various ways in the cell wall, usually more regularly in the secondary wall. In the primary wall the microfibrils are often orientated in a direction transverse to the long axis. They become arranged more longitudinally during growth of the cell as the subsequent wall layers are formed. This transition is gradual and the change in direction of the microfibrils in successive wall layers may be about 120°. For example in cotton root hairs there is a gradual transition from an approximately axial orientation of microfibrils on the outer surface through a central region of crossed microfibrils to the transverse orientation of the inner layer. Also in the cells of different regions of a developing organ the wall may show differential orientations of the microfibrils. In the root of *Allium cepa* the cell walls of the apical initials show a loosely woven mesh of microfibrils; in slightly older cells the microfibrils are mainly aligned horizontally, and this is the case during active elongation.

Jensen revealed that in the onion root these changes in cell structure can apparently be correlated with changes in the selective amount of cell wall components both in cells of different stages of development, *i.e.* at different distances from the root tip and in cells of different tissues at any one level. In particular the transitional region between radial enlargement and rapid elongation of the root is characterised by changing relationships between the wall components.

Ben Hayyim and Ohad have suggested a possible model for the orientation of microfibrils in the cell wall. They introduced a synthetic cellulose called Ma-carboxymethyl cellulose (CMC) into the microfibrils. This, they considered, played a similar role to that of charged polysaccharides. The parallel orientation of fibrils containing CMC could be explained partially in terms of charged interactions and mechanical stress. Non-charged cellulose is considered to form a random mesh, the rigidity of which would depend on the number of interfibriller cross links, *i.e.* H bonds. Electrostatic repulsion between fibrils containing CMC can weaken the interfibriller links occurring in the vicinity of adjacent carboxyls. Stress or strain applied to this kind of mesh could result in an easier deformation of the matrix in the direction of the applied force. The most stable state would be achieved when the fibrils are parallel and the overall numbers of links formed between fibrils exceed the number of weak bonds due to electrostatic repulsion. From a similar consideration it is conceivable that a second layer of fibrils formed above the first would reach a more stable position when oriented at an angle with the first, when the repulsion between the layers would be minimal.

CELL WALL CONSTITUENTS

The principal constituents of the cell wall are cellulose, hemi-cellulose, pectic substances, lignin and proteins. Waxes, cutin, suberin and sporopollenin are also found.

STRUCTURAL COMPONENTS

Cellulose: This is the most abundant substance in the plant kingdom, and is a polymer of -b-glucose residues joined in long chains by 1-4 links. Over certain parts of their length these chains lie parallel to each other and are very regularly spaced, to form long crystalline microfibrils. These microfibrils may be arranged randomly or in a regular fashion. Meyer and Misch proposed the 'unit cell' of cellulose. They considered that the cellulose chains lie antiparallel in such a way that alternate chains point in opposite directions. Within the microfibrils themselves are smaller units, the micelles, which are small aggregations of cellulose molecules that lie parallel to one another and thus confer a crystalline structure upon the microfibrils. More recently it has been claimed that the ultimate structural units of the cell wall are elementary fibrils about 35 A in diameter which are not aggregated into larger strands. The microfibrils are necessary to bear the stress in the wall due to turgor pressure.

AMORPHOUS COMPONENTS

The Pectic Substances: These consist of polymers of d-galacturonic acid, l-arabinose, d-galactose and l-rhamnose. These substances are found mainly in the middle lamella of primary walls. Northcote has studied the metabolic changes which occur in the pectic substances deposited in the wall. He showed that the strongly acidic polygalacturonic acid was formed first. He found that pectin in the cell walls of a mature tissue such as in apple fruit was composed mainly of neutral and weakly acidic material and only traces of the strongly acidic polymer could be detected.

Hemi Celluloses: These are amorphous and consist of linear or branched polymers of d-xyloses, d-galactose, d-mannose, l-arabinose, and l-rhamnose. In contrast to cellulose the hemicelluloses are not crystalline in their natural condition although they have been found in a crystalline state after extraction.

The distinction between the hemicelluloses and the pectins is not absolutely one of chemical constituents, but rather of physical properties, mainly solubility, which depends on their degree of methylation, cross-linkage and the extent to which different sugars are joined in the same molecule.

Lignin: This is an amorphous substance which occurs as an incrustation between cellulose microfibrils. The concentration is highest in the middle lamella and falls off towards the lumen. It is an important structural material and it is this substance that gives strength to wood. Because of this deposition of lignin between the existing cellulose framework there is always a swelling of the cell wall during lignification.

Protein: Recent work has demonstrated the occurrence of a group of proteins containing hydroxyproline in the primary walls of various tissues. The amount present increases during growth and it is thought that the protein may serve enzymatic as well as structural functions.

The primary and secondary walls differ in their chemical composition and in fine structure. In most cases the secondary walls have a higher percentage of cellulose and lignin while the pectic substances are present only in trace quantities as compared to the primary walls.

Thus the general cell wall consists of the structural components of the wall, the cellulose microfibrils, orientated in the amorphous component comprising the pectic substances, hemicelluloses, lignin and the protein substances.

THE BIOSYNTHESIS OF CELL CONSTITUENTS

Biologists once depended on the light microscope to study the morphology of cells found in higher plants and animals. The functioning of cells in unicellular and in multicellular organisms was then postulated from observation of the structure; the discovery of the chloroplastids in the cell, for example, led to the investigation of the process of photosynthesis. With the invention of the electron microscope, the fine organization of the plastids could be utilized for further quantitative studies of the different parts of this process.

The energy derived from the breakdown of organic molecules (catabolism) is used to drive the synthesis of other required components of the cell. Most catabolic pathways involve the oxidation of organic molecules coupled to the generation of both energy (ATP) and reducing power (NADH). In contrast, biosynthetic (anabolic) pathways generally involve the use of both ATP and reducing power (usually in the form of NADPH) for the production of new organic compounds. One major biosynthetic pathway, the synthesis of carbohydrates from CO_2 and H_2O during the dark reactions of photosynthesis. Additional pathways leading to the biosynthesis of major cellular constituents (carbohydrates, lipids, proteins, and nucleic acids) are reviewed in the sections that follow.

Carbohydrates

In addition to being obtained directly from food or generated by photosynthesis, glucose can be synthesized from other organic molecules. In animal cells, glucose synthesis (gluconeogenesis) usually starts with lactate (produced by anaerobic glycolysis), amino acids (derived from the breakdown of proteins), or glycerol (produced by the breakdown of lipids). Plants (but not animals) are also able to synthesize glucose from fatty acids—a process that is particularly important during the germination of seeds, when energy stored as fats must be converted to carbohydrates to support growth of the plant. In both animal and plant cells, simple sugars are polymerized and stored as polysaccharides.

Gluconeogenesis involves the conversion of pyruvate to glucose—essentially the reverse of glycolysis. However, the glycolytic conversion of glucose to pyruvate is an energy-yielding pathway, generating two molecules each of ATP and NADH. Although some reactions of glycolysis are readily reversible, others will proceed only in the direction of glucose breakdown, because they are associated with a large decrease in free energy. These energetically favourable reactions of glycolysis are bypassed during gluconeogenesis by other reactions (catalysed by different enzymes) that are coupled to the expenditure of ATP and NADH in order to drive them in the direction of glucose synthesis.

Overall, the generation of glucose from two molecules of pyruvate requires four molecules of ATP, two of GTP, and two of NADH. This process is considerably more costly than the simple reversal of glycolysis (which would require two molecules of ATP and two of NADH), illustrating the additional energy required to drive the pathway in the direction of biosynthesis.

In both plant and animal cells, glucose is stored in the form of polysaccharides (starch and glycogen, respectively). The synthesis of polysaccharides, like that of all other macromolecules, is an energy-requiring reaction. As noted earlier, the linkage of two sugars by a glycosidic bond can be written as a dehydration reaction, in which H_2O is removed. Such a reaction, however, is energetically unfavourable and therefore unable to proceed in the forward direction. Consequently, the formation of a glycosidic bond must be coupled to an energy-yielding reaction, which is accomplished by the use of nucleotide sugars as intermediates in polysaccharide synthesis. Glucose is first phosphorylated in an ATP-driven reaction to glucose-6-phosphate, which is then converted to glucose-1-phosphate. Glucose-1-phosphate reacts with UTP (uridine triphosphate), yielding UDP-glucose plus pyrophosphate, which is hydrolyzed to phosphate with the release of additional free energy. UDP-glucose is an activated intermediate that then donates its glucose residue to a growing polysaccharide chain in an energetically favourable reaction. Thus, chemical energy in the form of ATP and UTP drives the synthesis of polysaccharides from simple sugars.

Proteins

In contrast to carbohydrates and lipids, proteins (as well as nucleic acids) contain nitrogen in addition to carbon, hydrogen, and oxygen. Nitrogen is incorporated into organic compounds from different sources in different organisms. Some bacteria can use atmospheric N_2 by a process called nitrogen fixation, in which N_2 is reduced to NH_3 at the expense of energy in the form of ATP. Although relatively few species of bacteria are capable of nitrogen fixation, most bacteria, fungi, and plants can use nitrate (NO_3^-), which is a common constituent of soil, by reducing it to NH_3 via electrons derived from NADH or NADPH. Finally, all organisms are able to incorporate ammonia (NH_3) into organic compounds.

NH_3 is incorporated into organic molecules primarily during the synthesis of the amino acids glutamate and glutamine, which are derived from the citric acid cycle intermediate α-ketoglutarate. These amino acids then serve as donors of amino groups during the synthesis of the other amino acids, which are also derived from central metabolic pathways, such as glycolysis and the citric acid cycle. The raw material for amino acid synthesis is thus obtained from glucose, and the amino acids are synthesized at the cost of both energy (ATP) and reducing power (NADPH). Many bacteria and plants can synthesize all 20 amino acids. Humans and other mammals, however, can synthesize only about half of the required amino acids; the remainder must be obtained from dietary sources.

The polymerization of amino acids to form proteins also requires energy. Like the synthesis of polysaccharides, the formation of the peptide bond can be considered a dehydration reaction, which must be driven in the direction of protein synthesis by being coupled to another source of metabolic energy. In the biosynthesis of polysaccharides, this coupling is accomplished through the conversion of sugars to activated intermediates, such as UDP-glucose. Amino acids are similarly activated before being used for protein synthesis.

A critical difference between the synthesis of proteins and that of polysaccharides is that the amino acids are incorporated into proteins in a unique order, specified by a gene. The order of nucleotides in a gene specifies

GENERATION OF ENERGY

Unlike eucaryotes, bacteria don't have intracellular organelles for energy producing processes such as respiration or photosynthesis. Instead, the cytoplasmic membrane carries out these functions. The membrane is the location of electron transport systems (ETS) used to produce energy during photosynthesis and respiration, and it is the location of an enzyme called ATP synthetase (ATPase) which is used to synthesize ATP.

When the electron transport system operates, it establishes a pH gradient across of the membrane due to an accumulation of protons (H^+) outside and hydroxyl ion (OH^-) inside. Thus the outside is acidic and the inside is alkaline. Operation of the ETS also establishes a charge on the membrane called proton motive force (pmf). The outer face of the membrane becomes charged positive while inner face is charged negative, so the membrane has a positive side and a negative side, like a battery. The pmf can be used to do various types of work including the rotation of the flagellum, or active transport as described above. The pmf can also be used to make ATP by the membrane ATPase enzyme which consumes protons when it synthesizes ATP from ADP and phosphate. The connection between electron transport, establishment of pmf, and ATP synthesis during respiration is known as oxidative phosphorylation; during photosynthesis, it is called phosphorylation.

Figure below illustrates the membrane of *E. coli*. The topographical features of the membrane from top to bottom are

1. Lactose transport system;
2. The flagellar motor coupled to the hook and filament;
3. Na^+ transport (export) system;
4. Ca^{++} transport (export) system;
5. Electron transport system;
6. ATPase enzyme;
7. Proline transport system.

The operation ot the electron transport system during respiration produces the H^+ charge on the membrane (pmf). The pmf (H^+) is used by the transport systems to move molecules from one side of the membrane to the other; by the flagellar motor ring to rotate the flagellar filament; and by the ATPase enzyme to synthesize ATP.

THE CYTOPLASMIC MEMBRANE

The cytoplasmic membrane of bacterial cells is a delicate and plastic structure that completely encloses the cell cytoplasm (or protoplasm). The bacterial membrane is composed of 40 per cent phospholipid and 60 per cent protein. The phospholipids are amphoteric molecules, meaning they have a water-soluble hydrophilic region (the glycerol "head") attached to two insoluble hydrophobic fatty acid "tails". In water, such molecules naturally form the molecular bilayer characteristic of membranes. The fatty acid tails from one layer face towards the fatty acid tails of the second layer ("likes dissolve like"), and the glycerol heads naturally turn towards the water.

Dispersed throughout the bilayer are various structural and enzymatic proteins which carry out most membrane functions. This arrangement of proteins and phospholipids forms what is called the fluid mosaic membrane as illustrated in Figure.

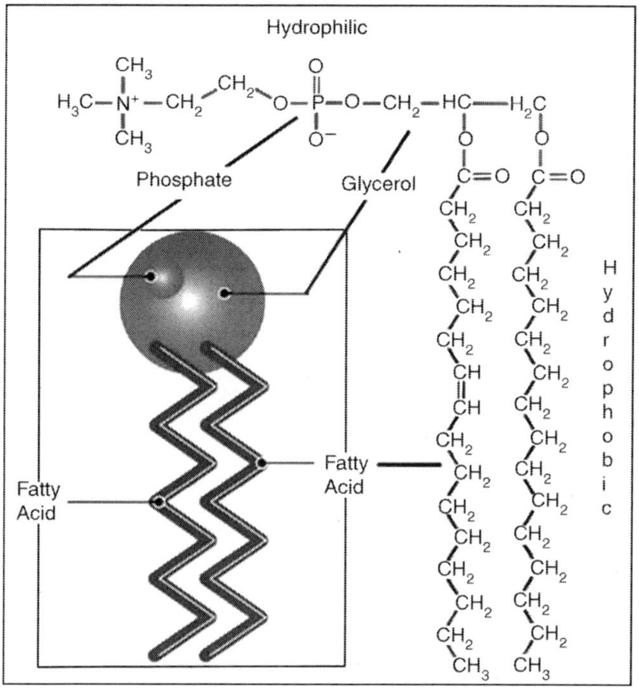

Fig. Molecular Structure of a Phospholipid, the Building block of Membranes. Inset - the usual Depiction of a Membrane Phospholipid Containing a Phosphatidyl-glycerol "head" Attached to two Fatty acid "tails".

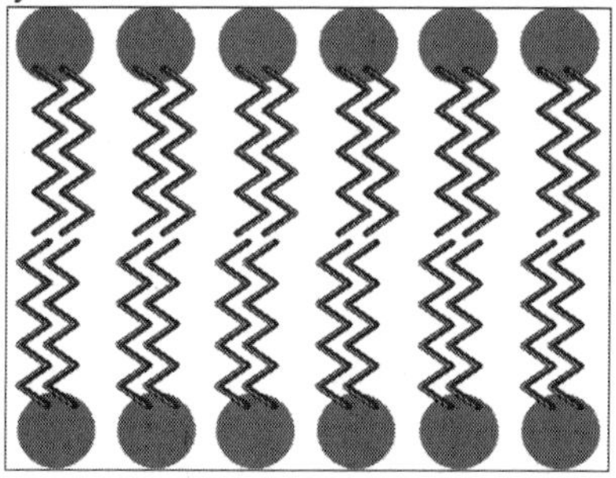

Fig. Organization of Phospholipids in Aqueous Solution to form a Bilayer. The Hydrophilic Phosphatidyl Glycerols form the inner and outer Faces of the Membrane. The Fatty Acids Orient Themselves Towards one Another to form the Hydrophobic Interior of the Membrane.

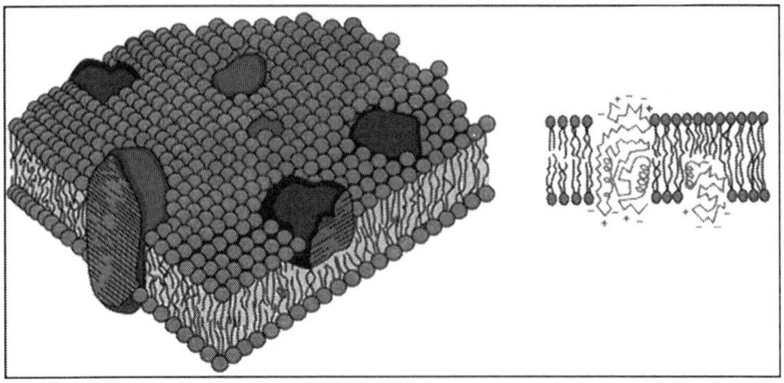

Fig. Fluid Mosaic Model of a Biological Membrane. In Aqueous Environments Membrane Phospholipids Arrange themselves in Such a way that they Spontaneously form a Fluid Bilayer. Membrane Protein may be Either Structural or Functional. Proteins may be Permanently or Transiently Associated with one Side or the other of the Membrane or built into the Bilayer, or they may Span the Bilayer Forming Transport Channels through the Membrane.

PROCARYOTIC CELL ARCHITECTURE

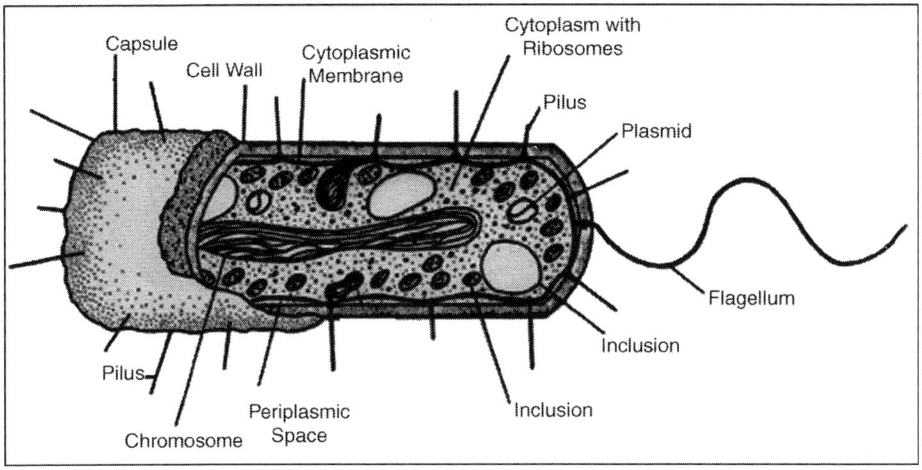

Fig. Schematic Drawing of a Typical Bacterial Cell.

At one time it was thought that bacteria were essentially "bags of enzymes" with no inherent cellular architecture. The development of the electron microscope in the 1950s revealed the distinct anatomical features of bacteria and confirmed the suspicion that they lacked a nuclear membrane. Structurally, a bacterial cell has three architectural regions: appendages (attachments to the cell surface) in the form of flagella and pili (or fimbriae); a cell envelope consisting of acapsule, cell wall and plasma membrane; and a cytoplasmic region that contains the cell chromosome (DNA) and ribosomes and various sorts of inclusions.

In this lecture, we will discuss the anatomical structures of procaryotic cells in relation to their adaptation, function and behaviour in natural environments.

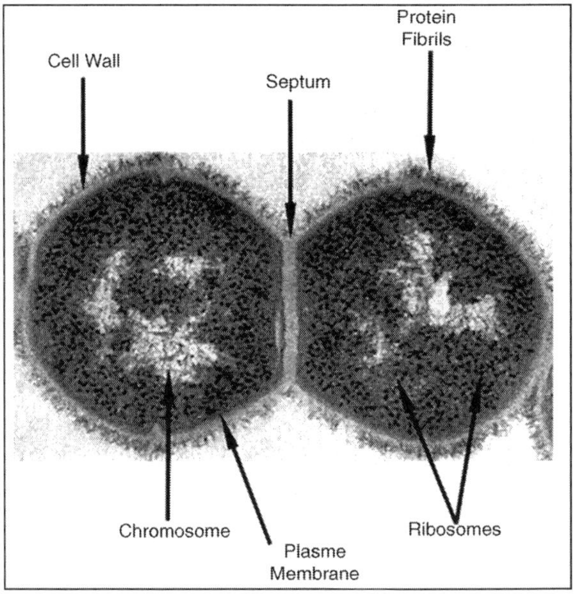

Fig. Electron Micrograph of an Ultra-thin Section of a Dividing Pair of Group A Streptococci (20,000X).

The Cell Surface Fibrils, Consisting Primarily of Protein, are Evident. The Bacterial Cell Wall, to which the Fibrils are Attached, is also Clearly seen as the Light Staining Region between the Fibrils and the Dark Staining cell Interior.

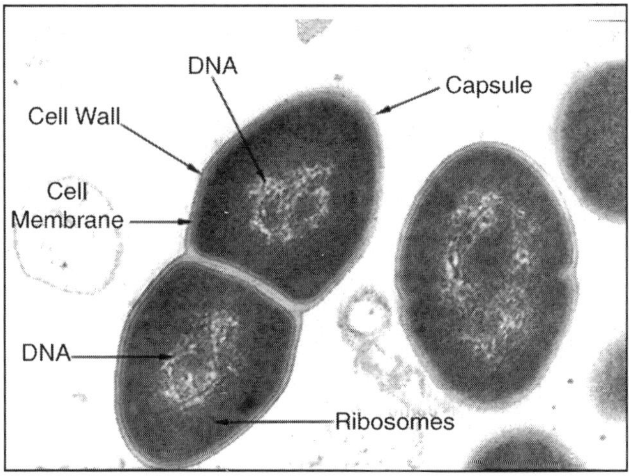

Fig. Electron Micrograph of an Ultra-thin Section of a Dividing Pair of Enterococci

Table. Characteristics of Typical Bacterial Cell Structures.

Structure	Function(s)	Predominant Chemical Composition
Flagella	Swimming movement	Protein
Pili		
Sex pilus	Mediates DNA transfer during conjugation	Protein
Common pili or fimbriae	Attachment to surfaces; protection against phagotrophic engulfment	Protein
Capsules (includes "slime layers" and glycocalyx)	Attachment to surfaces; protection against phagocytic engulfment, occasionally killing or digestion; reserve of nutrients or protection against desiccation	Usually polysaccharide; occasionally polypeptide
Cell wall		
Gram-positive bacteria	Prevents osmotic lysis of cell protoplast and confers rigidity and shape on cells	Peptidoglycan (murein) complexed with teichoic acids
Gram-negative bacteria	Peptidoglycan prevents osmotic lysis and confers rigidity and shape; outer membrane is permeability barrier; associated LPS and proteins have various functions	Peptidoglycan (murein) surrounded by phospholipid protein-lipopolysaccharide "outer membrane"
Plasma membrane	Permeability barrier; transport of solutes; energy generation; location of numerous enzyme systems	Phospholipid and protein
Ribosomes	Sites of translation (protein synthesis)	RNA and protein
Inclusions	Often reserves of nutrients; additional specialized functions	Highly variable; carbohydrate, lipid, protein or inorganic
Chromosome	Genetic material of cell	DNA
Plasmid	Extrachromosomal genetic material	DNA

APPENDAGES

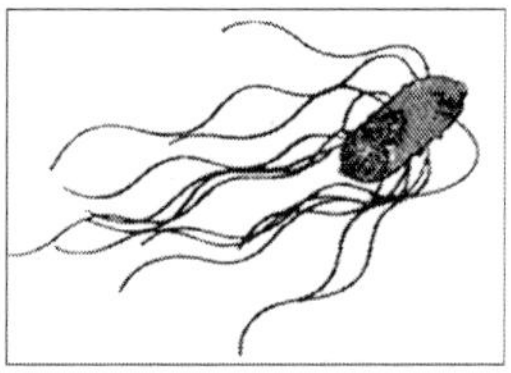

Fig. *Salmonella Enterica. Salmonella* is an Enteric Bacterium Related to *E. coli.* The Enterics are Motile by Means of Peritrichous Flagella.

FLAGELLA

Flagella are filamentous protein structures attached to the cell surface that provide the swimming movement for most motile procaryotes. Procaryotic flagella are much thinner than eucaryotic flagella; the diameter of a procaryotic flagellum is about 20 nanometers, well-below the resolving power of the light microscope. The flagellar filament is rotated by a motor apparatus in the plasma membrane allowing the cell to swim in fluid environments.

The ultrastructure of the flagellum of *E. coli* is illustrated in Figure below. The flagellar apparatus consists of several distinct proteins: a system of rings imbedded in the cell envelope (the basal body), a hook-like structure near the cell surface, and the flagellar filament.

The innermost rings, the M and S rings, located in the plasma membrane, comprise the motor apparatus. As the M ring turns, powered by an influx of protons, the rotary motion is transferred to the filament which rotates thereby propelling the bacterium.

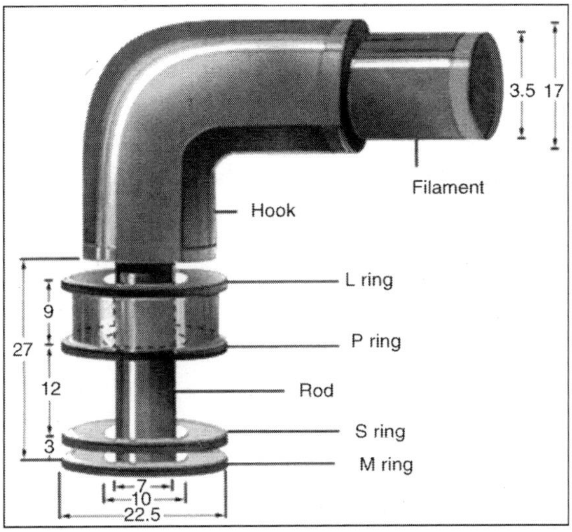

Fig. The Ultrastructure of a Bacterial Flagellum.

Measurements are in Nanometers. The Flagellum of *E. coli* Consists of three Parts, Filament, hook and Basal body, all composed of Different Proteins. The Basal body and hook Anchor the Whip-like Filament to the cell Surface. The Basal body Consists of four ring-shaped Proteins Stacked like Donuts around a Central rod in the cell Envelope.

Flagella may be variously distributed over the surface of bacterial cells in distinguishing patterns, but basically flagella are either polar (one or more flagella arising from one or both poles of the cell) or peritrichous (lateral flagella distributed over the entire cell surface). Flagellar distribution is a genetically-distinct trait that is occasionally used to characterize or distinguish bacteria. For example, among Gram-negative rods, pseudomonads have polar flagella to distinguish them from enteric bacteria, which have peritrichous flagella.

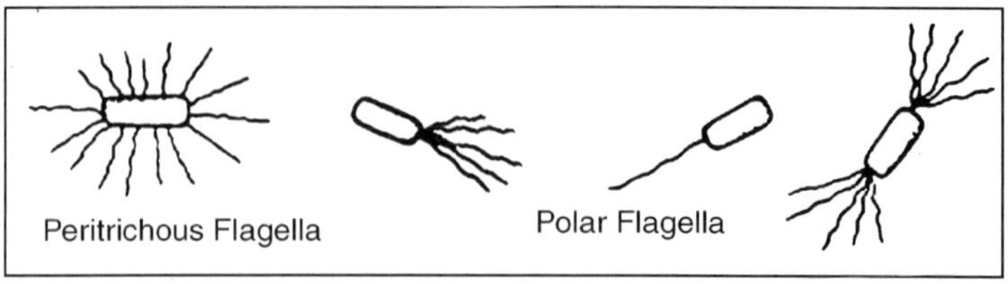

Fig. Different Arrangements of Bacterial Flagella. Swimming Motility, Powered by Flagella, Occurs in Half the Bacilli and Most of the Spirilla.

Procaryotes are known to exhibit a variety of types of tactic behaviour, *i.e.*, the ability to move (swim) in response to environmental stimuli. For example, duringchemotaxis a bacterium can sense the quality and quantity of certain chemicals in its environment and swim towards them (if they are useful nutrients) or away from them (if they are harmful substances). Other types of tactic response in procaryotes include phototaxis, aerotaxis and magnetotaxis. The occurrence of tactic behaviour provides evidence for the ecological (survival) advantage of flagella in bacteria and other procaryotes.

DETECTING BACTERIAL MOTILITY

Since motility is a primary criterion for the diagnosis and identification of bacteria, several techniques have been developed to demonstrate bacterial motility, directly or indirectly. Flagellar stains outline flagella and show their pattern of distribution. If a bacterium possesses flagella, it is presumed to be motile.

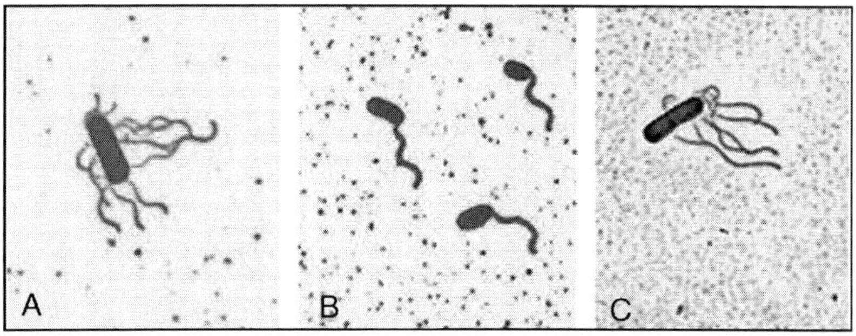

Fig. Flagellar Stains of Three Bacteria a. *Bacillus Cereus* b. *Vibrio Cholerae* c. *Bacillus brevis*. CDC. Since the Bacterial Flagellum is below the Resolving power of the Light Microscope, although Bacteria can be seen Swimming in a Microscope Field, the Organelles of Movement cannot be Detected.

Motility test medium demonstrates if cells can swim in a semisolid medium. A semisolid medium such as O.75 per cent agar is inoculated with the bacteria in a straight-line stab with a needle. After incubation, if turbidity (cloudiness) due to bacterial growth can be observed away from the line of the stab, it is evidence that the bacteria were able to swim through the medium.

Fig. Bacterial Cultures Grown in Motility test medium. The tube on left is a non Motile Organism; the tube on Right is a Motile Organism. Motility test medium is a semi-soft medium that is Inoculated with a Straight Needle.

Direct microscopic observation of living bacteria in a wet mount. One must look for transient movement of swimming bacteria. Most unicellular bacteria, because of their small size, will shake back and forth in a wet mount observed at 400X or 1000X. This is Brownian movement, due to random collisions between water molecules and bacterial cells. True motility is confirmed by observing a bacterium swim from one side of the microscope field to the other side.

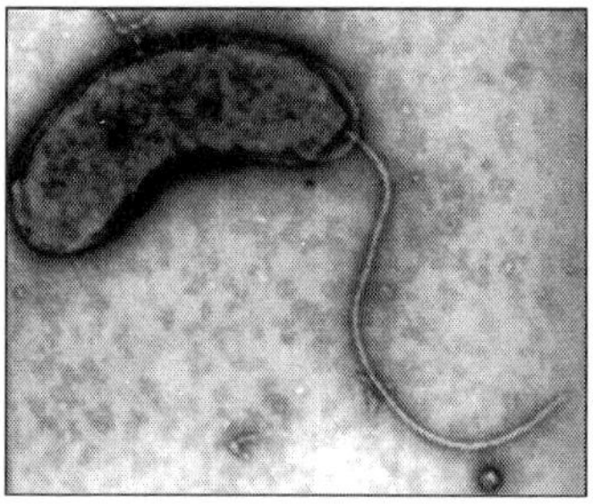

Fig. A *Desulfovibrio* Species. TEM. About 15,000X. The Bacterium is Motile by Means of a Single Polar Flagellum. Of Course, one can Detect Flagella by Means of Electron Microscopy. Perhaps this is an Alternative way to Determine Bacterial Motility, if you happen to have an Electron Microscope.

FIMBRIAE AND PILI

Fimbriae and pili are interchangeable terms used to designate short, hair-like structures on the surfaces of procaryotic cells. Like flagella, they are composed of protein. Fimbriae are shorter and stiffer than flagella, and slightly smaller in diameter. Generally, fimbriae have nothing to do with bacterial movement (there are exceptions, *e.g.* twitching movement on *Pseudomonas*). Fimbriae are very common in Gram-negative bacteria, but occur in some archaea and Gram-positive bacteria as well. Fimbriae are most often involved in adherence of bacteria to surfaces, substrates and other cells or tissues in nature. In *E. coli,* a specialized type of pilus, the F or sex pilus, apparently stabilizes mating bacteria during the process of conjugation, but the function of the smaller, more numerous common pili is quite different Common pili (often called fimbriae) are usually involved in specific adherence (attachment) of procaryotes to surfaces in nature. In medical situations, they are major determinants of bacterial virulence because they allow pathogens to attach to (colonize) tissues and/or to resist attack by phagocytic white blood cells. For example, pathogenic Neisseria gonorrhoeae adheres specifically to the human cervical or urethral epithelium by means of its fimbriae; enterotoxigenic strains of E. coli adhere to the mucosal epithelium of the intestine by means of specific fimbriae; the M-protein and associated fimbriae of *Streptococcus pyogenes* are involved in adherence and to resistance to engulfment by phagocytes.

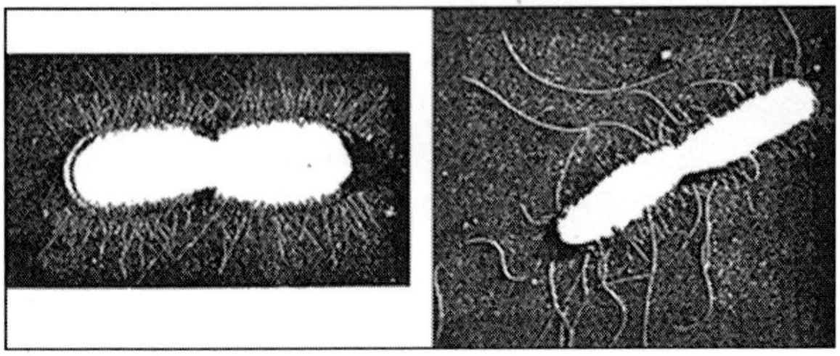

Fig. Fimbriae (Common Pili) and Flagella on the Surface of Bacterial cells. Left: Dividing *Shigella* Enclosed in Fimbriae. The Structures are Probably Involved in the Bacterium's Ability to Adhere to the Intestinal Surface.

Table. Some Properties of Pili and Fimbriae.

Bacterial Species Where Observed	Typical Number on Cell	Distribution on Cell Surface	Function
Escherichia coli (F or sex pilus)	1-4	Uniform	Stabilizes mating during conjugation
Escherichia coli (common pili or Type 1 fimbriae)	100-200	Uniform	Surface adherence to epithelial cells of the GI tract
Neisseria gonorrhoeae	100-200	Uniform	Surface adherence to epithelial cells of the urogenital tract
Streptococcus pyogenes (fimbriae plus the M-protein)	?	Uniform	Adherence, resistance to phagocytosis; antigenic variability
Pseudomonas aeruginosa	10-20	Polar	Surface adherence

MOLECULAR MODELS OF THE CELL MEMBRANE

On the molecular structure of the membrane before the isolation of plasma membranes, theories were generally based on indirect information. Since substances soluble in lipid solvents penetrate the plasma membrane easily, Overton postulated in. 1902 that the plasma membrane is composed of a thin layer of lipid. In 1926 Gorter and Grendell found that the lipid content of hemolyzed erythrocytes was sufficient to form a double layer of lipid molecules over the entire cell surface. This theory was also supported by electrical measurements that indicated a high impedance at the plasma membrane. The high impedance is because of the faex fact that it is difficult for ions to penetrate a lipid layer.

Tension at a water-oil interface is about 10 to 15 dynes per centimetre, whereas surface tension of cells is almost nil. It has been postulated that the low tension is due to the presence of protein layers on the lipid components. In fact, when a very small amount of protein is added to a model lipid-water system, the surface tension is lowered comparably. In 1935 Danielli and Davson explained about all these properties. They proposed that the plasma membrane contained a *lipid bilayer,* with protein adhering to both lipid-aqueous interfaces. The different membrane models based on the concept of a lipid bilayer, but which differ in the type of protein (*i.e.,* globular, a or b-helix) or the penetration of proteins within the bilayer. Other models containing globular lipid micelles, globular proteins, or a combination of both have been proposed a globular-bilayer transition has also been postulated. These globular models do not account satisfactorily for the high electrical impedance.

On the fine structure of the plasma membrane, electron microscopy has thrown some light and has revealed the numerous structural differentiations that this membrane and the underlying cell cytoplasm have in different cell types. To resolve the structure of the plasma membrane, extremely thin sections (~ 20 nm) must be used, otherwise one would observe different orientations of the plasma membrane with respect to the plane of the section. The membrane appears most thin when it is exactly perpendicular to the plane of the section. Membranes of 6 to 10 mm have been observed at the surface of all cells. The membranes of two cells that are in close contact appear as dense lines separated by a space of 11 to 15 nm, which is strikingly uniform and contains a material of low electron density. This intercellular component can be considered as a kind of cementing substance.

Re evaluation of the EM Image

There are three layer in cell membrane outer dense layers of 2.0 nm and a middle clear one of about 3.5 nm led Robertson to postulate the so-called *unit membrane* model. The electron microscopic image was interpreted along the lines of the Danielli-Davson model, so that the clear layer corresponded to the hydrocarbon chains of the lipids, and the dense layers, to the proteins. This unit membrane model was also extended to all kinds of intracellular membranes. That this concept is an oversimplication which does not account for the many proteins traversing the membrane can be readily seen by observing some high resolution micrographs in which fine bridges cross the unit membrane.

On the basis of some experiments the unit membrane has been re-evaluated recently methods designed to avoid the removal of proteins from membranes during the preparation for conventional electron microscopy or freeze-fracturing. With both methods a granular appearance of the membrane is obtained, which is more in keeping with the fact that a high proportion of the proteins traverse the membrane. According to these findings the unit membrane image is, to a great extent, artifactual.

THE FLUID MOSAIC MODEL

The main source of the knowledge of the molecular organization of biological membrane comes from an integration of the data on chemical analysis and from the application of several biophysical techniques. The important concepts are (1) that the lipid and integral proteins are disposed in a kind of mosaic arrangement; and (2) that biological membranes are quasi-fluid structures in which both the lipids and the integral proteins are able to perform translational movements within the overall bilayer. The concept of fluidity implies that the main components of the membrane are held in place only by means of non-covalent interactions. For better study of the molecular organization of the membrane, it is necessary to remember that, not only the lipids, but also many of the intrinsic proteins and glycoproteins of the membrane, are amphipathic molecules. The term *amphipathy,* coined by Hartley in 1936, refers to the presence, within the same molecule, of hydrophilic and hydrophobic groups these amphipathic molecules constitute liquid crystalline aggregates in which the polar groups are directed toward the water phase and the non-polar groups are situated inside the bilayer. The integral proteins of the membrane are intercalated to a greater or lesser extent into a rather continuous lipid bilayer. This arrangement is based on the fact that these integral proteins are also amphipathic, with polar regions protruding from the surface and non-polar regions embedded in the hydrophobic interior of the membrane.

It is well recognized that a protein of appropriate size or a cluster of protein subunits may pass across the entire membrane (transmembrane proteins). Such traversing proteins could be in contact with the aqueous solvent on both sides of the membrane. A similar model has been postulated for the cholinergic receptor. This model also reflects the generally accepted view that the major proportion of the phospholipids in the membrane are arranged in a bilayer form. This mosaic model of the membrane, with intercalating proteins, comes from the use of freeze-fracturing techniques in erythrocytes and other cell membranes. The red cell ghosts show a large number of particles, about 8 nm in diameter, that have been interpreted as representing proteins embedded within the plane of cleavage which passes through the middle of the lipid bilayer.

There are between 500,000 and 600,000 such particles in a cell and more are attached to the inner half of the membrane (PF) than to the outer half (EF). We mentioned earlier that these particles represent dimers of the polypeptide found in *band 3*, and that they may also correspond to anionic permeation channels. The degree of dispersion or aggregation of these particles within the membrane appears to be influenced by the *spectrin-actin system.* This suggests that some link between this system and the dimers of band 3 could exist in the living membrane. The mosaic model allows for the characteristic *asymmetry* of the membrane, in which specific components predominate in the outer half or the inner half of the membrane. This implies that there are constraints on the transmembrane rotation of those macromolecules that are well oriented for the carrying of information across the cell membrane.

In this model the fluidity of the membrane also implies that both the lipids and the proteins have considerable freedom of movement within the bilayer. The fluidity of the lipids depends (1) on the degree of saturation of the hydrocarbon chains and (2) on the ambient temperature. A considerable proportion of the lipids in the membrane are unsaturated, so that the melting point of the bilayer is below body temperature.

Membrane Fluidity

For studying purpose, it can be classified as physical and biological. The *physical techniques* are of two main classes (1) those that involve a minimal perturbation of the membrane, such as *X-ray diffraction* and *nuclear magnetic resonance (NMR) spectroscopy,* and (2) those that use certain added molecules to monitor specific sites of the membrane. Into this second class fall *fluorescence microscopy,* which uses fluorescent probes and *electron spin resonance (ESR)* spectroscopy. In practice, these physical techniques can be subdivided on the basis of their ability to examine mainly either the lipid or the protein components, or both. ESR spectroscopy uses paramagnetic probes (*e.g.*, nitroxide-containing amphipathic molecules) which are introduced into the lipid bilayer. This can be done with natural and artificial membranes The data obtained provide information on lipid-protein packing, lateral diffusion of lipids, lipid-protein interactions, the fluidity of the membrane, and the rate of transmembrane rotation (so-called flip-flop) of molecules across the bilayer. The *biological techniques* involve light and fluorescence microscopy and electron microscopy, including freeze-fracturing and radioisotope labelling methods. One of the simplest methods consists of binding gold or carbon particles to the cell surface and observing under the light microscope the movement of those particles on the surface. More information has come from studies in which different ligands, such as antibodies and plant lectins, interact with cell surface receptors. If these ligands are labelled with fluorescent dyes their movement can be followed by fluorescence microscopy.

For example, in a classical experiment by Frye and Edidin[23] mouse L cells and human transformed cells which have different surface antigens, were marked with the corresponding antibodies labelled with two distinct